ADVANCES IN ELECTRONICS AND ELECTRON PHYSICS

VOLUME 27

CONTRIBUTORS TO THIS VOLUME

G. J. Béné
R. W. Crompton
P. H. Dawson
E. Geneux
E. B. Hooper, Jr.
Predhiman Krishan Kaw
J. Perrenoud
Mahendra Singh Sodha
N. R. Whetten

Advances in Electronics and Electron Physics

EDITED BY
L. MARTON

National Bureau of Standards, Washington, D.C.

Assistant Editor

CLAIRE MARTON

VOLUME 27

1969

ACADEMIC PRESS　　　New York and London

CONTRIBUTORS TO VOLUME 27

Numbers in parentheses indicate the pages on which the authors' contributions begin.

G. J. BÉNÉ (19), Institute of Experimental Physics, University of Geneva, Geneva, Switzerland

R. W. CROMPTON (1), Electron and Ion Diffusion Unit, The Australian National University, Canberra, Australia

P. H. DAWSON (59), General Electric Research and Development Center, Schenectady, New York*

E. GENEUX (19), Institute of Experimental Physics, University of Geneva, Geneva, Switzerland†

E. B. HOOPER, JR. (295), Department of Engineering and Applied Science, Yale University, New Haven, Connecticut

PREDHIMAN KRISHAN KAW (187), Physics Department, Indian Institute of Technology, New Delhi, India‡

J. PERRENOUD (19), Institute of Experimental Physics, University of Geneva, Geneva, Switzerland

MAHENDRA SINGH SODHA (187), Physics Department, Indian Institute of Technology, New Delhi, India

N. R. WHETTEN (59), General Electric Research and Development Center, Schenectady, New York

* *Present address*: Centre de Recherches sur les Atomes et les Molécules, Université Laval, Quèbec, Canada.

† *Temporary address*: Lawrence Radiation Laboratory, University of California, Berkeley, California.

‡ *Present address*: Plasma Physics Laboratory, Princeton University, Princeton, New Jersey.

FOREWORD

In the foreword to our last volume I mentioned that Volume 26 happened to be largely device-oriented. To show that no trend is implied, the present volume is largely gaseous-discharge-oriented; three out of five of the contributions represent different aspects of this branch of physical electronics. The first of these three is the revised version of an invited talk given by Dr. Crompton before the 1968 Gaseous Electronics Conference. Reflex and Penning discharges are the subject of Dr. Hooper's review. The third review, belonging more loosely in the same category, is by Drs. Sodha and Kaw, and treats the theory of the generation of harmonics and combination frequencies in a plasma.

In the past not much space was devoted in these volumes to magnetic resonance phenomena. The contribution of Drs. Geneux, Béné, and Perrenoud at least partially fills this gap by reviewing such resonances and transitions at zero frequency.

Different aspects of mass spectroscopy were the subject of earlier reviews (Volumes 1 and 8). Relatively new is the use of RF quadrupole fields for such purposes, reviewed here by Drs. Dawson and Whetten.

As usual, I would like to list here expected future reviews, with the names of their authors.

Light Interaction with Plasma   Heinz Raether
Superconducting Magnets   P. F. Smith
Recent Advances in Field Emission   L. Swanson and F. Charbonnier
Microfabrication Using Electron Beams   A. N. Broers
The Measurement of Lifetimes of Free Atoms,
   Molecules, and Ions   A. Corney
Energy Distribution in Thermionically Emitted
   Electron Beams   B. W. Zimmermann
Information Storage in Microspace   S. Newberry
Frequency FET Noise Parameters and Approxi-
   mation of the Optimum Source Admittance   M. Strutt
The Effects of Radiation in MIS Structures   Karl Zaininger
Research in Solid State Electronics with Electron
   Microprobes   David Wittry
Recent Advances in Biological Temperature
   Measurement   Hardy W. Trolander
Recent Progress on Fluidics   H. Burke Horton
Network Theory   L. Weinberg
The Formation of Cluster Ions in Gaseous Dis-
   charges and in the Ionosphere   W. Roth and R. Narcissi

L. MARTON

Washington, D.C.
August, 1969

Contents

The Contribution of Swarm Techniques to the Solution of Some Problems in Low Energy Electron Physics

R. W. CROMPTON

Magnetic Coherence Resonances and Transitions at Zero Frequency

E. GENEUX, G. J. BÉNÉ, and J. PERRENOUD

Mass Spectroscopy Using RF Quadrupole Fields

P. H. DAWSON and N. R. WHETTEN

Theory of the Generation of Harmonics and Combination Frequencies in a Plasma

MAHENDRA SINGH SODHA and PREDHIMAN KRISHAN KAW

A Review of Reflex and Penning Discharges

E. B. Hooper, Jr.

The Contribution of Swarm Techniques to the Solution of Some Problems in Low Energy Electron Physics

R. W. CROMPTON

Electron and Ion Diffusion Unit
The Australian National University
Canberra, Australia

I. Introduction

The measurement of the transport properties of electron swarms drifting and diffusing through gases and the interpretation of the experimental results in terms of fundamental collision processes began with the pioneering work of J. S. Townsend in the early years of this century. Two factors are primarily responsible for the increased activity in this field in recent years. The first is the need for quantitative data at very low energies for electron-loss processes, such as attachment and recombination. These processes are important in many branches of science ranging from biology to astrophysics. The second factor is the success recently achieved in analyzing the data from swarm experiments to derive detailed information about elastic and inelastic collision processes at low energies. Two recent reviews by Phelps (*1*) have covered many aspects of the application of swarm measurements to these problems. It is primarily the purpose of this paper to review some experimental methods that have been developed to make precise measurements of electron transport coefficients and to show how the results of such experiments can be used in special cases to obtain accurate cross sections for very low energy electrons.

The experimental data for the energy dependence of electron-neutral cross sections below about 1 eV are scarce and often conflicting. While the accurate measurement of these cross sections by an electron beam method of adequate energy resolution is in principle the most desirable goal, the difficulties associated with measurements of this kind at very low energies have not yet been solved (2). On the other hand, there has been some reluctance to accept the data deduced from swarm measurements, partly because of the complexity of the analysis involved and partly because the full capability of these methods has not often been appreciated or applied.

The fundamental principles and characteristics of swarm measurements and their analysis will first be outlined in order to show where the method would be expected to make significant contributions. The question then discussed is whether the reliability of the experimental techniques themselves is adequate or whether there are still good reasons for mistrusting the experimental data in common with the results of some other methods of investigating very low energy collisions. Finally, a brief survey will be given of the areas in which successes have been achieved by this technique, together with a more detailed account of two specific applications which will serve to illustrate these points.

II. A Comparison of Beam and Swarm Experiments

In swarm measurements of the type that is the subject of this paper, measurements are first made of one or more properties that characterize the motion of electrons drifting and diffusing through a gas at relatively high number density, after which the data are related to the collision processes between the electrons and neutrals. At first sight the method appears to be an unattractive alternative to a suitably designed beam experiment, since an electron beam of well-defined energy is replaced by a swarm in which the energy distribution is at best never much narrower than a Maxwellian distribution. This disadvantage of swarm techniques is fundamental, and it generally precludes the application of the method to the examination of fine structure in energy-dependent cross sections, particularly at higher energies. Nevertheless there are some features of swarm methods which, for certain applications, give them a particular advantage at all energies.

Although the techniques have found some important applications at higher energies, their unique contributions have been made in the energy range below a few tenths of an electron volt. Here many factors combine to make swarm measurements easier to perform, whereas the reverse is true of beam experiments. Figure 1 shows the energy distribution of an electron swarm whose most probable energy is about 0.010 eV. The curve is the calculated energy distribution for an electron swarm in parahydrogen at 77°K, with the experimental conditions chosen to ensure that the swarm is almost in thermal equilibrium with the gas. Since the width of the distribution at "half-maximum"

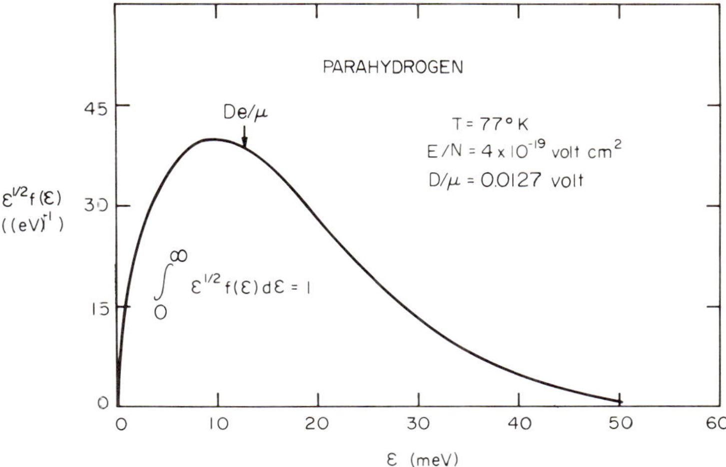

FIG. 1. The energy distribution of an electron swarm of low mean energy.

is a few tens of millielectron volts, it can be seen that the energy spread of the swarm is not much worse than that achieved in many beam experiments. It may be added that many swarm measurements have been made at energies as low as 0.01 to 0.02 eV but that, so far, energies as low as this have been outside the range of beam experiments.

The difficulties associated with extremely low energy beam experiments are well known. Not only are there difficulties in establishing an absolute energy scale for energies of a fraction of an electron volt, but other difficulties arise from the influence of small, stray electric and magnetic fields and from the fact that beam intensities are usually low at these energies. In addition, the requirement of low gas target number densities in the collision chamber (necessitating pressures of the order of 0.1 to 10 mT) or the use of a crossed molecular beam makes difficult the determination of absolute cross sections. Some normalizing procedure is therefore often necessary.

The procedure for determining cross sections through swarm measurements differs in almost every detail from that used in beam experiments. For this reason the method is not faced with the difficulties already enumerated for beam experiments although there are, of course, other difficulties; the advantages and disadvantages of each method are such, in fact, that accurate knowledge of low energy cross sections can often be obtained only by combining the results obtained from both types of measurements. The first essential difference between the two methods is that the field-free collision chamber of a beam experiment is replaced by one in which a relatively strong electric field is established, the field strength being typically of some tens of volts per centimeter. The second main difference is that the number density of the target gas is high so that, typically, each electron makes of the order of 10^6 collisions in crossing the collision chamber rather than a maximum of one collision. The energy of the electrons prior to entering the collision chamber, and within it, is now controlled by the balance established between the power fed into the swarm by the electric field and the rate at which energy is lost through collision with the gas molecules, rather than by an energy selection technique. The controlling factors are therefore the gas temperature T, the nature of the gas, and the ratio E/N of the electric field strength to the gas number density. The lower energy limit of the experiments is thus set by the gas temperature, while the upper limit is of the order of 10 eV.

The experiments consist of the measurement of a number of transport coefficients as the ratio E/N (and hence the energy distribution of the electrons in the swarm) is varied. Those of chief significance for the determination of cross sections in the region below electronic excitation and ionization are the drift velocity W and the diffusion coefficient D. The drift velocity is measured in effect by pulsing the electron swarm at the entry to the collision chamber and subsequently measuring the arrival time of the electrons at a plane a known distance from the pulsed source. A well-known technique for measurements of this kind, and one that is capable of high precision, is the Bradbury-Nielsen method (3) in which one electrical shutter produces the pulses while a second, a known distance away from the first, samples the arrival times. Although methods have recently been devised for measuring the diffusion coefficient directly (4) the method more often used has depended on the measurement of the ratio W/D by determining the distribution of current over an electrode a known distance away from the source. In the Townsend-Huxley method (5), the source (at the entry of the collision chamber) consists of a small hole that acts as a point source, while the anode terminating the chamber is divided to form a central disk and surrounding annulus. From the geometry of the apparatus and the measured ratio of the currents received by the insulated segments of the anode, the ratio W/D can be determined, from which D can be calculated by combining these results with those for the drift velocity.

The final procedure in the cross-section determination is the analysis of the experimental results to find a set of energy-dependent cross sections that are consistent with these data. This analysis will be briefly described subsequently.

In this summary of the swarm method for studying low energy electrons, two important points should be noted:

1. The low energy electrons are controlled without the need to resort to excessively weak electric fields. Consequently the effects of stray electric and magnetic fields are greatly reduced and the accuracy of the experiments remains good even at ultralow energies. This is the reason why swarm techniques acquire an increasing advantage over beam techniques as zero energy is approached.

2. The experiments are carried out under static conditions at high gas pressures—typically from a few torr to an atmosphere or more. As a consequence the calculation of the target gas number density presents no problem so that absolute cross section determinations can be made without difficulty. Swarm measurements are therefore of great value in certain circumstances in the normalization of cross sections determined by other means.

III. Analysis of the Results of Swarm Experiments

From what has been said it can be seen that swarm techniques can lead to a precise control of low energy electrons, but the problem of interpreting the measured transport data in terms of fundamental collision processes remains to be discussed. The difficulty arises not only from the fact that energy-dependent cross sections are being examined by a probe of poor energy resolution, but also from the fact that the peak energy of the probe and its energy distribution are themselves determined by the interaction of the swarm and the target gas, and therefore have to be calculated for a given set of experimental conditions.

As is well known, the effect of an electric field applied to an electron swarm in a gas at high number density is to establish a new steady-state energy distribution in which the mean energy may lie anywhere between that corresponding to the gas temperature and a value many hundred times that value. In addition to raising the average electron energy, the electric field causes the velocity distribution to be no longer spherically symmetrical in velocity space. As a consequence there is a resultant drift motion, which is related to the momentum transfer cross section q_m and the energy distribution function $f(\varepsilon)$ through the equation

$$W = \frac{-eE(2/m)^{1/2}}{3N} \int_0^\infty \frac{\varepsilon}{q_m(\varepsilon)} \frac{df}{d\varepsilon} \, d\varepsilon \tag{1}$$

where e and m are the electronic charge and mass, ε is the electron energy, and the distribution function $f(\varepsilon)$ is normalized through $\int_0^\infty \varepsilon^{1/2} f(\varepsilon) = 1$.

The diffusive motion of the swarm is described quantitatively through the diffusion coefficient D for which the formula analogous to Eq. (1) is

$$D = \frac{(2/m)^{1/2}}{3N} \int_0^\infty \frac{\varepsilon f(\varepsilon)}{q_m(\varepsilon)} \, d\varepsilon \tag{2}$$

It should be pointed out that these formulas are developed for the case of elastic collisions only and that they are therefore, strictly speaking, applicable only to that situation. However, their application is in fact a good deal wider than this. In many cases the energy distribution $f(\varepsilon)$ is largely controlled by one or more inelastic processes even though the combined inelastic collision frequency is very much less than the elastic collision frequency. In these situations, of which there are many examples, the use of Eqs. (1) and (2) is still justified. In less clear-cut situations where it is no longer true that elastic collisions greatly outnumber inelastic collisions, the problem requires further investigation. Cavalleri and Sesta (6) have recently examined this problem analytically, but their work so far is incomplete. An alternative approach to the problem would be through the application of Monte Carlo techniques, as in the work of Itoh and Musha (7) and of Bell and Kostin (8).

In addition, we note that the ratio W/D, the quantity that can be measured directly in the Townsend-Huxley experiment, can be expressed as a ratio of the integrals in Eqs. (1) and (2) in a way that shows the physical significance of this transport coefficient ratio. If the reciprocal of the ratio is written in terms of the electron mobility $\mu = W/E$, it becomes

$$D/\mu = -e \int_0^\infty \frac{\varepsilon f(\varepsilon)\, d\varepsilon}{q_m(\varepsilon)} \Big/ \int_0^\infty \frac{\varepsilon}{q_m(\varepsilon)} \frac{df}{d\varepsilon}\, d\varepsilon$$
$$= e F \bar{\varepsilon} \tag{3}$$

where $F \cong 1$. Thus the measurement of D/μ (i.e., of W/D) gives an approximate measure of the mean energy of the swarm.

In order to evaluate the transport integrals of Eqs. (1) and (2), it is necessary to determine the energy distribution function $f(\varepsilon)$. In earlier work, estimates of the function were made by first assuming a functional form for the distribution in order to calculate F (9). The function $f(\varepsilon)$ was then calculated, using Eq. (3) and the measured values of D/μ. The energy dependence of the momentum transfer cross section was subsequently determined in effect by using Eq. (1) with the now known $f(\varepsilon)$, although an approximate treatment had to be used because of the difficulty of carrying out the integration unless $q_m(\varepsilon)$ was assumed to be of the form $q_m(\varepsilon) = q_0 \varepsilon^n$. This procedure was obviously limited, since it enabled only approximate estimates to be made of the energy dependence of the momentum transfer cross section, while practically no information could be obtained about inelastic cross sections.

Further progress could not be made until it was possible to apply high speed numerical methods to the solution of the Boltzmann equation and to the evaluation of the transport integrals. In a form appropriate to the conditions of the experiments described in this paper, the Boltzmann equation reads (10)

$$\frac{E^2 e^2}{3} \frac{d}{d\varepsilon}\left(\frac{\varepsilon}{N q_m(\varepsilon)} \frac{df}{d\varepsilon}\right) + \frac{2m}{M} \frac{d}{d\varepsilon}(\varepsilon^2 N q_m(\varepsilon) f) + \frac{2mkT}{M} \frac{d}{d\varepsilon}\left(\varepsilon^2 N q_m(\varepsilon) \frac{df}{d\varepsilon}\right)$$
$$+ \sum_j [(\varepsilon + \varepsilon_j) f(\varepsilon + \varepsilon_j) N q_j(\varepsilon + \varepsilon_j) - \varepsilon f(\varepsilon) N q_j(\varepsilon)]$$
$$+ \sum_j [(\varepsilon - \varepsilon_j) f(\varepsilon - \varepsilon_j) N q_{-j}(\varepsilon - \varepsilon_j) - \varepsilon f(\varepsilon) N q_{-j}(\varepsilon)] = 0 \tag{4}$$

where M is the molecular mass of the gas, k is Boltzmann's constant, $q_j(\varepsilon)$ is the cross section for the jth inelastic process for which the threshold is ε_j, and $q_{-j}(\varepsilon)$ is the cross section for collisions of the second kind of the jth process in which an electron receives energy ε_j from the molecule. The numerical solution of Eq. (4) was first successfully attempted by Frost and Phelps (*10*). Since the effects of both elastic and inelastic collisions are properly accounted for in calculating $f(\varepsilon)$ by the procedure developed by these authors, it is possible to derive in some detail the energy dependence of both elastic and inelastic cross sections. The technique now used, therefore, is to determine the set of cross sections that gives satisfactory agreement between calculated and measured values of the transport coefficients over the full range of values of E/N covered by the measurements. The cross sections are first used to calculate the energy distribution functions corresponding to the set of values of E/N, using Eq. (4), after which the transport integrals of Eqs. (1), (2), and (3) are evaluated by numerical integration. A comparison between calculated and measured values of the coefficients is then made, using all the experimental data, after which adjustments are made to the cross sections until the discrepancies lie within experimental error.

This brief summary has been given to show what can in principle be achieved with swarm techniques by employing sufficiently detailed analysis. It is now necessary to determine whether the experimental data themselves warrant the refinement of the analysis that is currently applied.

IV. THE ACCURACY OF TRANSPORT COEFFICIENT MEASUREMENTS

In applying the analysis that has just been described, the final accuracy and uniqueness of the data depend critically on the accuracy of the primary experimental data. Recent years have seen a marked improvement in the accuracy of many of these sets of data. Apart from the general improvement in experimental techniques that this field has shared with others, there has recently been an increasing understanding of factors that degrade the accuracy of the measurements apart from poor experimental techniques. The investigation of these factors arose from the experimentally observed dependence of the transport coefficients on experimental parameters, notably the geometry of the drift tube, or diffusion apparatus, and the number density of the gas. Since the steady state energy distribution is expected to be a function of E/N only, values of W and D/μ measured at the same E/N but at different N should be the same. While this was found to be generally true, significant dependence on the number density had been observed (*11*). Some dependence on the geometry of the apparatus (for example, the length of the drift tube) had also been reported (*11*). Where such departures were observed, however, the transport coefficients were usually found to approach asymptotic values as either the geometry of the apparatus was suitably varied (e.g., an increase in length of the drift tube) or the gas pressure increased.

These results initiated a series of theoretical investigations (*12*) into the interpretation of experiments of this kind, taking account of the effects of longitudinal density gradients, (for example, gradients introduced by the boundary condition $n = 0$ at the electrodes). In this work, which was applied both to lateral diffusion experiments of the Townsend-Huxley type and to time-of-flight experiments, it was assumed that $f(\varepsilon)$ was spatially independent. With this assumption, both the steady state and the time-dependent continuity equations were solved, allowing for the appropriate boundary conditions.

At about the same time an examination was made of the problem of the spatial variation of $f(\varepsilon)$ as a result of density gradients in the diffusing stream of the Townsend-Huxley experiment (*13*). Very recently a similar analysis has been made of the effect of spatial variation of $f(\varepsilon)$ caused by the density gradients in the drifting and diffusing pulses of time-of-flight experiments (*14*). This work not only accounts for most of the pressure dependence observed in drift velocity measurements, but also provides a satisfactory explanation of the longitudinal diffusion coefficient measurements first reported by Hurst and Parks (*15*) and by Wagner *et al.* (*15*). Using a time-of-flight technique, these workers found that the longitudinal spread of the electron pulses was very much less than that calculated by using the lateral diffusion coefficient. Their results in argon, for example, required a diffusion coefficient less than 20% of the accepted value from lateral diffusion experiments. These results have now been satisfactorily explained.

The situation can be summarized by saying that analyses of diffusive effects and of the spatial variation of the electron energy distribution have satisfactorily accounted for the experimentally observed asymptotic behavior of the measured transport coefficient as the experimental parameters are varied. The results of this work can be used to design experiments in which errors from these effects can be reduced to any desired level. The correct design of experiments (*3, 5*), coupled with modern techniques for the precise measurement of voltage, frequency, pressure, and low level currents, has resulted in data for W and D/μ for a large number of gases over an extended range of E/N for which error limits of $\pm 1\%$ can be claimed. This point is illustrated in Fig. 2, which shows results for W and D/μ in helium at two temperatures. The data are plotted as functions of E/N expressed in terms of a unit named the Townsend (*16*), 1 Townsend (Td) being defined as 10^{-17} volt cm^2. The results are taken at many values of E/N, using a large range of pressures, but the data points are not shown because it is not possible to record the scatter on a scale as small as that used in the figure. The low temperature drift measurements may be noted particularly. Measurements were taken at the values of E/N represented by the points and with the number of pressures shown.

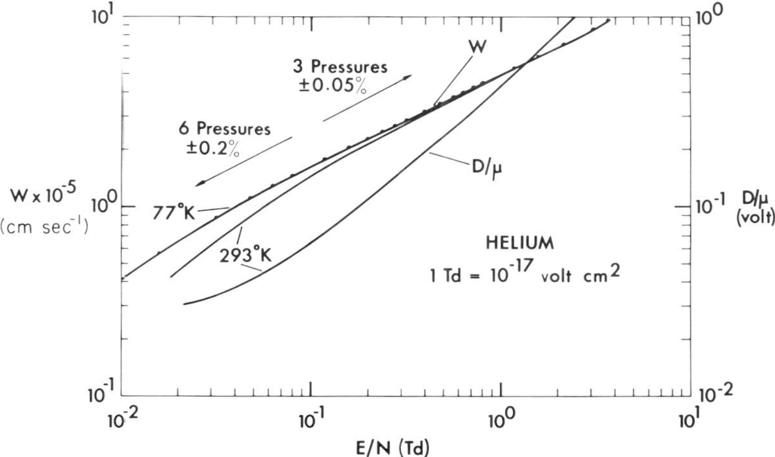

FIG. 2. Typical results of transport coefficient measurements; experimentally determined values of W and D/μ in helium at 77 and 293°K.

Below an E/N of about 10^{-1} Td, at least six pressures were used, and all values lay within $\pm 0.2\%$ of the line of best fit. Above this value, only three pressures were used, but the agreement was within $\pm 0.05\%$. The claimed accuracy of the experimental results follows from the small scatter in the data and the accuracy with which voltage, frequency, and pressure can be determined over the ranges used in these experiments.

V. Low Energy Collision Cross Sections from Swarm Measurements

The application of the analysis described in Section III to transport coefficient measurements of adequate accuracy provides a method of determining elastic and inelastic cross section that are difficult to determine by other methods. The remainder of this paper is devoted to a discussion of some examples of this method.

A. Elastic Scattering; the Momentum Transfer Cross Section in Helium

When swarm experiments are performed in a monatomic gas under conditions that ensure that a negligible fraction of the electrons cause excitation or ionization, a considerable simplification of Eq. (4) results. The solution of the equation without the inelastic terms is

$$f(\varepsilon) = A \exp\left[- \int_0^\varepsilon \left(\frac{ME^2 e^2}{6mN^2 q_m^{\,2}(\varepsilon)\varepsilon} + kT \right)^{-1} d\varepsilon \right]$$

where the constant A is found from the normalizing equation $\int_0^\infty \varepsilon^{1/2} f(\varepsilon)\, d\varepsilon = 1$.

For a given value of E/N and T, therefore, $f(\varepsilon)$ and hence the transport integrals are specified by specifying the energy dependence of q_m. Furthermore, an examination of the energy distribution for any particular value of E/N shows that only a relatively small section of the cross section curve plays a significant part in determining $f(\varepsilon)$ and hence the transport coefficients at that value of E/N. It is therefore possible to examine the whole curve section by section by displacing the energy distribution along the energy axis and comparing calculated and experimental values of one transport coefficient. In practice this is done by comparing the calculated and experimental curves for the transport coefficient plotted as a function of E/N. The accuracy and u-niqueness of the cross section determined in this way is clearly dependent on the accuracy of the initial data.

Figure 3 shows the results of applying this technique to the determination

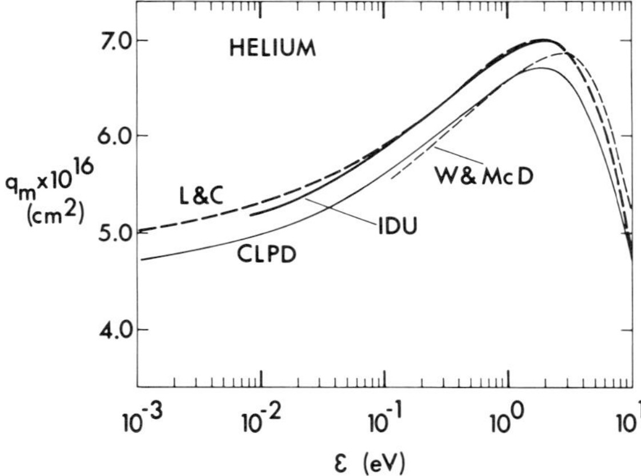

FIG. 3. Experimental and theoretical determination of the momentum transfer cross section for low energy electrons in helium: IDU, experimental determination from drift velocity data; W and McD, Williamson and McDowell (17); L & C, La Bahn and Callaway (18); CLPD, Callaway et al. (19).

of the momentum transfer cross section in helium (3) from the drift velocity results shown in Fig. 2. The cross section that has been derived leads to calculated values of W at the two temperatures that agree with the experimental results to within 0.5%. An independent check on the accuracy of the cross section is afforded by comparing the values of D/μ calculated by using this cross section with the experimental values shown in Fig. 2. Again the agreement is to within 1%, the accuracy claimed for these measurements. An analysis of the uniqueness of the cross section has been made based on the scatter in the drift data and its overall accuracy. The analysis showed that an error limit of $\pm 2\%$ can be placed on the cross section from 0.009 to 3 eV.

Despite the apparent indirectness of this method of obtaining an elastic scattering cross section, it may be claimed that, first, the accuracy compares more than favorably with that achieved by beam techniques at any energy, and that, secondly, this method provides the only data currently available below 0.1 eV. Moreover, in this energy range there is no reduction in the experimental accuracy; in fact, for energies down to those approaching the lower limit, the accuracy tends to increase rather than decrease in this case. These claims arise directly from the points discussed earlier, namely, the use of high number densities for the target gas and the ability to control low energy electrons without the need to resort to correspondingly small accelerating potentials. For example, the lowest field used in the low temperature helium measurements was 20 volt/cm.

The results of recent theoretical calculations of the cross section in helium are also shown in Fig. 3 in order to compare them with the experimental results. The theoretical curves shown are those of Williamson and McDowell (*17*) who used an adiabatic approximation, La Bahn and Callaway (*18*) who used the so-called dynamic exchange approximation, and the latest work of Callaway *et al.* (*19*) who derived an extended polarization potential. It can be seen that there is now a satisfactory convergence of experiment and theory. Nevertheless the remaining discrepancies appear to be significant, owing to the accuracy that can be claimed for the experimentally determined cross section.

B. *Elastic and Inelastic Scattering in Molecular Gases*

The analysis of transport data for low energy electron swarms in the majority of molecular gases represents the opposite extreme to the case just considered. In addition to elastic energy losses there are now, typically, many inelastic processes through which the swarm interacts with the gas. Figure 4 (*20*) illustrates this point for the case of nitrogen. Here the fraction of the total power supplied by the field to the swarm that is dissipated either in elastic collisions or in one of a number of inelastic processes is plotted as a function of E/N. It can be seen that at any value of E/N, several inelastic processes must usually be considered although the relative importance of each process changes as E/N, and hence the energy distribution of the swarm, is varied. Since data are available for a limited number of transport coefficients only (usually no more than two), it is clear that unique cross sections cannot in general be unfolded by the procedure that has been described. In this situation, where there are competing inelastic processes, swarm methods can still provide useful results.

First, if all but one of the inelastic cross sections are known accurately, the remaining inelastic cross section can be determined. The region in which the

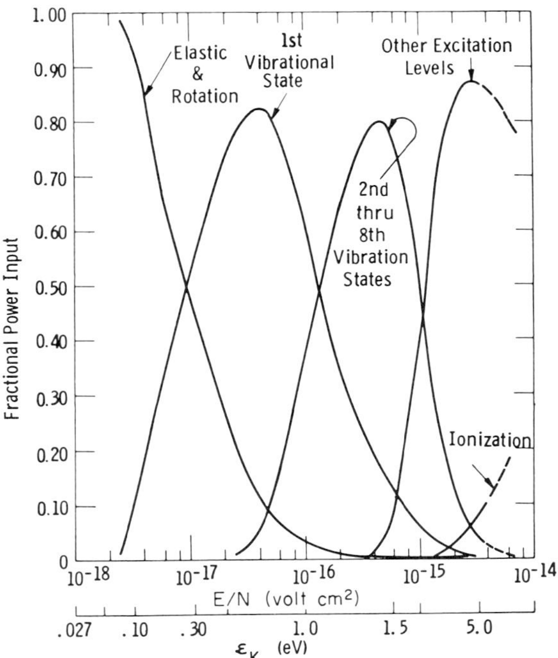

FIG. 4. Showing the distribution between the elastic and inelastic collision processes of the total power input into an electron swarm in nitrogen [from A. G. Engelhardt, A. V. Phelps, and C. G. Risk, *Phys. Rev.* **135**, A1566 (1964)].

data from this procedure are most precise is near the threshold of the process, since an examination of the energy dependence of the cross section well above threshold rests on an analysis of data for electron swarms for which the energy losses from an inelastic process of higher threshold are likely to predominate.

Secondly, if the only significant inelastic losses are due to one type of excitation process (e.g., rotational excitation), swarm methods can be used either to normalize the amplitudes of the cross sections measured by a beam experiment or, alternatively, if the cross sections have been calculated from theory, both the theory and the physical parameters used in the theory (for example, electric dipole and quadrupole moments) can be checked by the measure of agreement that is found with the results of swarm experiments. The important contributions that have been made to the body of data on molecular rotational and vibrational excitation as a result of the application of this method have been reviewed by Phelps (*1*) in a recent paper.

Finally it is to be noted that the momentum transfer cross section can be obtained with reasonable accuracy at all energies, provided the inelastic collision frequency is small compared with that for elastic scattering. Again the most valuable contributions from swarm measurements are those in the energy range below about 1 eV.

C. Elastic Scattering and Rotational Excitation in Hydrogen

While it is generally true that it is difficult to derive unique inelastic cross section data from the analysis of swarm data, for the reasons outlined above, it has been found possible to determine the cross section for the $J = 0 \to 2$ rotational transition in hydrogen (*21*), for which there are also several theoretical calculations (*22*).

In normal hydrogen at room temperature (293°K) the populations of the rotational levels are as follows:

J	0	1	2	3	4
population (%)	13.5	67	11.2	7.9	0.3

The thresholds of the four rotational cross sections of significance are 0.044, 0.073, 0.101, and 0.128 eV, while the threshold of the vibrational excitation cross section, $v = 0 \to 1$, is about 0.52 eV. To analyze transport data for electrons in hydrogen at this temperature, it is therefore necessary to take account of each of these processes and to include the effect of collisions of the second kind with molecules in the $J = 2$ and $J = 3$ rotational states, each of which has an appreciable population.

The situation becomes simpler to analyze if the experiments are carried out at low temperatures. At 77°K the populations become:

J	0	1	2
population (%)	24.87	75.00	0.13

and because the $J = 2$ state is now so thinly populated, collisions of the second kind can be neglected for all cases except those in which the swarm is almost in thermal equilibrium with the gas, that is, for very low values of E/N. The neglect of collisions of the second kind greatly simplifies the analysis (*10*). An analysis of the transport data for normal hydrogen at this temperature has been carried out by Engelhardt and Phelps (*23*).

A still further simplification of the problem can be achieved by analyzing the data for parahydrogen at 77°K (*24*). In this case 99.5% of the molecules occupy the $J = 0$ state, and 0.5% the $J = 2$ state, while the odd states are, of course, unpopulated. Thus, at sufficiently low values of E/N, only elastic collisions and those exciting the $J = 0 \to 2$ transition need to be considered; it is to be expected that the transport data could be analyzed to get unique data for the energy dependence of these processes without resorting to the

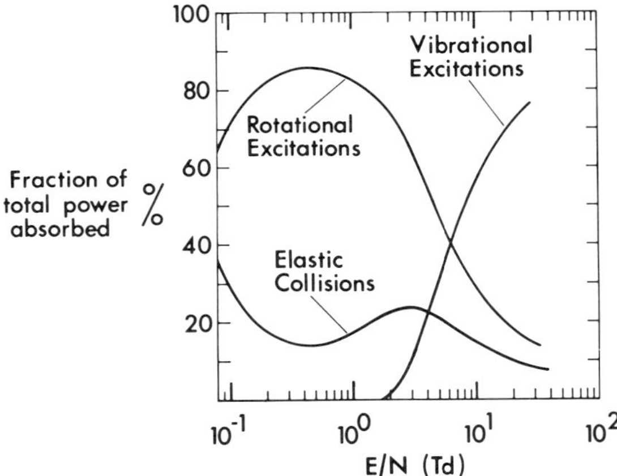

FIG. 5. The distribution of the total power input into an electron swarm in parahydrogen between elastic, rotational, and vibrational collision processes.

results of other experiments or theory. From Fig. 5 it is possible to determine the range over which the analysis is expected to yield unique results. Below $E/N \sim 2$ Td it can be seen that a negligible fraction of the total energy loss goes toward exciting vibrational transitions. In this region, therefore, the energy distribution of the swarm and hence the transport coefficients are controlled almost entirely by elastic collisions and the single rotational excitation process, and the analysis should therefore lead to unique cross sections for these processes.

Figure 6 shows the results of applying the technique developed by Frost and Phelps (*10*) to the analysis of the transport data obtained in parahydrogen (*25*). The momentum transfer cross section can be obtained with the same accuracy over the whole energy range and is therefore indicated everywhere as a full line. An examination of the uniqueness of the rotational cross section showed that it could be determined to $\pm 5\%$ up to about 0.3 eV, that is, just below the threshold of vibrational excitation, as is to be expected. Above this limit the cross section, shown as a broken line, was based on a reasonable extrapolation of Henry and Lane's theoretical cross section (*22*). The vibrational cross section is shown everywhere as a broken line, since it is nowhere possible to determine it uniquely. With the rotational cross section determined in the way described above, the vibrational cross section is adjusted to give results that are in agreement with the measured values of the transport coefficients. In this way agreement between calculated and measured values to within 1% has been achieved.

Although the vibrational cross section cannot be uniquely determined, the degree of its uniqueness is much higher than might be expected at first sight.

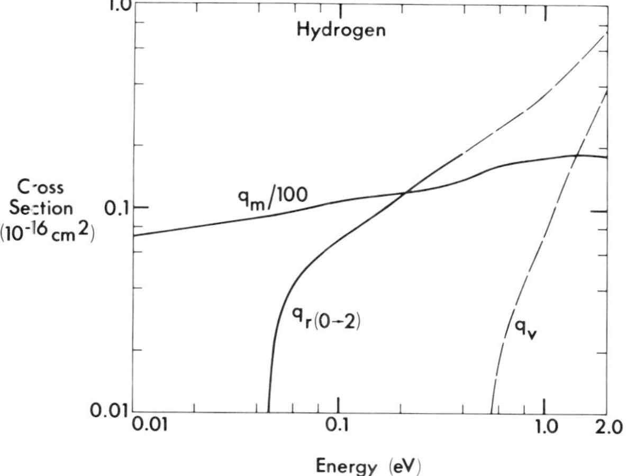

FIG. 6. Elastic and inelastic collision cross sections derived from the analysis of transport coefficient measurements in parahydrogen.

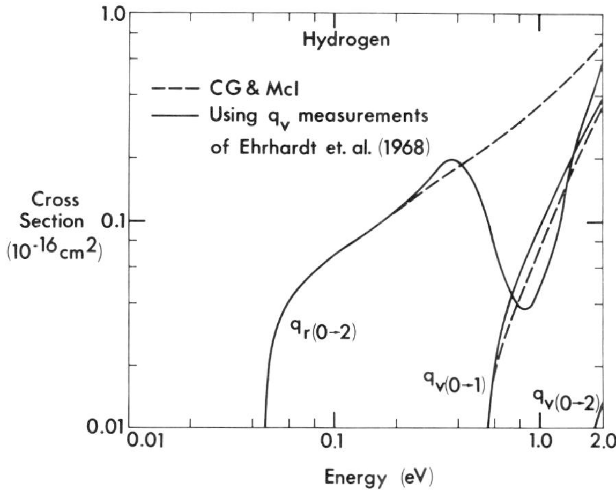

FIG. 7. Redetermination of the high energy portion of the rotational excitation cross section using Ehrhardt *et al.* (*26*) vibrational excitation cross sections: CG & McI, Crompton *et al.* (*21*).

This point is illustrated in Fig. 7 in which an attempt has been made to fit the transport data, using the vibrational excitation cross sections recently published by Ehrhardt *et al.* (*26*). Their curves are shown as full lines. Although there is generally good agreement with the result of Crompton *et al.* (*21*), which is shown dashed in this figure, it is not possible to fit the swarm data to the accuracy warranted by the experimental accuracy without making the some-what unrealistic adjustment to the rotational cross section shown by the full line. This is because at higher values of E/N the contribution to the energy losses from rotational excitation is much less than from vibrational excitation so that large adjustments to the rotational cross section are required to compensate for small adjustments to the vibrational cross section. It should also be noted that below 0.3 eV the rotational cross section is essentially unaffected, thus substantiating the claims for uniqueness made earlier. The results of this analysis suggest that the near-threshold behavior of the vibrational excitation cross section is perhaps best investigated by the swarm technique. However, it should be emphasized that the results obtained in this way become increasingly inaccurate as the energy is increased well beyond threshold, while the reverse is true of the results from a beam experiment. This illustrates the complementary nature of the two experimental methods and suggests that considerably more progress can be made by synthesizing the results from both (27).

Since the main aim of this investigation was the determination of the rotational excitation cross section, the region of the curve in which a unique result was obtained is shown in more detail in Fig. 8 together with other

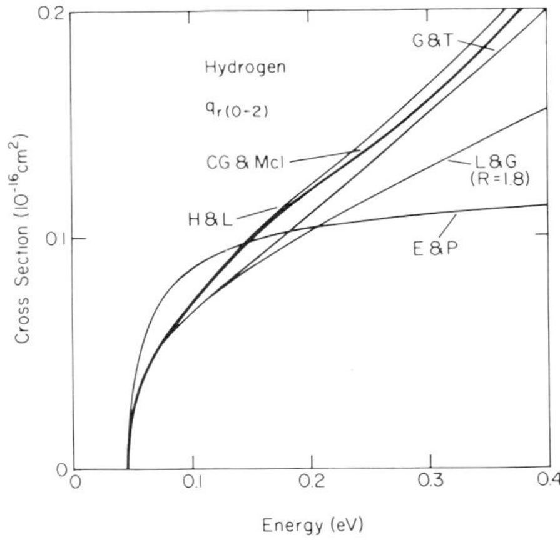

Fig. 8. A comparison of experimental and theoretical $J = 0 \rightarrow 2$ rotational cross sections in hydrogen: E & P, Engelhardt and Phelps (23); CG & McI, Crompton et al. (21); G & T, Geltman and Takayanagi (22); L & G, Lane and Geltman (22); H & L, Henry and Lane (22).

recent experimental and theoretical determinations. Engelhardt and Phelp's cross section (23), obtained from an analysis of the data in normal hydrogen at 77°K, is based on Gerjuoy and Stein's theory of rotational excitation as extended by Dalgarno and Moffett (22). In order to get an acceptable fit between calculated and measured transport coefficients, Engelhardt and Phelps found it necessary to multiply the theoretical cross section by a factor of approximately 1.5 which, as they pointed out, implies an error of about 25 % in the accepted value of the quadrupole moment. On the other hand, the results of the analysis of the parahydrogen data indicate a rather different form of the energy dependence and one that cannot be fitted by any simple adjustment to the Dalgarno and Moffett cross section. The difference between these two results from swarm measurements is almost certainly due to some discrepancy in the experimental results rather than to the additional complexity of the analysis required in the case of normal hydrogen.

The comparison between the latest experimental result and the recent theoretical work (22) of Geltman and Takayanagi, Lane and Geltman, and Henry and Lane provides a further example of the convergence between theory and experiment which was noted earlier for helium. The most complete of these analyses is that of Henry and Lane, who modified the earlier close-coupling calculation of Lane and Geltman to include both polarization and exchange. The discrepancies that remain between theory and experiment in this range are scarcely significant.

VI. Conclusion

The application of high speed numerical methods to the problem of unfolding elastic and inelastic cross sections from swarm measurements has removed a serious obstacle which has previously limited the usefulness of measurements of this kind. While the nature of swarm techniques is such that there are fundamental limitations to their applicability, there are situations in which cross sections can be obtained in this way with an accuracy that is difficult to match by other methods. Moreover, the techniques are particularly valuable at energies below a few tenths of an electron volt, where the data from them are currently the only data available.

The analytical methods now applied are sufficiently precise that the limitation to the accuracy of the cross sections is frequently the accuracy of the transport measurements themselves. The realization of this fact has led to a critical examination of the experimental methods and their interpretation. As a consequence there has been a more exact formulation of the macroscopic behavior of electron swarms in gases under the conditions of the experiments, and this work has led to the design of experiments from which adequately precise transport data have been obtained. The analysis of these results has produced cross section data for very low energy electrons which are sufficiently precise to serve as a critical test of current theories of low energy electron scattering by simple atoms and molecules.

REFERENCES

1. A. V. Phelps, *Rev. Mod. Phys.* **40**, 399 (1968); *Can. J. Chem.* **47**, 1783 (1969).
2. R. H. Bullis, T. L. Churchill, and W. J. Wiegand, *Bull. Am. Phys. Soc.* **14**, 251 (1969).
3. J. J. Lowke, *Australian J. Phys.* **16**, 115 (1963); M. T. Elford, *ibid.* **19**, 629 (1966); R. W. Crompton, M. T. Elford, and R. L. Jory, *ibid.* **20**, 369 (1967).
4. G. Cavalleri, *Phys. Rev.* **179**, 186 (1969).
5. R. W. Crompton, and R. L. Jory, *Australian J. Phys.* **15**, 451 (1962); R. W. Crompton, M. T. Elford, and J. Gascoigne, *ibid.* **18**, 409 (1965); R. W. Crompton, M. T. Elford, and R. L. Jory, *ibid.* **20**, 369 (1967).
6. G. Cavalleri and G. Sesta, *Phys. Rev.* **170**, 286 (1968).
7. T. Itoh and T. Musha, *J. Phys. Soc. Japan* **15**, 1675 (1960).
8. M. J. Bell and M. D. Kostin, *Phys. Rev.* **169**, 150 (1968).
9. L. G. H. Huxley and R. W. Crompton in "Atomic and Molecular Processes" (D. R. Bates, ed), p. 335. Academic Press, New York, 1962.
10. L. S. Frost and A. V. Phelps, *Phys. Rev.* **127**, 1621 (1962).
11. J. J. Lowke, *Australian J. Phys.* **15**, 39 (1962); R. W. Crompton and R. L. Jory, *ibid.* **15**, 451 (1962).
12. See, for example, the following papers and references therein: L. G. H. Huxley, *Australian J. Phys.* **12**, 171 (1959); J. J. Lowke, *ibid.* **15**, 39 (1962); J. Lucas, *J. Electron Control* **17**, 43 (1964); C. A. Hurst and B. S. Liley, *Australian J. Phys.* **18**, 521 (1965). D. S. Burch (personal communication, 1968).
13. J. H. Parker, *Phys. Rev.* **132**, 2096 (1963).
14. J. H. Parker and J. J. Lowke, *Phys. Rev.* **181**, 290 (1969); J. J. Lowke and J. H. Parker, **181**, 302 (1969); H. R. Skullerud, *J. Phys. B* **2**, 696 (1969).
15. G. S. Hurst and J. E. Parks, *J. Chem. Phys.* **45**, 282 (1966); E. B. Wagner, F. J. Davis, and G. S. Hurst, *ibid.* **47**, 3138 (1967).
16. L. G. H. Huxley, R. W. Crompton, and M. T. Elford, *Brit. J. Appl. Phys.* **17**, 1237 (1966).
17. J. H. Williamson and M. R. C. McDowell, *Proc. Phys. Soc.* **85**, 719 (1965).
18. R. W. La Bahn and J. Callaway, *Phys. Rev.* **147**, 28 (1966).
19. J. Callaway, R. W. La Bahn, R. T. Pu, and W. M. Duxler, *Phys. Rev.* **168**, 12 (1968).
20. A. G. Engelhardt, A. V. Phelps, and C. G. Risk, *Phys. Rev.* **135**, A1566 (1964).
21. R. W. Crompton, D. K. Gibson, and A. I. McIntosh, *Australian J. Phys.* (to be published 1969).
22. See, for example: E. Gerjuoy and S. Stein, *Phys. Rev.* **97**, 1671 (1955); A. Dalgarno and R. J. Moffett, *Proc. Natl. Acad. Sci. India* **A33**, 511 (1963); S. Geltman and K. Takayanagi (*Phys. Rev.* **143**, 25 (1966); N. F. Lane and S. Geltman, *ibid.* **160**, 53 (1967); R. J. W. Henry and N. F. Lane, *Bull. Am. Phys. Soc.* **14**, 255 (1969).
23. A. G. Engelhardt and A. V. Phelps, *Phys. Rev.* **131**, 2115 (1963).
24. R. W. Crompton and A. I. McIntosh, *Phys. Rev. Letters* **18**, 527 (1967).
25. R. W. Crompton and A. I. McIntosh, *Australian J. Phys.* **21**, 637 (1968).
26. H. Ehrhardt, L. Langhans, F. Linder, and H. S. Taylor, *Phys. Rev.* **173**, 222 (1968).
27. See also L. G. Christophorou, R. N. Compton, G. S. Hurst, and P. W. Reinhardt, *J. Chem. Phys.* **43**, 4273 (1965).

Magnetic Coherence Resonances and Transitions at Zero Frequency

E. GENEUX*

G. J. BÉNÉ

AND

J. PERRENOUD

*Institute of Experimental Physics, University of Geneva
Geneva, Switzerland*

I. Introduction

In this article, we have summarized theoretical bases, fundamental experiments and possible applications of two phenomena that seem to be unrelated: coherence magnetic resonance and zero frequency transitions.

Both phenomena, recently discovered, are further development of the method of optical detection of magnetic transitions between Zeeman sublevels. We wish to point out that results in this area of research, and the most important ones, were first obtained by Alfred Kastler, Nobel Prize 1966. Numerous publications concerning this field have appeared, so details of the basic procedure will not be repeated here.

We shall present a more general point of view by showing that these phenemona have analogs in magnetic resonances by condensed matter. Examples will be given especially in nuclear magnetic resonance (NMR) by nonparamagnetic liquids. Use of such liquids will permit a simpler interpretation of these phenomena. First of all, let us remark that well-known NMR phenomena discovered before these two developments can be simply interpreted in terms of the formalism utilized in atomic physics. So Minneman (4)

---

* *Temporary Address:* Lawrence Radiation Laboratory, University of California, Berkeley, California.

and Hanuise (*2, 5*) experiments can be explained with the help of "coherence resonance" in NMR. The well-known Packard and Varian (*30*) free precession experiment (after nuclear prepolarization) can be seen as zero frequency transitions by transient state.

These two effects have two common features:

1. Both can be interpreted by the common formalism used in NMR and atomic physics.

2. In the laboratory frame or in the rotating frame, coherence resonance has been induced on degenerate levels; conversely, zero frequency transitions have been observed utilizing coherence resonance.

At the end of the article we shall describe in some detail the possibility of an experiment of spin echoes without rf irradiation.

II. Magnetic Coherence Resonances

A. Theory of Coherence Resonance.

1. Phenomenological Theory

a. Principle of the NMR (nuclear magnetic resonance) (*1, 2*). Let us consider an atom subjected to a constant magnetic field $\mathbf{H}_0$ along the Oz axis of the laboratory reference system in which $\mathbf{I}$ = angular momentum of the nucleus and $\mathbf{m}$ = magnetic momentum. $\mathbf{I}$ and $\mathbf{m}$ are bound by the relation

$$\mathbf{m} = \gamma \mathbf{I} \tag{1}$$

where γ is the gyromagnetic ratio. Assuming a nucleus where $\gamma \neq 0$, the nucleus is subject in a field H_0 to a couple $\mathbf{C}$

$$\mathbf{C} = \mathbf{m} \times \mathbf{H}_0 \tag{2}$$

where $H_0 = B_0$ in the Gauss unit system and

$$\mathbf{C} = d\mathbf{I}/dt \tag{3}$$

Combining Eqs. (1), (2), and (3), one obtains

$$d\mathbf{m}/dt = \gamma \mathbf{m} \times \mathbf{H}_0 \tag{4}$$

the solution of which shows that $\mathbf{m}$ is rotating round the Oz axis at an angular displacement equal to ω_0 (Fig. 1):

$$\omega_0 = -\gamma \mathbf{H}_0$$

where ω_0 corresponds to the Larmor frequency.

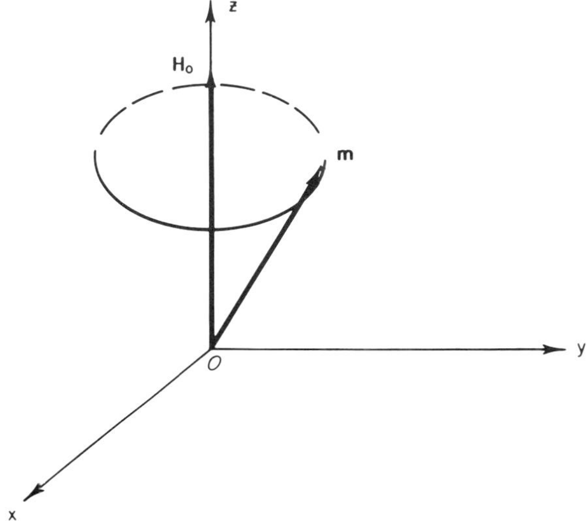

FIG. 1. Classical **m** precession around $\mathbf{H}_0$.

Add to the $\mathbf{H}_0$ field an $\mathbf{H}_1$ field rotating around Oz at a frequency ω in an (Ox, Oy) plane with

$$H_1 \ll H_0$$

Introduce a reference system (Ox', Oy', Oz') revolving around Oz at frequency ω and let Oz' and Oz be parallel. The Ox' axis is taken along $\mathbf{H}_1$.

In the rotating vectors, the evolution is given by

$$\frac{d\mathbf{m}}{dt} = \gamma \mathbf{m} \times \mathbf{H} = \gamma \mathbf{m} \times \left(\mathbf{H}_0 + \mathbf{H}_1 + \frac{\boldsymbol{\omega}}{\gamma} \right)$$

$$\mathbf{H} = \mathbf{H}_0 + \mathbf{H}_1 + \frac{\boldsymbol{\omega}}{\gamma}$$

Magnetic momentum, **m** precesses around **H**, and the angle θ between $\mathbf{H}_0$ and **H** is given by

$$\tan \theta = \frac{H_1}{H_0 + (\omega/\gamma)} = \frac{\omega_1}{\omega_0 - \omega}$$

with $\gamma H_0 = -\omega_0$ and $\gamma H_1 = -\omega_1$.

When ω strongly differs from ω_0, $\theta \cong 0$. The **m** components along Ox and Oy can be neglected (Fig. 2a).

When ω nearly corresponds to ω_0, $\theta \cong \pi/2$. The **m** components along Ox and Oy are then important (Fig. 2b).

                    E. GENEUX, G. J. BÉNÉ, AND J. PERRENOUD

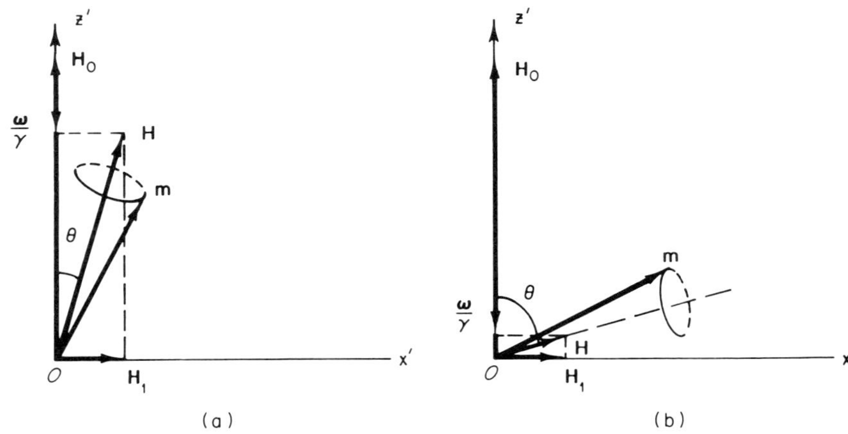

FIG. 2. **m** precession in classical NMR: (a) far from resonance; (b) near resonance.

There now appears a resonance phenomenon, the system's own frequency corresponding to the Larmor frequency. The moment turns in the Oz direction on the Ox, Oy plane. Such is the classical NMR principle.

b. Resonance for a group of atoms (1). Both spin/lattice interaction (thermal agitation) and spin/spin interaction must be taken into account.

Moment **m** of a nucleus is substituted by the density of momentum **M**, a volume unit. **M** is often called magnetic polarization density, or magnetization.

The form of Eq. (4) is then

$$dM/dt = \gamma M \times H \tag{4'}$$

whatever the H value of the magnetic field may be.

When equilibrated, the magnetization of the system placed in a field $\mathbf{H_0}$ is equal to $\mathbf{M_0}$:

$$M_0 = X_m H_0 \tag{5}$$

where X_m = magnetic susceptibility.

The evolution of $\mathbf{M}_z$ toward a state of equilibrium is given by

$$dM_z/dt = -(M_z - M_0)/T_1 \tag{6}$$

where T_1 is the thermal or longitudinal relaxation time.

On the other hand, nuclear interactions create a transversal perturbation field H_1', giving rise to a transversal relaxation time T_2:

$$T_2 = 1/|\gamma| H'$$

In such a transversal magnetic field, the evolution of the transversal magnetization component $\mathbf{M}_T$ is given by

$$dM_T/dt = -(M_T/T_2) \tag{7}$$

If the group of atoms is a gas, the relation is

$$T_1 \cong T_2$$

If the group of atoms is a liquid or a solid, the corresponding relation is

$$T_1 \cong T_2 \qquad \text{or} \qquad T_1 \gg T_2$$

The result of combining Eqs. (4′), (6), and (7) in normal Bloch equations is

$$\frac{d\mathbf{M}}{dt} = \gamma \mathbf{M} \times \mathbf{H} - \frac{M_x}{T_2}\mathbf{i} - \frac{M_y}{T_2}\mathbf{j} - \frac{M_z - M_0}{T_1}\mathbf{k}$$

in which the explicit components are

$$\frac{dM_x}{dt} = -\frac{M_x}{T_2} + \gamma H_z M_y - \gamma H_y M_z$$

$$\frac{dM_y}{dt} = -\gamma H_z M_x - \frac{M_y}{T_2} + \gamma H_x M_z$$

$$\frac{dM_z}{dt} = \gamma H_y M_x - \gamma H_x M_y - \frac{M_z - M_0}{T_1}$$

In a weak magnetic field, and when $T_1 = T_2 = T$, Bloch equations are altered to

$$\frac{d\mathbf{M}}{dt} = \gamma \mathbf{M} \times \mathbf{H} - \frac{\mathbf{M} - X_m \mathbf{H}}{T}$$

and are often applied.

The foregoing equations are more rigorously exact.

c. Coherence resonance principle (3). In a classical resonance, the appearance of a transversal component of moment **m** corresponds to a diminution of the longitudinal component.

The case of a nucleus placed in a constant $\mathbf{H}_0$ field along the Oz axis can now be reexamined. The transversal moment component precesses around Oz at Larmor frequency. For all atoms combined, the phases occur haphazardly and the resulting magnetization is equal to null (Fig. 3a).

When phases are synchronized (whatever the cause may be; for instance, the intervention of an rf field), it will be noticed that transversal moments group together. A transversal magnetization then appears without noticeable modification of the longitudinal magnetization. The appearance of such magnetization corresponds to a resonance phenomenon called "coherence resonance" (Fig. 3b).

As a rule, both types of resonances (NMR and coherence resonance) occur simultaneously. The theoretical study of resonance phenomena leads to solution of Bloch equations.

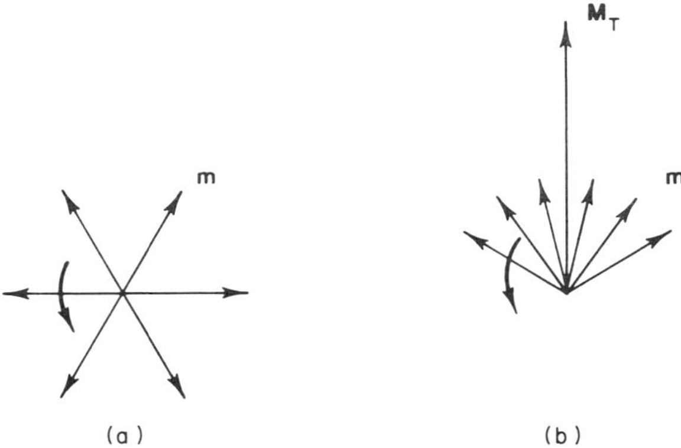

(a)   (b)

FIG. 3. (a) Distribution of transversal moments without phase synchronization. (b) Distribution of transversal moments with phase synchronization.

d. Principal systems to the study of resonance. The system applied by Minneman *et al.* (*4*) (Fig. 4a) is

$$H_x = \frac{\omega}{\gamma} = H_T, \qquad H_y = 0, \qquad H_z = H_0 + H_1 \cos \Omega t$$

This system revolves round the Ox axis at an angular velocity ω, producing a magnetic field $\mathbf{H}_T$ along Ox (Fig. 4b).

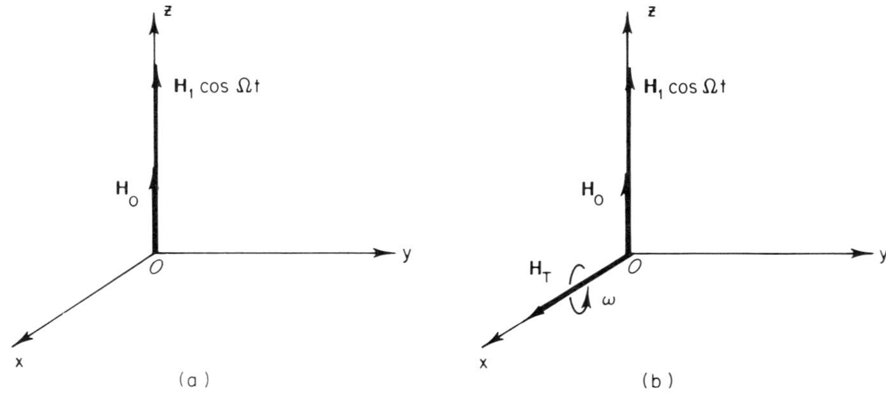

(a)   (b)

FIG. 4. Minneman's system. (a) Static system. (b) Revolving system.

The system applied by Hanuise (*5*) (Fig. 5) is

$$H_x = H_T, \qquad H_y = 0, \qquad H_z = H_0 + H_1 \cos \omega t$$

From a phenomenological point of view, this system is equivalent to Minneman's.

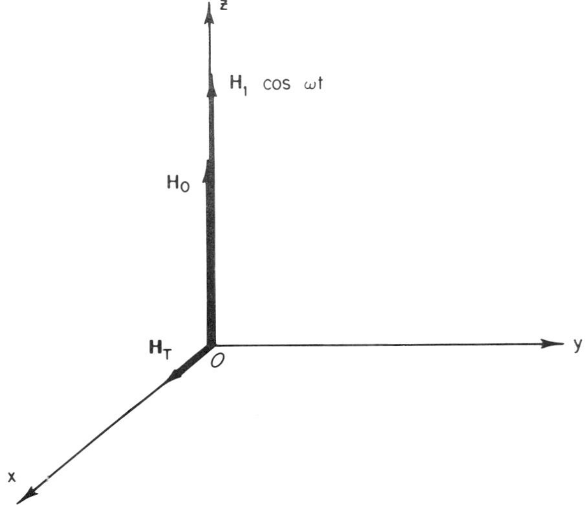

FIG. 5. Hanuise's system.

The system by Béné and Geneux (6) (Fig. 6) is

$$H_x = (H_1 + h_1 \cos \Omega t)\cos \omega_0 t$$
$$H_y = (H_1 + h_1 \cos \Omega t)\sin \omega_0 t \qquad H_z = H_0$$

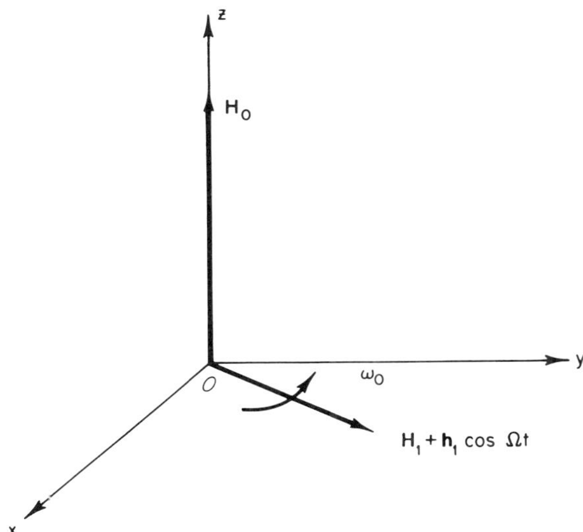

FIG. 6. Béné's and Geneux' system.

Observed in a reference system rotating around the Oz axis (interaction reference) at a Larmor frequency ω_0, this system gives

$$H_{x'} = H_1 + h_1 \cos \Omega t, \qquad H_{y'} = 0, \qquad H_{z'} = H_0 + (\omega_0/\gamma) = 0$$

In a revolving reference system, this system results in a particular case of Minneman's and Hanuise's systems.

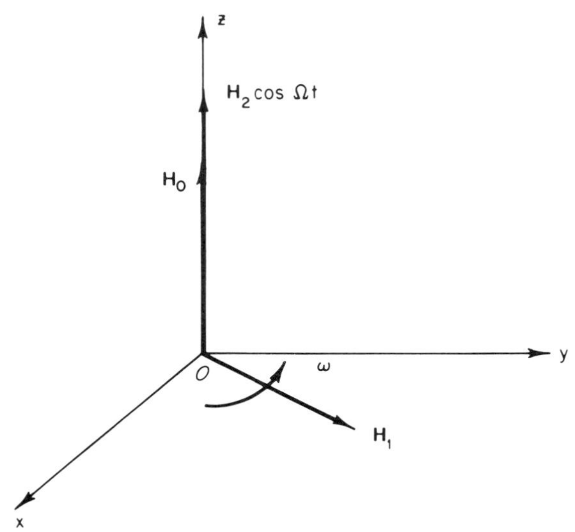

FIG. 7. Novikov's system in laboratory frame.

The system applied by Novikov *et al.* (7) (Fig. 7) is

$$H_x = H_1 \cos \omega t, \qquad H_y = H_1 \sin \omega t, \qquad H_z = H_0 + H_2 \cos \Omega t$$

In a reference revolving around Oz at frequency ω, this system also corresponds to those of Minneman's and Hanuise's (Fig. 8):

$$H_x = H_1, \qquad H_y = 0, \qquad H_z = H_0 + (\omega/\gamma)$$

A comparison between Béné and Geneux' system (6) and those of Hanuise (5) and Novikov *et al.* (7) follows.

     e. *Béné and Geneux system* (6). In a reference system revolving around Oz at Larmor frequency, Bloch normal equations become

$$dM_x/dt = -(M_x/T_2)$$
$$dM_y/dt = -(M_y/T_2) + \gamma(H_1 + h_1 \cos \Omega t)M_z$$
$$dM_z/dt = -\gamma(H_1 + h_1 \cos \Omega t)M_y - (M_z - M_0)/T_1$$

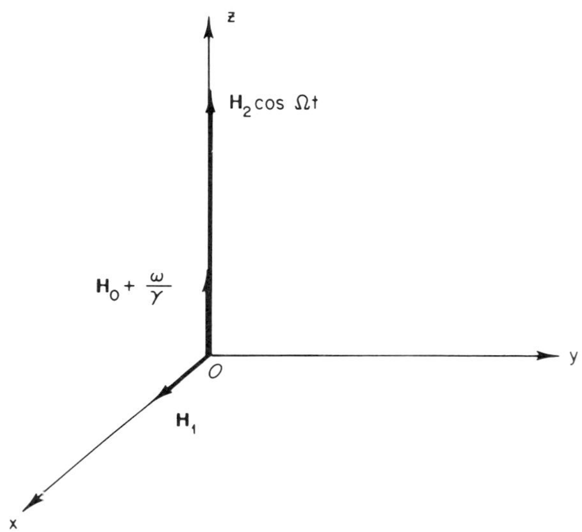

Fig. 8. Novikov's system in rotating frame.

When $T_1 \neq T_2$, there is no solution in simple analytical form. When $T_1 = T_2 = T$ (more frequent case, corresponding in particular to the experiments made), the system can easily be integrated. The solution is sought only in the permanent regime.

Deprived of a second member, we have

$$M_{1y} = A\, e^{j\alpha(t)}, \qquad M_{1z} = B\, e^{j\alpha(t)}$$

For M_{1y} and M_{1z} this gives

$$M_{1y} = j\, e^{-t/T}[a\, \exp(-j(\gamma H_1 t + (\gamma h_1/\Omega)\sin \Omega t))$$
$$+ b\, \exp(j(\gamma H_1 t + (\gamma h_1/\Omega)\sin \Omega t))]$$
$$M_{1z} = e^{-t/T}[a\, \exp(-j(\gamma H_1 t + (\gamma h_1/\Omega)\sin \Omega t))$$
$$- b\, \exp(j(\gamma H_1 t + (\gamma h_1/\Omega)\sin \Omega t))]$$

where a and b are the arbitrary integration constants.

The solution with a second member can be obtained with the aid of the constants variation method coupled with a serial development:

$$e^{jz \sin \theta} = \sum_{n=-\infty}^{+\infty} J_n(z)\, e^{jn\theta}$$

where $J_n(z)$ is the Bessel function.

As a permanent solution:

$$M_x = 0$$

$$M_y = \frac{M_0}{T} \sum_{p=-\infty}^{+\infty} \sum_{n=-\infty}^{+\infty} \frac{J_n(\gamma h_1/\Omega)J_{n+p}(\gamma h_1/\Omega)}{((1/T^2) + (\gamma H_1 + n\Omega)^2)^{1/2}} \sin(p\Omega t + \phi_n)$$

$$M_z = \frac{M_0}{T} \sum_{p=-\infty}^{+\infty} \sum_{n=-\infty}^{+\infty} \frac{J_n(\gamma h_1/\Omega)J_{n+p}(\gamma h_1/\Omega)}{((1/T^2) + (\gamma H_1 + n\Omega)^2)^{1/2}} \cos(p\Omega t + \phi_n)$$

with

$$\tan \phi_n = (\gamma H_1 + n\Omega)T,$$

where $\tan \phi_n = \phi_n = 0$ at resonance.

These are characteristic formulas of coherence resonance. In particular, the width of the line is independent of the amplitude of the field that produces the resonance.

f. Hanuise's solution (5). The Bloch normal equations appear here as follows:

$$dM_x/dt = -(M_x/T_2) + \gamma(H_0 + H_1 \cos \omega t)M_y$$
$$dM_y/dt = -\gamma(H_0 + H_1 \cos \omega t)M_x - (M_y/T_2) + \gamma H_T M_z$$
$$dM_z/dt = -\gamma H_T M_y - (M_z - M_0)/T_1$$

No solution in simple analytical form is found.

In his **GIN** theory (nuclear induction gyrometer), Hanuise gives a most interesting method for an approximate solution in the permanent regime.

In a laboratory system, the magnetic field components are

$$H_x = H_T, \qquad H_y = 0, \qquad H_z = H_0 + H_1 \cos \omega t$$

In a reference system revolving around Oz at a variable angular velocity $d\theta/dt$, the efficient fields components are

$$H_{x'} = H_T \cos \theta, \qquad H_{y'} = H_T \sin \theta, \qquad H_{z'} = H_0 + H_1 \cos \omega t + (1/\gamma)(d\theta/dt)$$

The $d\theta/dt$ angular speed is chosen so as to ensure that the component be static, i.e.,

$$d\theta/dt = -\gamma(H + H_1 \cos \omega t)$$
$$\theta = -(\gamma Ht + (\gamma H_1/\omega)\sin \omega t)$$

With the serial developments of $\sin \theta$ and $\cos \theta$,

$$\sin \theta = \sum_{n=-\infty}^{+\infty} J_n(\gamma H_1/\omega)\sin(\gamma H + n\omega)t$$

$$\cos \theta = \sum_{n=-\infty}^{+\infty} J_n(\gamma H_1/\omega)\cos(\gamma H + n\omega)t$$

where

$$H_{x'} = H_T \sum_{n=-\infty}^{+\infty} J_n(\gamma H_1/\omega)\cos(\gamma H + n\omega)t$$

$$H_{y'} = H_T \sum_{n=-\infty}^{+\infty} J_n(\gamma H_1/\omega)\sin(\gamma H + n\omega)t$$

$$H_{z'} = H_0 - H$$

The rf field can be split into two counter-rotating fields. Only the Larmor frequency component is noticeably affected. In the revolving reference system, the influence of the nonstatic components of the field is ignored (second-hand effect such as Bloch–Siegert effects are not taken into account).

If the transversal field must have a static component, one of the serial development cosines must have an argument equal to null,

$$\gamma H + n\omega = 0$$

whereby the H values satisfying the above conditions are satisfied, i.e.,

$$H^{(n)} = -n(\omega/\gamma)$$

Consequently, there is an infinite number of revolving reference systems.

In the n potential revolving reference, the efficient field is equal to

$$H_{x'}^{(n)} = J_n(\gamma H_1/\omega)H_T$$

$$H_{y'}^{(n)} = 0$$

$$H_{z'}^{(n)} = H_0 - H^{(n)} = (1/\gamma)(n\omega - \omega_0)$$

where

$$\omega_0 = -\gamma H_0 = \text{Larmor frequency}$$

Bloch equations are then as follows:

$$\frac{dM_{x'}^{(n)}}{dt} = -\frac{M_{x'}^{(n)}}{T_2} + (n\omega - \omega_0)M_{y'}^{(n)}$$

$$\frac{dM_{y'}^{(n)}}{dt} = -(n\omega - \omega_0)M_{x'}^{(n)} - \frac{M_{y'}^{(n)}}{T_2} + \gamma H_T J_n\left(\frac{\gamma H_1}{\omega}\right)M_{z'}^{(n)}$$

$$\frac{dM_{z'}^{(n)}}{dt} = -\gamma H_T J_n\left(\frac{\gamma H_1}{\omega}\right)M_{y'}^{(n)} - \frac{M_{z'}^{(n)} - M_0}{T_1}$$

with permanent regime $d/dt = 0$ for linear differential equations of the first order with constant coefficient and constant second number, where from:

$$M_{x'}^{(n)} = M_0 \frac{(n\omega - \omega_0)\gamma H_T J_n(\gamma H_1/\omega)T_2^2}{1 + (n\omega - \omega_0)^2 T_2^2 + \gamma^2 H_T^2 T_1 T_2 J_n^2(\gamma H_1/\omega)}$$

$$M_{y'}^{(n)} = M_0 \frac{\gamma H_T T_2 J_n(\gamma H_1/\omega)}{1 + (n\omega - \omega_0)^2 T_2^2 + \gamma^2 H_T^2 T_1 T_2 J_n^2(\gamma H_1/\omega)}$$

$$M_{z'}^{(n)} = M_0 \frac{1 + (n\omega - \omega_0)^2 T_2^2}{1 + (n\omega - \omega_0)^2 T_2^2 + \gamma^2 H_T^2 T_1 T_2 J_n^2(\gamma H_1/\omega)}$$

Dividing the result by $1 + (n\omega - \omega_0)^2 T_2^2$ under serial development,

$$(1 + n)^{-1} = 1 - n + n^2 \cdots$$

we get for the two first terms of the development:

$$M_{x'}^{(n)} = \frac{\gamma H_T(n\omega - \omega_0)T_2^2 J_n(\gamma H_1/\omega)}{1 + (n\omega - \omega_0)^2 T_2^2} M_0 \left[1 - \frac{\gamma^2 H_T^2 T_1 T_2 J_n^2(\gamma H_1/\omega)}{1 + (n\omega - \omega_0)^2 T_2^2}\right]$$

$$M_{y'}^{(n)} = \frac{\gamma H_T T_2 J_n(\gamma H_1/\omega)}{1 + (n\omega - \omega_0)^2 T_2^2} M_0 \left[1 - \frac{\gamma^2 H_T^2 T_1 T_2 J_n^2(\gamma H_1/\omega)}{1 + (n\omega - \omega_0)^2 T_2^2}\right]$$

$$M_{z'}^{(n)} = M_0 \left[1 - \frac{\gamma^2 H_T^2 T_1 T_2 J_n^2(\gamma H_1/\omega)}{1 + (n\omega - \omega_0)^2 T_2^2}\right]$$

The approximate analytic resolution of Bloch equations gives in the third order identical results (as above). In the second the line is widening due to saturation. When its result is not too big, it serves as a verification of the GIN theory.

Hanuise has also given an approximate solution of altered Bloch equations. If the saturation term is not too big, the result obtained in the first order is the same as that obtained by the application of the GIN theory to these equations.

The GIN theory is corroborated also by the results obtained by Winter multiquanta transition theory applied to the particular case of a resonance line width derived from an excitation field when $H_1 \ll H_0$.

The GIN theory applied to line widths gives

$$\delta H_n = (2/\gamma T_1)[1 + \gamma^2 H_T^2 T_1 T_2 J_n^2(\gamma H_1/\omega)]^{1/2}$$

When $T_1 \cong T_2 = T$ (T being a high relaxation time), we have

$$\delta H_n \cong 2 H_T J_n(\gamma H_1/\omega)$$

$$\gamma H_1 \ll \omega \Rightarrow J_n(\gamma H_1/\omega) \cong (1/n!)(\gamma H_1/2\omega)^n$$

$$\delta H_n \cong (2 H_T/n!)(\gamma H_1/2\omega)^n$$

However, in the revolving reference system of n order,

$$\omega/\gamma = \omega_0/n\gamma = H_0/n$$
$$\delta H_n \cong (n^n/2^{n-1}n!)H_T(H_1/H_0)^n$$

which is the result given by Winter multiquanta transition theory.

g. Novikov et al. (7). In an (Ox', Oy', Oz') reference system revolving around Oz at an angular speed ω, with the Oz' axis in the direction of efficient field H_e (Fig. 9).

$$H_{x'} = -H_2 \sin\theta \cos\Omega t = (\omega_2/\gamma)\sin\theta \cos\Omega t$$
$$H_{y'} = 0$$
$$H_{z'} = H_e + H_2 \cos\theta \cos\Omega t$$
$$= -(1/\gamma)(\omega_e + \omega_2 \cos\theta \cos\Omega t)$$

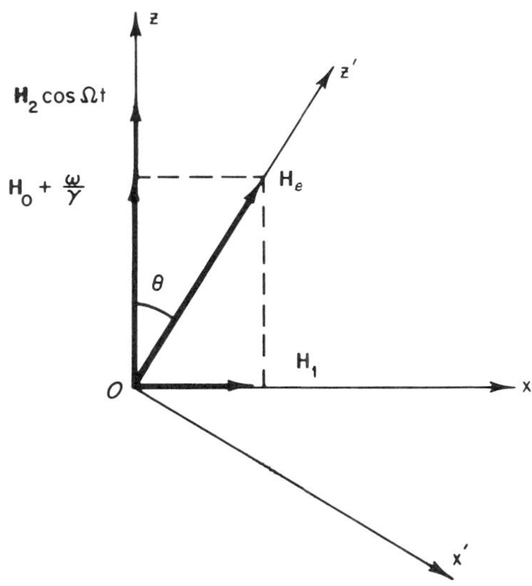

FIG. 9. Novikov's system in rotating frame after θ rotation.

If M_+ and M_- are introduced,

$$M_+ = M_{x'} + jM_{y'}, \qquad M_- = M_{x'} - jM_{y'}$$

Bloch equations give

$$M_\pm = \pm j(\omega_e + \omega_2 \cos\theta \cos\Omega t)M_\pm \pm j\omega_2 \sin\theta \cos\Omega t M_{z'}$$
$$- (M_\pm/T) - (M_0/T)\sin\theta$$
$$M_{z'} = j(\omega_2 \sin\theta \cos\Omega t/2)(M_+ - M_-) - (M_{z'}/T) + (M_0/T)\cos\theta$$

32　　　　　　E. GENEUX, G. J. BÉNÉ, AND J. PERRENOUD

In the vicinity of an n order resonance, M_z, in the first equation can be regarded as constant, which gives

$$M_+ = - \sum_{m,\,p=-\infty}^{+\infty} \frac{J_p(\omega_2/\Omega \cos \theta)J_{p+m}(\omega_2/\Omega \cos \theta)}{(1/T) - j(\omega_e + p\Omega)}$$

$$\times\, e^{jm\Omega t}\{(M_0/T)\sin \theta - jp\Omega M_z^{(n)} - \tan \theta\}$$

$$M_- = \overline{M}_+$$

If the value above is introduced in the second equation $M_z^{(n)}$ in the vicinity of the n order resonance,

$$M_z^{(n)} = M_0 \frac{(1/T^2) + (\omega_e + n\Omega)^2}{(1/T^2) + (\omega_e + n\Omega)^2 + (n\Omega)^2(\tan^2\theta)J_n^{\,2}(\omega_2/\Omega \cos \theta)} \cos \theta$$

Reverting to the (Ox, Oy, Oz) revolving coordinate system,

$$M_z = M_{z'} \cos \theta - M_{x'} \sin \theta$$

the final solution becomes

$$M_z^{(n)} = M_0 \cos^2 \theta \frac{(1/T^2) + (\omega_e + n\Omega)^2}{(1/T^2) + (\omega_e + n\Omega)^2 + (n\Omega)^2(\tan^2 \theta)\, J_n^{\,2}(\omega_2/\Omega \cos \theta)}$$

$$+\, M_0 \sin^2 \theta \sum_{m=-\infty}^{+\infty} \{A_m^{(n)} \cos m\Omega t + B_m^{(n)} \sin m\Omega t\}$$

where

$$A_{(m)}^{(n)} = \sum_{p=-\infty}^{+\infty} \frac{J_p(\omega_2/\Omega \cos \theta)J_{p+m}(\omega_2/\Omega \cos \theta)}{(1/T^2) + (\omega_e + p\Omega)^2} \left\{ \frac{1}{T^2} + p\Omega(\omega_e + p\Omega) \right.$$

$$\times\, \left. \frac{(1/T^2) + (\omega_e + n\Omega)^2}{(1/T^2) + (\omega_e + n\Omega)^2 + (n\Omega)^2(\tan^2 \theta)\, J_n^{\,2}(\omega_2/\Omega \cos \theta)} \right\}$$

$$B_{(m)}^{(n)} = \frac{1}{T} \sum_{p=-\infty}^{+\infty} \frac{J_p(\omega_2/\Omega \cos \theta)J_{p+n}(\omega_2/\Omega \cos \theta)}{(1/T^2) + (\omega_e + p\Omega)^2} \left\{ \omega_e + p\Omega \right.$$

$$+\, p\Omega \left. \frac{(1/T^2) + (\omega_e + n\Omega)^2}{(1/T^2) + (\omega_e + n\Omega)^2 + (n\Omega)^2(\tan^2 \theta)\, J_n^{\,2}(\omega_2/\Omega \cos \theta)} \right\}$$

This result compared with that of the GIN theory is interesting:

$$\sin \theta = H_1/H_e, \qquad \tan \theta = H_1/H_0$$

The equivalents of Novikov's and Hanuise's notations are shown in the tabulation.

Novikov	$\longrightarrow$ Hanuise
$H_0 + \omega/\gamma$	$H_0 = -(n\omega/\gamma)$
H_1	H_T
H_2	H_1
Ω	ω
n	$-n$

where from

$$\sin\theta = H_T/(H_0^2 + H_T^2)^{1/2} \ll 1, \qquad \sin^2\theta \cong 0, \qquad \cos\theta \cong \cos^2\theta \cong 1$$

$$\tan\theta = H_T/H_0, \qquad H_e \cong H_0, \qquad \omega_e \cong \omega_0$$

Adding the following approximations, $M_z^{(n)}$ gives

$$M_z^{(n)} \cong M_0 \frac{(1/T^2) + (\omega_0 - n\omega)^2}{(1/T^2) + (\omega_0 - n\omega)^2 + (\gamma H_T)^2 J_n^2(\gamma H_1/\omega)}$$

which is the result obtained by the GIN theory.

The method of Novikov *et al.* thus gives a better approximation of $M_z^{(n)}$ than does the GIN theory.

2. Use of Density Matrix Formalism

When each compatible observable of a system has a well-defined value at a time t, the system is said to be in pure state. Most of the time, this state is not fully defined or known. The prediction of its evolution is, however, given by statistical mechanics (8).

In this case, one says that the system has a defined probability $p_1, \ldots, p_n$ of being in one of the possible states given by the kets $|1\rangle, \ldots, |n\rangle$. These expressions have been normalized and do not need to be orthonormalized. The p_i are normalized. Such a system is called a *mixture* state.

The mean value of an observable A is then

$$\langle A \rangle = \sum_n p_n \langle n| A |n\rangle$$

We introduce a statistical operator ρ (density operator) defined by

$$\rho = \sum_n p_n |n\rangle\langle n|$$

The mean value of A can also be given by

$$\langle A \rangle = \mathrm{Tr}(\rho A)$$

Two systems having the same statistical operator are considered as identical, because one obtains the same mean value A for this observable. Each system is fully defined by its statistical operator. One advantage over wave functions is the elimination of phase indetermination.

Time evolution of ρ (8). Introduction of evolution operator $u(t, t_0)$ gives

$$\rho(t) = u(t, t_0)\rho(t_0)u^*(t, t_0)$$

The Schroedinger equation for $u(t, t_0)$ is

$$j\hbar \, du(t, t_0)/dt = \mathscr{H}u(t, t_0)$$

For ρ, this equation is

$$j\hbar \, d\rho/dt = [\mathscr{H}, \rho] \tag{8}$$

Let us consider matrix representation of the statistical operator in a basis of compatible observables. Diagonal elements give information about population. Off-diagonal elements are coupled with noncompatible observables.

For example, if eigenstates of J_z are taken, diagonal elements of ρ are populations of the eigenstates of J_z. Off-diagonal elements correspond to J_x and J_y values.

3. Classical NMR and Coherence Resonance

Let us examine the spin-$\frac{1}{2}$ problem. The eigenstates are those of J_z:

$$J_z \left| \pm \frac{1}{2} \right\rangle = \pm \frac{\hbar}{2} \left| \pm \frac{1}{2} \right\rangle$$

a. Classical NMR. The system is in a magnetic field $\mathbf{H}_0$ parallel to the Oz axis and in a field $\mathbf{H}_1$ rotating with angular frequency ω in a plane perpendicular to $\mathbf{H}_0$.

The total Hamiltonian is split into two identities:

$$\mathscr{H} = \mathscr{H}_0 + \mathscr{H}'$$

where

$$\mathscr{H}_0 = -\gamma H_0 J_z = \omega_0 J_z$$
$$\mathscr{H}' = -\gamma H_1(J_x \cos \omega t + J_y \sin \omega t) = \omega_1(J_x \cos \omega t + J_y \sin \omega t)$$

Evolution operator for $\mathscr{H}_0$ is known:

$$u_0(t) = \exp(-j(\omega_0/\hbar)J_z t)$$

Time-dependent eigenstates of $\mathscr{H}_0$ are

$$|\,\rangle_t = u_0(t)|\,\rangle_{t=0}$$

In the interaction representation, which corresponds to a system rotating at the Larmor angular frequency ω_0, $\mathscr{H}'$ becomes

$$\mathscr{H}_I = u_0^* \mathscr{H}' u_0$$

and the matrix elements of $\mathscr{H}_1$ are

$$\langle +\tfrac{1}{2}|\,\mathscr{H}_1\,|+\tfrac{1}{2}\rangle = \langle -\tfrac{1}{2}|\,\mathscr{H}_1\,|-\tfrac{1}{2}\rangle = 0$$
$$\langle -\tfrac{1}{2}|\,\mathscr{H}_1\,|+\tfrac{1}{2}\rangle = \tfrac{1}{2}\hbar\omega_1\,\exp(j(\omega - \omega_0)t)$$
$$\langle +\tfrac{1}{2}|\,\mathscr{H}_1\,|-\tfrac{1}{2}\rangle = \tfrac{1}{2}\hbar\omega_1\,\exp(-j(\omega - \omega_0)t)$$

Let us introduce following notations:

$$\langle -\tfrac{1}{2}|\,\rho\,|-\tfrac{1}{2}\rangle = \rho_{11} \qquad \langle +\tfrac{1}{2}|\,\rho\,|-\tfrac{1}{2}\rangle = \rho_{21}$$
$$\langle -\tfrac{1}{2}|\,\rho\,|+\tfrac{1}{2}\rangle = \rho_{12} \qquad \langle +\tfrac{1}{2}|\,\rho\,|+\tfrac{1}{2}\rangle = \rho_{22}$$

Evolution equations of the matrix elements of ρ are

$$\frac{d\rho_{11}}{dt} = +j\,\frac{\omega_1}{2}\,(\exp(-j(\omega - \omega_0)t))\rho_{12} - j\,\frac{\omega_1}{2}\,(\exp(j(\omega - \omega_0)t))\rho_{21}$$

$$\frac{d\rho_{12}}{dt} = +j\,\frac{\omega_1}{2}\,(\exp(j(\omega - \omega_0)t))(\rho_{11} - \rho_{22})$$

$$\frac{d\rho_{21}}{dt} = -j\,\frac{\omega_1}{2}\,(\exp(-j(\omega - \omega_0)t))(\rho_{11} - \rho_{22})$$

$$\frac{d\rho_{22}}{dt} = -j\,\frac{\omega_1}{2}\,(\exp(-j(\omega - \omega_0)t))\rho_{12} + j\,\frac{\omega_1}{2}\,(\exp(j(\omega - \omega_0)t))\rho_{21}$$

If at $t = 0$ one has the situation

$$\rho_{11}(t = 0) = 1, \qquad \rho_{12}(t = 0) = \rho_{21}(t = 0) = \rho_{22}(t = 0) = 0$$

the solutions are as follows:

$$\rho_{11} = 1 - \frac{\omega_1{}^2}{\omega_1{}^2 + (\omega - \omega_0)^2}\,\sin^2\frac{1}{2}(\omega_1{}^2 + (\omega - \omega_0)^2)^{1/2}t$$

$$\rho_{22} = \frac{\omega_1{}^2}{\omega_1{}^2 + (\omega - \omega_0)^2}\,\sin^2\frac{1}{2}(\omega_1{}^2 + (\omega - \omega_0)^2)^{1/2}t$$

$$\rho_{12} = \bar{\rho}_{21} = \frac{\omega_1(\omega - \omega_0)}{\omega_1{}^2 + (\omega - \omega_0)^2}\,\sin^2\frac{1}{2}(\omega_1{}^2 + (\omega - \omega_0)^2)^{1/2}t$$

$$+ \frac{1}{2}\,j(\omega_1/(\omega_1{}^2 + (\omega - \omega_0)^2)^{1/2})\sin(\omega_1{}^2 + (\omega - \omega_0)^2)^{1/2}t$$

For $\omega = \omega_0$, resonance conditions ρ_{12} and ρ_{21} are

$$\rho_{12} = \tfrac{1}{2}j\,\sin\omega_1 t, \qquad \rho_{21} = -\tfrac{1}{2}j\,\sin\omega_1 t$$

These two terms correspond to a precession of $\mathbf{m}$ around $\mathbf{H}_1$ with angular velocity ω_1. This precession takes place in the (Oy, Oz) plane, $m_x = 0$ following the initial conditions.

b. Pure coherence resonance. Here the system is in a field $\mathbf{H} = \mathbf{H}_0 + \mathbf{H}_1 \cos \Omega t$ parallel to the Oz axis. If initially the system is in the $|-\tfrac{1}{2}\rangle$ or $|+\tfrac{1}{2}\rangle$ state, nothing happens. We must have a superposition state; for example,

$$| \rangle_{t=0} = (1/2^{1/2})(|-\tfrac{1}{2}\rangle + |+\tfrac{1}{2}\rangle)$$

The corresponding ρ_0 is

$$\rho_0 = \begin{pmatrix} \tfrac{1}{2} & \tfrac{1}{2} \\ \tfrac{1}{2} & \tfrac{1}{2} \end{pmatrix}$$

The Hamiltonian has the following form:

$$\mathscr{H} = -\gamma(H_0 + H_1 \cos \Omega t)J_z$$

This Hamiltonian is diagonal:

$$\langle -\tfrac{1}{2}|\mathscr{H}|-\tfrac{1}{2}\rangle = \tfrac{1}{2}\hbar\gamma(H_0 + H_1 \cos \Omega t)$$

$$\langle +\tfrac{1}{2}|\mathscr{H}|+\tfrac{1}{2}\rangle = -\tfrac{1}{2}\hbar\gamma(H_0 + H_1 \cos \Omega t)$$

Therefore the system appears as

$$d\rho_{11}/dt = 0, \qquad d\rho_{22}/dt = 0$$

$$d\rho_{12}/dt = j\gamma(H_0 + H_1 \cos \Omega t)\rho_{12}$$

$$d\rho_{21}/dt = -j\gamma(H_0 + H_1 \cos \Omega t)\rho_{21}$$

Integration of the equation system gives

$$\rho_{11} = \rho_{22} = \tfrac{1}{2}$$

$$\rho_{12} = \tfrac{1}{2} \sum_{n=-\infty}^{+\infty} J_n(\gamma H_1/\Omega)\exp(j(\gamma H_0 + n\Omega)t)$$

$$\rho_{21} = \tfrac{1}{2} \sum_{n=-\infty}^{+\infty} J_n(\gamma H_1/\Omega)\exp(-j(\gamma H_0 + n\Omega)t)$$

Coherence resonance appear when one off-diagonal matrix element of ρ has a static component. The resonance condition is given by

$$\gamma H_0 + n\Omega = 0, \qquad \Omega^{(n)} = -(\gamma H_0/n)$$

c. System utilized by Béné and Geneux (6). The system is placed in a constant magnetic field $\mathbf{H}_0$ parallel to the Oz axis and in a field $\mathbf{H}_1 + \mathbf{h}_1 \cos \Omega t$, rotating at angular frequency ω_0 in a plane perpendicular to $\mathbf{H}_0$.

Magnetization $\mathbf{M}_0$ lies along Oz. At $t = 0$, all atoms are in the $|-\tfrac{1}{2}\rangle$ state. So at $t = 0$,

$$\rho_0 = \begin{pmatrix} 1 & 0 \\ 0 & 0 \end{pmatrix}$$

In interaction representation, the Hamiltonian matrix elements are

$$\langle -\tfrac{1}{2}|\mathcal{H}_{\mathrm{I}}|-\tfrac{1}{2}\rangle = \langle +\tfrac{1}{2}|\mathcal{H}_{\mathrm{I}}|+\tfrac{1}{2}\rangle = 0$$
$$\langle -\tfrac{1}{2}|\mathcal{H}_{\mathrm{I}}|+\tfrac{1}{2}\rangle = -\tfrac{1}{2}\hbar\gamma(H_1 + h_1 \cos \Omega t)$$
$$\langle +\tfrac{1}{2}|\mathcal{H}_{\mathrm{I}}|-\tfrac{1}{2}\rangle = -\tfrac{1}{2}\hbar\gamma(H_1 + h_1 \cos \Omega t)$$

Evolution equations of the matrix elements of ρ are

$$d\rho_{11}/dt = -(d\rho_{22}/dt) = -\tfrac{1}{2}j\gamma(H_1 + h_1 \cos \Omega t)(\rho_{12} - \rho_{21})$$
$$d\rho_{12}/dt = -(d\rho_{21}/dt) = -\tfrac{1}{2}j\gamma(H_1 + h_1 \cos \Omega t)(\rho_{11} - \rho_{22})$$

Integration of the equation system gives

$$\rho_{11} = \tfrac{1}{2} + \tfrac{1}{2} \sum_{n=-\infty}^{+\infty} J_n(\gamma h_1/\Omega)\cos(\gamma H_1 + n\Omega)t$$

$$\rho_{12} = -\rho_{21} = -\tfrac{1}{2}j \sum_{n=-\infty}^{+\infty} J_n(\gamma h_1/\Omega)\sin(\gamma H_1 + n\Omega)t$$

$$\rho_{22} = \tfrac{1}{2} - \tfrac{1}{2} \sum_{n=-\infty}^{+\infty} J_n(\gamma h_1/\Omega)\cos(\gamma H_1 + n\Omega)t$$

Coherence resonance appears when magnetization along the Oz axis has a static component. The resonance condition is given by

$$\Omega^{(n)} = -(\gamma H_1/n)$$

Static population of the $|-\tfrac{1}{2}\rangle$ state is at maximum and of the $|+\tfrac{1}{2}\rangle$ state is at minimum.

4. Modified Density Matrix Evolution Equation for an Ensemble of Atoms

For one atom, evolution is given by Eq. (8). For an ensemble, one has to consider the following relations.

Initially, one tries to pump all atoms in a state given by density matrix ρ_0. On the other hand, the relaxation phenomenon or finite lifetime tends to destroy the state created ρ_0. The time constant of this relaxation is represented by $T = 1/\Gamma$.

Due to the two phenomena, pumping and relaxation, the following evolution equation (with I = pumping rate) is obtained:

$$d\rho/dt = I\rho_0 - \Gamma\rho$$

where steady state $\rho = I\rho_0/\Gamma$.

The preceding equation combined with Eq. (8) leads to

$$d\rho/dt = I\rho_0 - \Gamma\rho - (j/\hbar)[\mathcal{H}, \rho] \tag{9}$$

This equation couples the quantum equation with the phenomenological one. The outline is the same as the Bloch equation.

B. Experimental Developments

1. Discovery and Coherence Resonance in Atomic Physics

In this section the following situation is considered: there exists a macroscopic magnetization M_0 in a fixed direction, created by the application of a static magnetic field H_0. Oscillating rf fields, either perpendicular or parallel to this fixed direction, tend to modify such magnetization. This phenomenon can be detected by the induced tension of a properly placed coil.

In atomic physics of excited states, the presence of magnetic dipolar resonance can be detected by the changes appearing in the rate of polarization of spontaneous emitted light (Bitter, Brossel and Kastler (9)), a method that will be used here.

Coherence resonances are related with the transversal part (perpendicular to H_0) of the magnetization. An atom excited by an adequately polarized resonance optical line can be brought into a state of excitation, the kinetic (and consequently the magnetic) moment of which is perpendicular to H_0. Circularly polarized light (from the left or from the right) is actually composed of photons, the kinetic moment (value $\hbar$) of which is parallel to its vector of propagation k. It is assumed that this stack of rays is displacing itself perpendicularly to H_0. In the event of one atom absorbing one of these photons, the maintenance of the kinetic moment will result in the appearance of a transversal component of the kinetic moment of the atom parallel to **k**. This magnetic moment will precess around H_0 at Larmor frequency ω_0. According to the τ duration of the excited state, this moment will rotate along the short part of an arc only if $\omega_0\tau < 1$; otherwise, if $\omega_0\tau > 1$, it will perform several revolutions before the de-excitation of the atom. In the former case, the polarization of the emitted light ray will be identical with that of the absorbed light ray; in the latter case, the polarization of the light will be of any kind (emitted by a dipole that can assume any direction on a plane perpendicular to H_0).

Let us assume now that not one but a batch of atoms is absorbing light rays: The former case will result in a macroscopic magnetization parallel to **k**. In the latter case, the dipoles will be uniformly spread over a disc and the transversal magnetization will be equal to null. This disappearance of transversal magnetization with increasing $H_0(\omega_0)$, caused by the accompanying depolarization of the emission of rays by the atom, is typical of Hanle's effect (10).

The effect of coherence resonance consists of grouping these dipoles in one direction (generally $\mathbf{k}$) as far as possible, and of obtaining an emission of light evidencing polarization identical to that absorbed. Since this grouping by an oscillating field does not absorb any energy whatsoever, it does not amplify the resonance with an increasing hf field.

The superposition of a field $h_1 \cos \Omega t$ on the field H_0 will modulate the Larmor frequency: ω_0 becomes $\omega_0 + \gamma h_1 \cos \Omega t$. If atoms continue to be brought from their fundamental state to the excited state at constant pumping speed, the disc will be inhomogeneously filled: packed when ω is weak (i.e., $\omega_0 - \gamma h_1$) and widely spread when ω is at its maximum (i.e., $\omega_0 + \gamma h_1$). During a period $2\pi/\Omega$ of the modulation field, dipoles will have revolved around an angle of $2\pi/\Omega \omega_0$. If the effect is to be resonant, this angle must be a multiple of 2π and hence $n\Omega = \omega_0$.

This effect was observed for the first time by Alexandrov *et al.* (*11*) when modulating fluorescent light in the state of $5\ ^3P_1$ of cadmium, and independently by Favre and Geneux (*12*) in the same state. Polonsky and Cohen–Tannoudji (*13*) have observed the same effect at level $F = \frac{1}{2}$ of the fundamental state $^1S_0$ of mercury-199. As a function of h_1 they have studied the dependence of modulation amplitudes at $\Omega, \ldots, 5\Omega$ frequencies as well as the independence of width for $n = 1$. Figure 10 gives both results. An interpretation by quantification of the rf field has been given by the same authors (*14*). Considerations include: the proper states $|\pm, n\rangle(|+\rangle, |-\rangle$, proper states of $J = \frac{1}{2}$; $|n\rangle$, proper state of photons field). The coupling between field and atom superposes on state $|+, n\rangle$ a certain proportion of states $|+, n \pm 1$ (as well as for $|-, n\rangle$; this explains that the transverse magnetization, proportionate to σ_x or σ_y (matrix of Pauli) will not only precess at Larmor frequency ω_0, but will also take up components including $\omega_0 \pm \Omega$. The calculation of perturbations led as far as allowed frequencies of $\omega_0 \pm p\Omega$ to interfere.

Figure 11 shows the dependence of harmonic amplitude 2 on various values of n (*15*).

When an oscillating field $h_1 \cos \Omega t$ is adjoined to field H_0, the Larmor frequency and thereby the energies of Zeeman sublevels are modulated. Such a modulation can be obtained through the medium of an oscillating electric field. In fact, an electric field E is followed by a Stark effect that displaces the Zeeman sublevels m of a given state J with energy equal to $\Delta E_m = E^2$ (A and B are constant values depending on the electric dipolar moment of state J.) A field $E \sin \Omega t$ will call for a displacement of $(A + Bm^2)/(E^2/2)(1 + \sin 2\Omega t) - \Omega$ which is small compared with the optical transition frequency of atom $-$. Also applied is a sweeping magnetic field H_0 parallel to E. The total energy of sublevel m will then be

$$m\omega_0 + (E^2/2)(A + Bm^2)(\tfrac{1}{2} + \tfrac{1}{2} \sin \Omega t)$$

     E. GENEUX, G. J. BÉNÉ, AND J. PERRENOUD

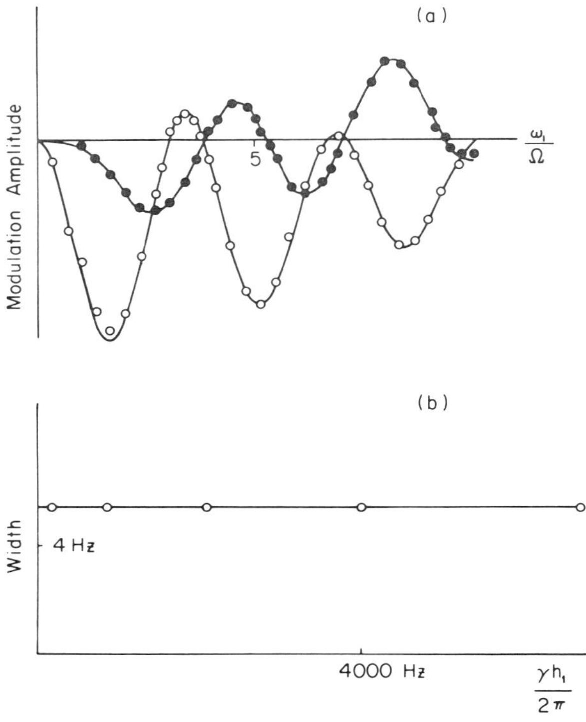

FIG. 10. (a) Field dependence of amplitude of resonance. Modulation at $p = 2$ and $p = 4$ for $n = 1$. Abcissa: $\omega_1/\Omega = \gamma h_1/\Omega$. (b) Independence of the width of resonance as a function of h_1. From Polonsky and Cohen–Tannoudji (14).

The stationary solution of Eq. (9) for the element ρ_{mm+1} will then be determined. A luminous coherent excitation to bring the atom into a state of superposition $\lambda|m\rangle + \mu|m + 1\rangle$ is necessary. The solution is

$$\rho^{\text{stat}}_{mm+1} = \frac{I}{\Gamma} \rho^0_{mm-1} \sum_p A_p \, e^{jp\Omega t}$$

$$A_p = \sum_n \frac{J_n(A^2 E^2/4\Omega) J_{n+p}(A^2 E^2/4\Omega)}{\Gamma^2 + (\omega_0 + (A^2 E^2/2)(2m + 1) - 2n\Omega)^2}$$

Figure 12 shows at (b) the static part ($p = 0$) of resonance $n = 1$ obtained with an oscillating electric field at 150 kHz and 20 kV/cm amplitude. Static part $(AE^2/2)(2m + 1)$ enables crossing of levels $|m\rangle$ and $|m + 2\rangle$, which gives the big left-hand signal. The experience is repeated on the even numbers of cadmium isotopes in the state $5\,^3P_1$ (16).

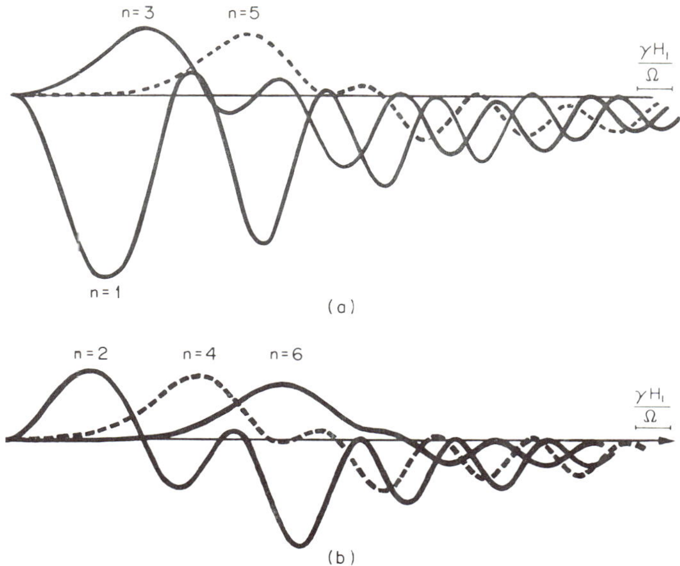

FIG. 11. Field dependence of modulation $p = 2$ for different n from $n = 1$ to $n = 6$.

All results described so far are exclusively based on nondiagonal elements of the density matrix corresponding to the transversal part of the magnetization. In the Geneux–Béné's case, the solution of the Bloch equation shows that M_z can be subject to coherence resonance with regard to field H_1, this magnetization is transversal, but can be described by diagonal elements of the density matrix when the quantification axis is parallel to H_0. The synchronization to the state of resonance, $\omega = \omega_0$, causes degeneration of energy levels in the revolving referential.

Such energy degeneracy occurs at the crossing of two Zeeman sublevels, either in hyperfine structure or on the occasion of a combined Stark–Zeeman effect. The application of a field $H_1 + h_1 \cos \Omega t$ coupling both states will create coherence resonances that can be observed not only on the nondiagonal elements but also on the whole population of said states. The experience is thus again realized on the $5\ ^3P_1$ cadmium state. A magnetic field H_0 and an electric field E parallel to H_0 enables intersection of levels $|0\rangle$ and $|1\rangle$ as shown in Fig. 12(a). An oscillating field $H_1 + h_1 \cos \Omega t$ perpendicular to this direction induces the transitions. Figure 13 shows the geometric field distribution together with the results obtained at a frequency $\Omega/2\pi$ of 225 kHz. Field H_1 appears as the abscissa (19).

Novikov et al. (7) have considered the case of a field $H_0 + h_1 \cos \Omega t$ together with a field H_1 revolving at the frequency ω in a plane perpendicular

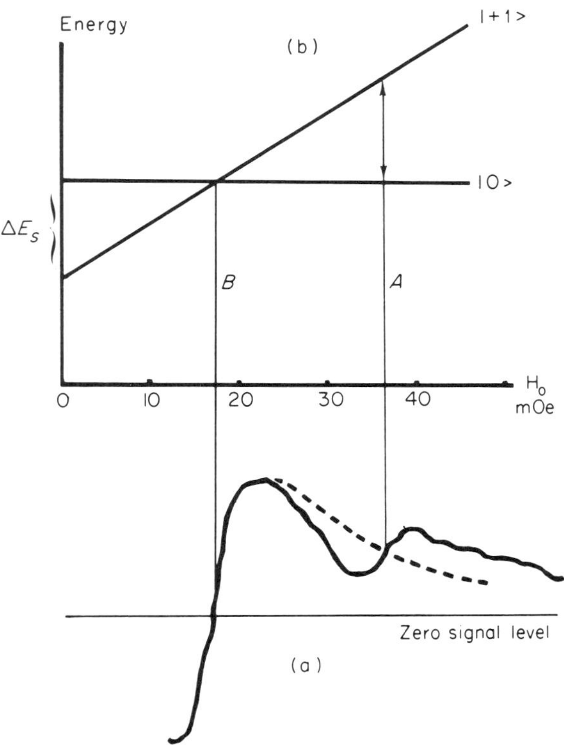

FIG. 12. (a) Energy level diagram for E fixed and H_0 variable. (b) Coherence resonance obtained with oscillating electric field (20 kV/cm, 150 kHz). The dashed line corresponds to the side of the dispersion curve detecting the crossing B due to the static part $\frac{1}{2}E^2$. This gives a static Stark effect separating the levels ± 1 from 0.

to H_0. They observed resonances when

$$n\Omega = \gamma H_{\text{eff}} = \left[\left(H_0 - \frac{\omega}{\gamma}\right)^2 + H_1{}^2\right]$$

This study calls for both types of resonances (the ordinary and the coherent resonance) according to H_{eff} orientation (rather perpendicular or rather parallel to H_0). These authors resolve the Bloch equation with an M_0 magnetization parallel to H_0. Such geometry had already been considered by Redfield (18), but he took into account only the actual resonances inducted by h_1.

Coherence resonances have also been observed in the case of oscillating fields perpendicular to H_0 (18). In such a case, the creation of a transversal moment is essential. On a spin-$\frac{1}{2}$ system these resonances can be discrimi-

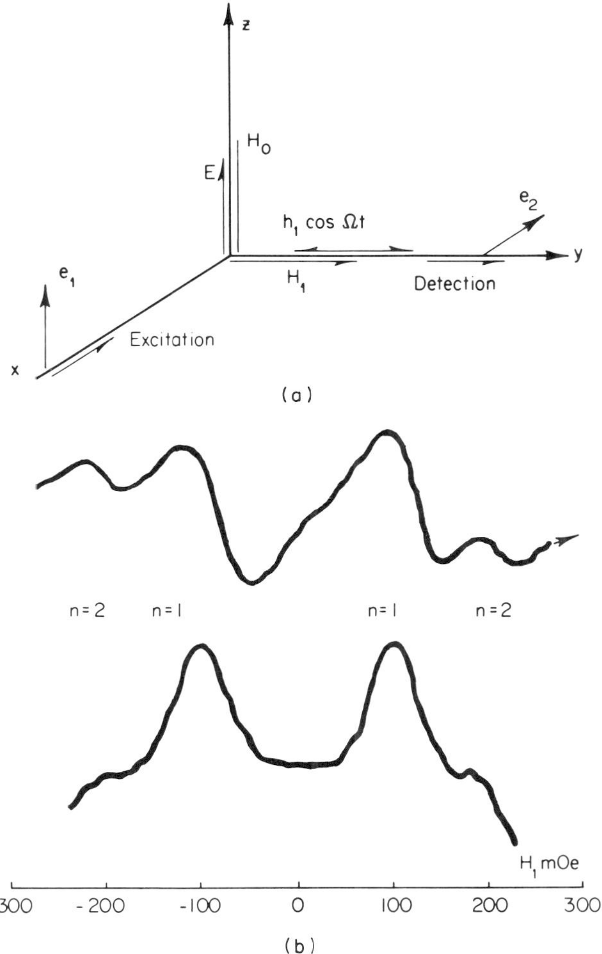

FIG. 13. Coherence resonance at crossing levels. (a) Geometrical setup of the fields and polarization vectors (e_1, excitation; e_2, detection). (b) Resonances $n = 1$ and $n = 2$. Static part ($p = 0$). Frequency, 225 kHz; $h_1 = 250$ mOe.

nated from ordinary multiquanta resonances as follows: The maintenance of the kinetic moment requires $2n + 1$ photons in the case of the usual resonance ($n\sigma^+$, $(n + 1)\sigma^-$, or inversely) and such resonance widths amplify with increasing rf power; conversely, parametric resonances require $2n$ photons, have a constant width, and show a comparatively important displacement when the rf power has increased.

An interpretation in terms of rf photons has also been given (20). In this case, the coupling field/atom superpose to the $|+, n\rangle$ state, with other states $|-, n \pm 1\rangle$. When calculating perturbations of p order, states $|+, n \pm p\rangle$ will be superposed when p corresponds to an even number, as will $|-, n \pm p\rangle$ when p corresponds to odd numbers. The transversal pumping results in superposition states: $\lambda |+, n\rangle + \mu |-, n\rangle$. These coherence resonances can be interpreted as crossings between states $|+, n\rangle$ and $|-, n + p\rangle$. This is possible by the superposition caused by the coupling only when p corresponds to even numbers. This qualitatively explains the appearance of these resonances for even numbers of hf photons.

Such transitions have been observed by the hyperfine structure of atomic excited states (21): $\Delta F = \pm 1$ transitions are induced in weak magnetic field H_0. The concept of hyperfine coherence $\rho_{F M F' M'}$ (22) has to be introduced. In order to excite such an element, the width of the optical exciting line has to be larger than the hyperfine separation. Excitation polarization is linear at 45 deg of the direction of H_0. Detection is made in a bridge: One takes the difference of two signals detected at $+45$ and -45 deg. (See Fig. 14a.) In this case, one is sensitive to $\rho_{F,M,F \pm 1,M \pm 1}$ elements. Figure 14(b) shows rf power dependence of the ordinary resonance (line with slope) and independence for the coherence resonance (horizontal line). Circles correspond to the rf field parallel to H_0; crosses, to the rf field perpendicular.

2. Coherence Effect in a Gyromagnetic Experiment (2)

In this experiment, the rf field $H_1 \cos \Omega t$ is oscillating parallel to the H_0 field and transitions are induced only if a small constant field H_T perpendicular to H_0 is applied or if the apparatus is rotating around an axis perpendicular to H_0 and creating an equivalent to the H_T field.

The detection method is that of Bloch, i.e., creation of an emf in a coil. To avoid the strong induced tension due to the particular direction of H_1, signals are recorded on the 2Ω frequency (each resonance of different n has terms oscillating at a multiple of Ω). However, a drastic filtering of frequency Ω is needed.

The frequency used is 26 kHz and the experiment is performed on water. This frequency is a compromise between achievement of high NMR signals (proportional to Ω^2) and difficulty of obtaining a high rf magnetic field (necessary for adequate values of Bessel function argument).

Signal amplitude and width in terms of the different parameters have been recorded. These dependences are not the same as these of ordinary NMR resonances, and leads to a proof of the theory. Measurements have been done for $n = 1$ and $n = 2$. A fairly good agreement is obtained. Resonance signals have been obtained by rotation, the angular velocity being 3 to 5 turns per second.

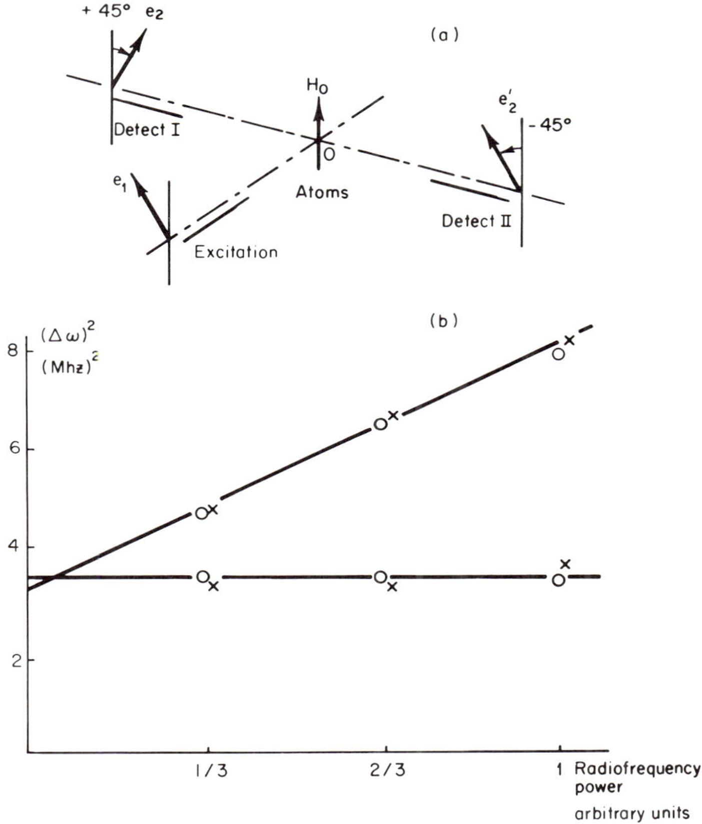

FIG. 14. Coherence resonance by hyperfine structure. (a) Geometrical disposition of polarization vectors at excitation and detection. (b) RF power dependence on width of ordinary resonance (steep line) and on coherence resonance (horizontal line). $\bigcirc$ is radio frequency field parallel to H_0. $\times$ is radio frequency field perpendicular to H_0.

3. Coherence Resonances in NMR

As stated above, the coherent magnetic resonances can be characterized by:

(a) a permanent polarization or alignment (e.g., in the Ox direction).

(b) the application of an H_0 field in an Oz direction perpendicular to the Ox direction. The field amplitude is given by the condition:

$$\omega_0 \gg 1/\tau$$

where

$$\omega_0 = \frac{\mu}{I} H_0 \qquad \text{(Larmor pulsation)}$$

where μ is the magnetic moment, I is the angular momentum of nuclei, and τ is the level lifetime or transition relaxation time.

(c) the modulation of field amplitude H_0 at a pulsation of

$$\omega_n = \omega_0/n \qquad (n \text{ integer positive})$$

giving a field $H_0 + H_1 \cos \omega_n t$ along the Oz axis.

The amplitude of these resonances is given by $J_n(\omega_1/\omega_n)$; the line width is independent of amplitude H_1 of the modulation ($H_1 = (I/\mu)\omega_1$) and enables the determination of the value of τ.

In order to obtain such resonances in RMN, the three preceding conditions must be fulfilled as given below (23):

(1) The transversal orientation of angular moments, to be obtained free of any external magnetic field whatsoever, may result from an internal field (for instance, nuclei aligned by the electrical gradient of a crystalline field).

(2) The H_0 field must be applied perpendicularly to the alignment axis and its intensity must correspond to

$$\omega_0 \gg 1/T_2$$

(T_2 is the transversal relaxation time of the nuclei).

(3) The field H_0 must be superposed by a modulation $H_1 \cos \omega_n t$. Such modulation reinstates a rotating transversal moment (previously destroyed by field H_0) at a pulsation of $\omega_n = \omega_0/n$.

III. Transitions at Zero Frequency

A. Steady Phenomenon

1. In Atomic Physics

The principle is based on the Hamiltonian of a classical resonance phenomenon expressed in the rotating referential:

$$(\gamma H_0 - \omega)J_z + \gamma H_1 J_x$$

This Hamiltonian is static and the nondiagonal elements of J_x induce transitions between Zeeman sublevels $|m\rangle$. These degenerate into energy when $\gamma H_0 = \omega$; the amplitude of the induced transitions then reaches its apex.

In the reference system of the laboratory, such degeneration into energy appears in fine or hyperfine structures at a level crossing or in the occurrence of a combined Stark–Zeeman effect.

In atomic physics, level crossings have been reported by Colegrove et al. (25). This requires the presence of an excitation in the course of which the atom comes in a state of superposition at both levels. The polarization of the emission of light is chosen so as to be affected by such a superposition state as well, which is possible only when the matrix elements of the electric dipolar operator between both levels and any one of the Zeeman sublevels of de-excitation state are simultaneously different from zero.

A static perturbation that couples both levels will induce transition, the amplitude of which will reach its apex at the very moment of energy degeneracy, i.e., at a crossing. The first instance of such a transition was observed by Eck *et al.* (*26*) when investigating an "anticrossing" phenomenon These co-workers were then examining the fine structure of the lithium doublet $2\ ^2P_{3/2}$, $^2P_{1/2}$, and in particular the crossings $|\frac{3}{2}, -\frac{3}{2}\rangle$ with $|\frac{1}{2}, \frac{1}{2}\rangle$ and $|\frac{1}{2} - \frac{1}{2}\rangle$.

The presence of a nuclear spin $1 = \frac{3}{2}$ for Li_7 introduces a hyperfine scalar interaction $A\mathbf{IJ}$. In accordance therewith, the levels $m_J = -\frac{1}{2}$, $m_I = -\frac{3}{2}$ and $m_J = -\frac{3}{2}$, $m_I = -\frac{1}{2}$ are coupled by such interaction, since they degenerate into $m_J + m_I$. If the atom is excited by a polarization that changes it to state $m_J = -\frac{3}{2}$ $m_I = -\frac{1}{2}$, the hyperfine interaction causes its transition toward $m_J = -\frac{1}{2}$ $m_I = -\frac{3}{2}$ and the fluorescent light will no longer be identical with that of excitation. Considerations that take into account this interaction influence on the nuclear polarization rate, which can be obtained by optical pumping, have been made by Lehman (*22*).

A second example of zero frequency transition, induced this time by a static magnetic field (*27*) is the following:

An electric field E and a parallel magnetic field H_0 are simultaneously applied to a state of kinetic moment $J = 1$. The diagram of the energy obtained is given in Fig. 12*b*. A static magnetic field placed perpendicularly to E and H_0 will induce transitions between level $|0\rangle$ and level $|1\rangle$. The differential equation resulting for a matrix density p can be solved. Atoms are continuously pumped in state $|1\rangle$ and the stationary solution for ρ_{00}, Re ρ_{01}, $Jm\rho_{01}$ is sought. Using the usual conventions given at the beginning of this article, we have

$$\rho_{00} = \frac{2\omega_1{}^2}{\Gamma^2 + (\Delta\omega)^2 + 4\omega_1{}^2}$$

$$\mathrm{Re}\ \rho_{01} = \frac{\Delta\omega\omega_1}{\Gamma^2 + (\Delta\omega)^2 + 4\omega_1{}^2}$$

$$J_m\ \rho_{01} = \frac{\Gamma\omega_1}{\Gamma^2 + (\Delta\omega)^2 + 4\omega_1{}^2}$$

The corresponding graphic curves appear in Fig. 15 and are similar to those appearing for NMR.

The final example of a zero frequency transition is that given by the simultaneous action of an electric field E placed perpendicularly to a magnetic field H_0 (*28*). Atoms are brought into the proper state $|m\rangle$ (z being selected parallel to H_0). As in any electric dipolar transition from the perpendicular field E, let the operators D_+ and D_- interact so that the atoms will be brought to transition from a state $|m\rangle$ to a state $m + 2$ through the medium of a

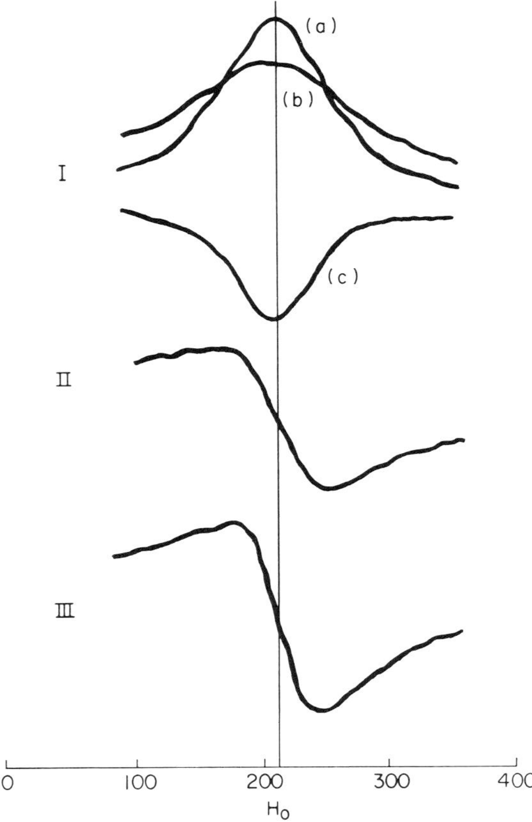

FIG. 15. I. Im(ρ_{01}). II. Re(ρ_{01}) for different H_1 amplitudes: (a) $H_1 = 35$ mOe; (b) $H_1 = 75$ mOe; (c) $H_1 = -20$mOe. III. Signal given by ordinary crossing level detection to compare relative sensitivity of the two methods.

Stark coupling via an intermediate state of opposed parity. State $5\ ^3P_1$ has been examined. Atoms are continuously brought into state $m = -1$, and the stationary solution of the equation for the density matrix should be sought. With the usual conventions and with ω_α being called *difference of energy* between state $5\ ^3P_1$ and 6^3S_1 (the latter being predominant with Stark coupling), we have

$$\rho_{11} = \frac{I}{\Gamma}\left(\frac{A^2E^2}{\hbar^2\omega_\alpha}\right)^2 \frac{1}{\Gamma^2 + 4\omega_0{}^2 + (4A^4E^4/\hbar^4\omega_\alpha{}^2)}$$

$$\text{Re}\ \rho_{1-1} = \frac{I}{\Gamma}\frac{A^2E^2}{\hbar^2\omega_\alpha} \frac{\omega_0}{\Gamma^2 + 4\omega_\alpha{}^2 + (4A^4E^4/\hbar^4\omega_\alpha{}^2)}$$

Particularly interesting is the fact that ρ_{1-1} is proportionate to $A^2 E^2/\hbar^2\omega_\alpha$ whereas ρ_{11} is proportionate to its square value. This phenomenon can thus be evidenced by a coherent detection called forth by electric fields of less intensity. Figure 16 shows the dependence of the graphic curve amplitude of

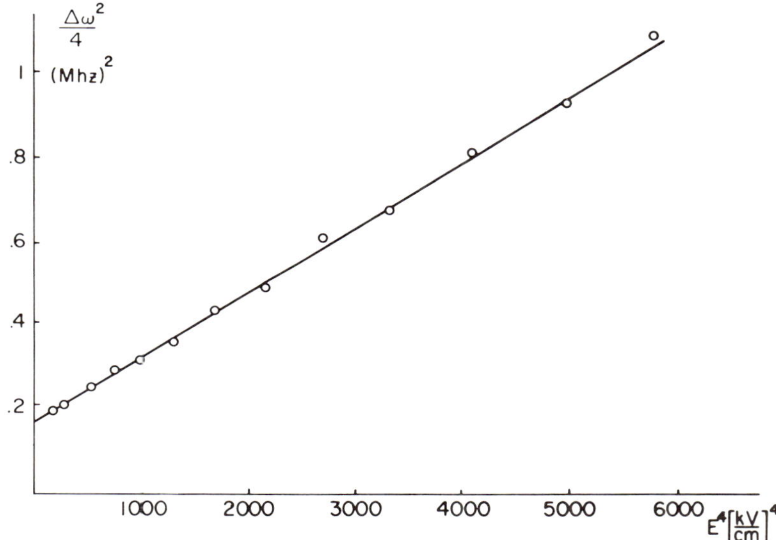

FIG. 16. Width dependence of the signal detecting ρ_{1-1} as a function of the amplitude of the E field.

the resonance in function with the electic field. Such dependence enables to measure the Stark constant of level $5\,{}^3P_1$.

2. RF Pumping of Nuclear Magnetic Levels

A phenomenon very similar to that described in the preceding text can be obtained by a real radioelectric pumping from one to another Zeeman sublevel when the same is subjected to a magnetic field.

In fact, as everybody knows, it is possible to modify the relative population of two levels E_1 and E_2 if a dipolar magnetic transition is allowed to take place between them by irradiation of the sample with a magnetic field at a frequency v given by the Einstein relation

$$hv = E_1 - E_2 \qquad (h = \text{Planck's constant})$$

If both levels have the same energy ($E_1 = E_2$) but distinct magnetic quantum numbers, the application of a steady field perpendicular to the quantification axis is likely to call forth an equalization of the populations of both E_1 and E_2 levels.

The experiment can be realized in NMR, for instance, in a system of two spins I_1 and I_2 ($I_1 = I_2 = \frac{1}{2}$) coupled with indirect spin interaction.

As in Erbeia (29) experiments, we can use the system of two coupled spins $H_1{}^1$ and P_{15}^{31} in phosphorous acid, HPO (OH)^2. Here $J = 692$ Hz.

In the absence of any external field, we can assume that each of the two nuclei H and P is in a field indirectly produced by the other via "scalar coupling."

When a low external field is introduced, it splits the resonance and changes the frequency. The Breit–Rabi diagram of the energy levels is given in Fig. 17 as a function of the magnetic field amplitude.

When $H_0 = H_c$, an inversion of both levels can be ascertained:

$$(F = 0; \quad m = 0) \quad \text{and} \quad (F = 1; \quad m = -1)$$

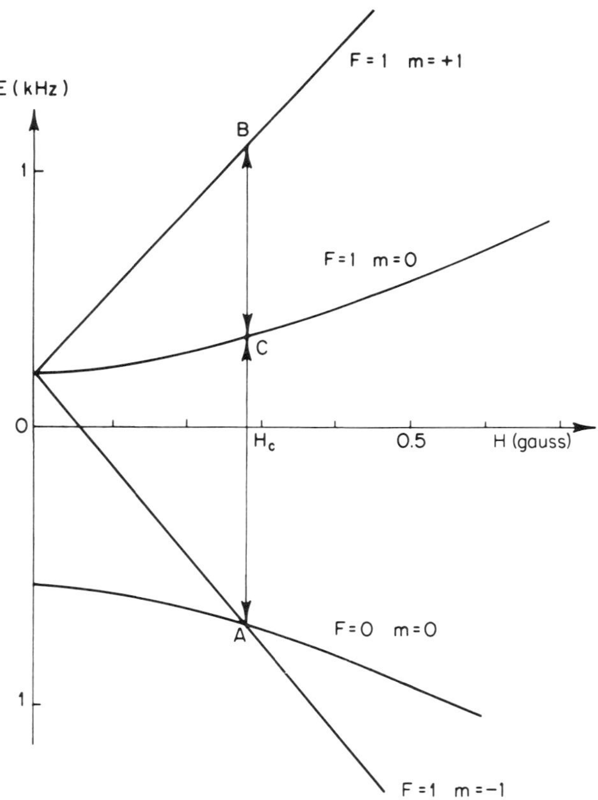

FIG. 17. The Breit–Rabi diagram for two coupled spins $H_1{}^1$ and P_{15}^{31} in low fields and transitions considered here.

If the sample is kept long enough in the field $H_0 = H_c$, the population of the two inverted levels will be the same, i.e., inferior to that of the level $(F = 1; m = 0)$. As a result of the Boltzmann factor, the population of level $(F = 1; m = +1)$ is naturally smaller.

A strong irradiation at a frequency corresponding to the AB transition is likely to equalize the population of both levels $(F = 0; m = 0)$ and $(F = +1; m = +1)$.

The population of level $(F = 1; m = -1)$ remains unchanged because $(F = 1; m = -1) \leftrightarrow (F = 1; m = +1)$ is forbidden.

The AC transition $(F = 1; m = -1) \leftrightarrow (F = 1; m = 0)$ can be observed by a classical NMR experiment, it being well-understood that the transition $(F = 0; m = 0) \leftrightarrow (F = 1; m = 0)$ is forbidden.

When applying a static magnetic field perpendicular to H_0, the population of the two levels $(F = 0; m = 0)$ and $(F = 1; m = -1)$, having the same energy (no forbidden transition), can be equalized. Such equalization will result in a diminution of the population of the $(F = 1; m = 1)$ level. In view of the strong irradiation of the AB transition, this can be ascertained along with a diminution in AC transition intensity (23).

Another procedure would be to irradiate the AC transition and to observe the AB transition, but in this case the saturation effect would be smaller.

A more suitable method for observing the difference of population arising between the $(F = 0; m = 0)$ and the $(F = 1; m = -1)$ levels at the crossing point A would be to detect the nutation of the resulting magnetic moment around a static magnetic cross field of arbitrary amplitude. Such nutation can be transformed into a coherence resonance and detected on the magnetization component z. This can be obtained by amplitude modulation of the magnetic cross field at a frequency that must be a submultiple of the nutation frequency. This effect on excited atomic states is described elsewhere.

B. Transient Phenomena

1. Free Precession of Nuclear Magnetization in a Steady Field

This experiment was actually carried out for the first time in 1954 when Packard and Varian (30) reported the free precession of proton magnetization around the terrestrial magnetic field after prepolarization in a steady magnetic field of high intensity, this field being perpendicular to the terrestrial magnetic field. Grivet and Molnar (31) explain the principle of this experiment in their paper. In the simple case of noninteracting nuclei, the resonance condition

$$\omega_0 = \gamma H_0$$

between a rf field pulsation, the gyromagnetic ratio, and the intensity of the applied magnetic field show the linear relation with the other two variables for a given nucleus. This linear relation can be extrapolated into null values of the two variables and will then give a simple interpretation of the well-known phenomenon of the free precession of nuclei in the terrestrial magnetic field after prepolarization in another high perpendicular magnetic field.

a. Free precession in the rotating frame (32). The sample containing resonant nuclei is subjected for a sufficiently long time, $t(t > T_1, T_2)$, to the main steady field H_0 and then suddenly applied to a rf field $H_1 \cos \omega t$ in which

$$\omega = \omega_0 = \gamma H_0$$

In the rotating frame, the useful rotating part of the rf field can be represented by the Coriolis vector (ω_0/γ) equal and of opposite sense, to H_0. The nuclear magnetization M_0, primitively parallel to H_0, precesses (in this frame) around the H_1 field with a pulsation $\omega_1 = \gamma H_1$.

In fact, this nutation is superposed in the laboratory frame to the precession of the M_0 vector around the applied steady field H_0.

b. Free precession in the laboratory frame. If the steady magnetic field amplitude tends toward null, the above description still applies, provided the pulsation of the H_1 field corresponds to $\omega = \omega_0$.

In the limiting case $H_0 = 0$, $\omega = \omega_0 = \gamma H_0 = 0$, and consequently the precession frequency around the H_0 field is suppressed, but the nutation around the H_1 field (now steady) is kept up. These conditions exactly correspond to those prevailing in the case of the free precession of the nuclear magnetization M_0 in the terrestrial magnetic field H_T after prepolarization in a perpendicular H_p field. This can easily be shown as follows:

(1) In the prepolarization phase, the terrestrial magnetic field H_T (very small as compared with the H_p polarization field) only intervenes in order to alter slightly the direction and amplitude of H_p.

(2) The sudden cutting off of H_p corresponds, in this picture, to sudden application of H_1 (here H_T) to the resonance frequency (here $v = 0$, since $H_p = 0$). The free nutation of M_0 magnetization can then be observed without superposed precession, since

$$\omega = \omega_p = \gamma H_p = 0$$

The condition of the presence of free precession in the terrestrial field, based on observation of the free precession in a rotating frame, can be deduced from these considerations. Rather than go into detail, we shall study an interesting application (*32*).

2. Spin Echoes without RF Excitation

a. Principle of realization. It has been seen that cutting off prepolarization field H_0 causes the resultant magnetization around direction $H_1 = H_T$ to start. Just as it is possible in the spin echo technique to stop the nutation and to orient the M_0 vector in an arbitrary direction of the plane perpendicular to H_1 by suddenly cutting off H_1, this can be done in this case by suddenly reinstating H_p.

If the time spent after cutting off H_p is equal to

$$t = \tfrac{1}{4}(2\pi/\gamma H_T)$$

at the time of receiving H_p, M_0 is in the x–y plane ($\pi/2$ nutation angle) and begins to precess freely around H_p at a frequency precession v_p. In this case it is possible to induce a signal of ω_p pulsation in a coil perpendicular to H_p. We can obtain the reversal of M_0 polarization by a π nutation angle. The well-known spin echo experiments, now without rf field, can then be easily described.

(1) *Extension of Hahn original method (34)*. The polarization of the nuclear moments along the H_p field during a time $t > T_1$, T_2 gives the magnetization M_0 parallel to H_p (Oz).

Note that the presence of the H_T field along the Ox axis has a negligible effect because

$$H_p \gg H_T$$

That is,

$$H_p \approx 100 \quad \text{Oe} \qquad H_T \approx 0.5 \quad \text{Oe}$$

(a) Just as H_p is cut off, M_p nutates around the H_T field perpendicular to H_p. This nutation is maintained during a quarter-cycle:

$$t = \frac{1}{4}\left(\frac{2\pi}{\gamma H_T}\right)$$

(b) Past this time t, the H_p field is reinstated. Then a precession of M_0 begins around it, but owing to the lack of homogeneity of said field, the resulting M_0 magnetization vanishes.

(c) After another time τ ($\tau > \omega_T^{-1}$), H_p is cut off again for a time $t = \tfrac{1}{4}(2\pi/\gamma H_T)$. The Hahn echo is obtained at the time of 2τ.

(2) *Carr and Purcell experiment (35)*. After a first $\pi/2$ pulse, which causes the H_0 vector to turn into the plane perpendicular to H_p, series of double timed pulses are applied after τ, 3τ, 5τ ... times and echoes are obtained at the times 2τ, 4τ ....

(3) *Meiboom and Gill (36)*. The phase of the first pulse is in quadrature in reference to other pulses. These echoes can be obtained with the following apparatus: two identical coils of rectangular axis in series give the fields $-H_T$ (equal and opposite to H_T) and $H_{T'}$, (equal and perpendicular to H_T). The first pulse is emitted by H_T' and all others by H_T (after cutting off $-H_T$ and $H_{T'}$).

b. Practical method to perform such experiments. In the simple transposition of spin echoes, we have to overcome some serious practical difficulties:

(1) It is necessary to cut off and to reinstate the H_p polarizing field with cutoff times smaller than the Larmor period, which is difficult to get if the H_p field is high and the Q factor of the coil has a high value.

(2) The sequence of operation in the Meiboom and Gill method is not easy to get because the first pulse must be out of phase by $\pi/2$ in comparison with that of other pulses. It is then necessary to cancel rigorously the terrestrial magnetic field and to realize another of the same amplitude and homogeneity but in a perpendicular direction.

We can eliminate both difficulties by proceeding as follows (*37*):

(a) As in the classical method of free precession, the signal is received in the polarization coil, but this coil is able to give the polarization field H_p parallel and in the same sense as that of the terrestrial field $H_{T'}$.

(b) Two pairs of identical coils with perpendicular axes, giving the same magnetic field amplitude as that of H_T but inhomogeneous (coils of small dimensions), are connected with equal current sources and oriented in such a way as to give, when the H_p field is cut off, a neutralization of the H_T field (field $-H_T$); and a production of an inhomogeneous H_{PT} field perpendicular to H_T and of approximately the same value, as shown in Fig. 18.

Fig. 18. Relative orientation of fields acting on the sample in the Meiboom and Gill sequence.

The Meiboom and Gill sequence can easily be obtained as follows:

(a) The sample is left in the H_p field (parallel to H_T and in the same sense) during a polarization time t ($t > T_1$), whereby a polarization M_p is obtained.

(b) After cutting off H_p, M_p slowly decreases at a constant time T_1 without change of direction.

(c) At the time θ ($\theta < T_1$), a circuit capable of giving a $-H_T$ and a H_{PT} field is connected. The effective field acting on M_p is inhomogeneous and the elementary contributions go rapidly out of phase.

(d) After a time τ, the elementary moments are distributed in a circular pattern perpendicular to the H_{PT} plane, whereupon the current is cut off (i.e., $\tau < T_1$ but $\tau > \omega_T^{-1}$).

(e) At this time the nuclear moments are subjected only to the H_T field; such condition is maintained during the time $t = \pi/\gamma H_T$. The circular pattern, in which the elementary moments are situated, is then turned at a π angle around the H_T axis. It is then necessary to have $t \ll \tau$.

(f) After the time t the current is connected (and the $-H_T$ and H_{PT} fields reinstated) and the echo can be observed at the time 2τ.

For the extension of the Hahn (*34*) and Carr–Purcell (*35*) method, the H_P field must be perpendicular to H_T and H_{PT} must be parallel to and in the same sense as H_p, as shown in Fig. 19.

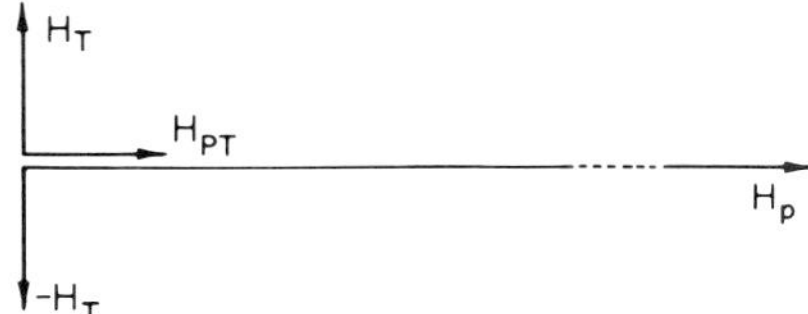

FIG. 19. Orientation of the various fields applied (for the Hahn and Carr–Purcell methods).

3. Rotary Echoes at Zero Frequency (*38*)

It is actually possible to overcome the necessity of using the terrestrial magnetic field as a source of pulsation at zero frequency and thus to be thoroughly independent of it. This can be obtained by rotary echoes at zero frequency by extrapolating from the first Solomon procedure (*39*).

As in the Packard and Varian (*30*) experiment of free precession, a prepolarizing field is used. The approximate cancellation of the terrestrial magnetic field is obtained by a suitable coil. In a perpendicular plane to H_T, a pair of small coils generates the feeble inhomogeneous reference field H_I.

Before the Larmor frequency of nuclei in the H_I field is reached, there must be the possibility of rapidly reversing the sense of the current in the coils.

The rotary echoes at zero frequency can be observed as follows:

(a) After cancellation of the terrestrial magnetic field, the sample is submitted to the polarizing field so as to get an extensive H_p magnetization. The current in the coils gives a H_I field obeying the following conditions:

$$H_I \perp H_P \qquad \text{and} \qquad H_I \ll H_P$$

(H_p and H_T are in line).

(b) H_p is cut off. The M_p magnetization plane, being perpendicular to $+H_I$ direction, precesses around that direction with Larmor pulsation ω_I. However, the $+H_I$ field is inhomogeneous and the elementary moments, subjected to slightly different fields, are soon out of phase and then spread over a circular path perpendicular to $+H_I$.

(c) After a time τ, where $\tau > (2\pi/\omega_I)$ but $\tau \ll T_1$, the current is reversed in the coils. The sample is then subjected to the inhomogeneous field $-H_I$, but the geometrical distribution of the inhomogeneity produced in the coils is symmetrical to that of $+H_I$. In each point of the sample the sense of Larmor precession is reversed, but the amplitude of the angular speed ω_I is kept up.

(d) At a time 2τ, an echo resulting from the reinstatement of the common phase of elementary moments can be observed. The magnitude of the echo is related to the inhomogeneity of field H_I.

(e) If at time 3τ the current is reversed once more, there will be another echo at time 4τ, and so on. Just as in the case of the Solomon or of the Meiboom and McGill methods, the errors due to small variations of τ or H_I are not cumulative.

Note Added in Proof

We were able to observe rotary echoes at zero frequency by the procedure outlined hereupon (B. Borcard, E. Hiltbrand, and G. J. Béné, submitted for publication to *Compt. Rend. Acad. Sci., Paris*, in May 1969).

IV. Conclusions

This article is limited to the description of the aspects of two particular magnetic resonances: atomic and nuclear. However, we have tried to show that the relatively old fields used in the past 20 to 30 years have still further applications. The description of planned experiments, described in more detail in other publications, shows that further developments of these techniques are possible.

References

1. J. P. Barrat, *J. Phys. Radium* **20**, 541, 633, 657 (1959).
2. G. Hanuise, *Office Natl. d'Etudes et de Rech. Aerospat.* No. 104, 49 (Jan.–Feb. 1965).
3. A. Kastler, *Proc. Zeeman Centennial Conf.* (Amsterdam 1965). Neth. Phys. Soc.; *Physica* **33**, 73 (1967).
4. M. J. Minneman *et al.*, Les gyroscopes avancés [*J. étude* **2**, 505 (Sept. 6, 1964)]. Centre Natl. Etudes Spat.
5. G. Hanuise, *Office Natl. d'Etudes et de Rech. Aerospat.*, Publ. No. 118 (1967).
6. G. J. Béné and E. Geneux, *Helv. Phys. Acta* **40**, 329 (1967).
7. L. N. Novikov, V. G. Pokazanev, and L. I. Yakub, *Soviet Phys. JETP* (*English Transl.*) **26**, 752 (1968).
8. A. Messiah, "Mécanique Quantique," Vol. I, p. 279. Dunod, Paris, 1959.
9. J. Brossel and F. Bitter, *Phys. Rev.* **86**, 308 (1952); J. Brossel and A. Kastler, *Compt. Rend.* **229**, 1213 (1949).
10. W. Hanle, *Ergeb. Exakt. Natur.* **4**, 214 (1925).

11. E. B. Alexandrov, O. V. Konstantinov, V. I. Perel', and V. A. Khodovoï, *J. Exptl. Theoret. Phys.* **45**, 503 (1963).

12. C. J. Favre and E. Geneux, *Phys. Letters*, **8**, 190 (1964).

13. C. Cohen-Tannoudji and N. Polonsky, *Compt. Rend.* **260**, 5231 (1965).

14. N. Polonsky and C. Cohen-Tannoudji, *J. Phys. Radium* **26**, 409 (1965).

15. E. Crettenand, Diploma Physics, Geneva (1968).

16. E. Geneux, *Proc. OPALS, Varsovie, 1968* (in press).

17. E. Geneux, and G. Béné, *Phys. Letters* **25A**, 199 (1967).

18. A. G. Redfield, *Phys. Rev.* **98**, 1787 (1955).

19. C. Cohen-Tannoudji and S. Haroche, *Compt. Rend.* **261**, 5400 (1965).

20. C. Cohen-Tannoudji and S. Haroche, *Compt. Rend.* **262**, 37 (1966).

21. P. Badan and E. Geneux, *Helv. Phys. Acta* **35**, 174 (1966).

22. J. C. Lehmann, *J. Phys.* **25**, 809 (1964).

23. G. J. Béné and E. Geneux, *Compt. Rend., SPHN, Geneva* **2**, 39 (1967).

24. H. Cottet, E. Geneux, and C. Rieben, *Helv. Phys. Acta* **38**, 343 (1965).

25. F. D. Colegrove, P. A. Franken, R. R. Lewis, and R. H. Sands, *Phys. Rev. Letters* **3**, 420 (1959).

26. T. G. Eck, L. L. Foldy, and H. Wieder, *Phys. Letters* **10**, 239 (1963).

27. E. Geneux, *Phys. Letters* **24A**, 295 (1967).

28. J. J. Forney and E. Geneux, *Phys. Letters* **20**, 632 (1966).

29. A. Erbeia, Thesis Paris, 1962, *Compt. Rend.* **251**, 1493–1495 (1960).

30. M. Packard and R. Varian, *Phys. Rev.* **93**, 941 (1954).

31. P. A. Grivet and L. Malnar, *Advan. Electron. Electron Phys.* **23**, 55–64 (1967).

32. H. C. Torrey, *Phys. Rev.* **76**, 145 (1949).

33. G. J. Béné, *Compt Rend.* **264**, 340 (1967).

34. E. L. Hahn, *Phys. Rev.* **76**, 145 (1949).

35. H. Y. Carr, and E. M. Purcell, *Phys. Rev.* **94**, 630 (1954).

36. S. Meiboom, and D. Gill, *Rev. Sci. Instr.* **29**, 688 (1958).

37. G. J. Béné, *Helv. Phys. Acta* **41**, 420 (1968).

38. G. J. Béné, *Compt. Rend. SPHN, Geneva* **3**, 40–42 (1968).

39. I. Solomon, *Phys. Rev. Letters* **2**, 301 (1959).

Mass Spectroscopy Using RF Quadrupole Fields

P. H. DAWSON* AND N. R. WHETTEN

General Electric Research and Development Center
Schenectady, New York

Present address: Centre de Recherche sur les Atomes et les Molecules, Université Laval, Quèbec, Canada.

I. Introduction

A. Organization and Scope

Two reviews of mass spectroscopy have appeared in earlier volumes of *Advances in Electronics and Electron Physics*. Volume I in 1948 included "Modern Mass Spectroscopy" by Mark Inghram. This was followed in Volume VIII of the series by Larkin Kerwin's "Mass Spectroscopy."

In the decade following the last review, significant changes have occurred in many areas of mass spectroscopy. In the present article we review a single class of instruments in which this change has been extremely rapid. At the time of the last review, quadrupole instruments using rf fields were not widely known and were hardly mentioned in that review. Yet today they perform a major role in partial pressure analysis, and their role appears likely to expand in other areas of mass spectroscopy in the near future. Much of the original work on quadrupole instruments is either not in the open literature or is not available in English. Several brief reviews (*1, 2*) have appeared, but these have been limited in scope.

Mass spectrometers can be grouped into two classes of instruments, static and dynamic. The latter are based on the time dependence of one or more parameters of the system. Dynamic spectrometers can be subdivided (*1*) into three classifications: energy balance spectrometers (omegatron, rf linear accelerators), time-of-flight spectrometers, and path stability spectrometers.

The three types of quadrupole instruments fall in the path stability class but differ significantly from each other in their method of operation. The quadrupole mass filter depends for its mass selectivity on path stability properties alone. The monopole spectrometer uses a combination of stability and focusing properties. The three-dimensional quadrupole spectrometer, the least known of the three, uses stability properties in three dimensions. It is an ion storage spectrometer and consequently has properties that are different from conventional mass spectrometers.

We first describe the general theoretical background for charged particle motion in rf quadrupole fields and include graphical illustrations of solutions to the Mathieu equations of motion of the particles. The application of this theory to each of the three quadrupole instruments is then considered in turn, together with a description of the performance that has been obtained to date. Typical uses that demonstrate characteristic features of these devices are included. The ion storage quadrupole appears to be a promising tool in ion and molecular physics in addition to its use as a mass spectrometer. A brief summary of the published work on ion storage has been included, even though strictly speaking it is not mass spectroscopy. The principles of operation are so closely connected with its use as a spectrometer that its treatment follows naturally.

The review concludes with a discussion of problems that can result from nonperfect quadrupole fields. Field faults can cause nonlinear resonances, with consequent peak distortion and loss of sensitivity. The problems are common to all quadrupole instruments and are presented in a single section.

A review of this kind provides an opportunity for crystal-ball gazing, and we have indulged ourselves in the last section by attempting to predict some future trends. As with most soothsayers, we are hopeful that only the successful predictions will be remembered.

B. Historical Introduction

The rf quadrupole mass spectrometers provide an interesting example of how scientific breakthroughs in one area sometimes provide the impetus necessary for breakthroughs in other areas that are seemingly unrelated.

A key breakthrough occurred in the field of accelerator physics in late 1952. Courant, Livingston, and Snyder (3) who were working at Brookhaven National Laboratory where the "cosmotron" accelerator was being constructed, announced the discovery of the strong-focusing, alternating gradient technique for focusing the proton beam. Quadrupole magnetic fields were used to "squeeze" the beam first in one direction and then alternatively in a direction at right angles to the first. The focusing thereby achieved was much stronger than that obtained with conventional techniques, making possible a reduction in the cross section of the beam and in the magnet cost.

Curiously, the strong-focusing technique had been discovered two years earlier by N. C. Christofilos, an electrical engineer in Athens, Greece, whose hobby was studying accelerators. He had sent reports of his work to the University of California Lawrence Radiation Laboratory and had applied for patents. His work was overlooked until after the strong-focusing, alternating gradient principle had been independently rediscovered at Brookhaven.

The quadrupole field strong-focusing principle was recognized by Wolfgang Paul (4) and his colleagues at the University of Bonn, Germany, as making possible quadrupole mass spectrometers using electric rf fields. Electric quadrupole field focusing had been shown to be feasible, and analogous to magnetic quadrupole focusing, by Blewett (5). Paul had been working with magnetic hexapole fields to focus paramagnetic atomic beams, and was able to transfer the new advance readily into the field of mass spectroscopy. Paul and Steinwedel applied for a patent (6) on the quadrupole mass filter in 1953, a year after the announcement of the strong-focusing, alternating gradient principle. Paul, his colleagues, and students pioneered most of the fundamental developments in quadrupole mass spectroscopy during the 1950s.

The quadrupole mass filter principle was independently proposed by R. F. Post at the University of California Radiation Laboratory. Post constructed a mass filter but did not publish his work except in Radiation Laboratory reports (7); consequently it is not widely known. The impetus for Post's work was also the discovery of the strong-focusing, alternating gradient principle.

The three-dimensional quadrupole ion trap was briefly described in the original patent of Paul and Steinwedel (6), and a description of an experimental device was given in little known reports in 1955 (8) and 1956 (9). The first formal publication was in 1959 (10) by Fischer. In the same year a similar apparatus for containing macroscopic charged particles was reported by Wuerker, Shelton, and Langmuir (11).

The monopole mass spectrometer was the last of the three quadrupole instruments to be proposed. It was first described by von Zahn (12) in 1963. von Zahn had participated with Paul in much of the earlier work on the quadrupole mass filter, but his initial work on the monopole was done at the University of Minnesota.

In the early 1960s, quadrupole mass filters began to be employed in upper atmosphere and space research, and this spurred further developmental efforts. A number of manufacturers now supply commercial versions of the quadrupole mass filter and the monopole mass spectrometer (see Appendix.) In the past few years the use of quadrupole spectrometers has rapidly increased. Their performance has steadily improved, and they have evolved from special devices into general purpose instruments.

C. List of Symbols

$a =$ Mathieu equation parameter, related to the applied dc voltage

$d =$ an integer, for exact focusing with quadrupole fields, $\beta = p/d$, where p and d are integers with no common factors

$e =$ electronic charge

$f =$ frequency of the applied voltage, in MHz

$h =$ number used in expressing the resolution of the monopole ($M/\Delta M = n^2/h$)

$m =$ ionic mass

$n =$ number of rf cycles an ion spends in the quadrupole field

$p =$ an integer (see symbol d)

$q =$ Mathieu equation parameter, related to the applied rf voltage

$r =$ radial direction

$r_0 =$ field radius

$t =$ time, in seconds

$u =$ distance parameter, representing x, y, z, or r

$z_0 =$ field size parameter for the ion trap, equal to $r_0/2^{1/2}$

A_N = weighting factor for the Nth order distortion term in the potential
D = diameter of entrance aperture
L = length of the quadrupole field
M = ionic mass, in amu
ΔM = width of mass peak in amu at half-height unless specified otherwise
N = number of ions in the trap
P = pressure in torr
U = dc voltage applied between opposite sets of electrodes
V = rf voltage (zero to peak) applied between opposite sets of electrodes
β = parameter characteristic of the frequency of ion motion obtained from solution of the Mathieu equation
ξ = time, defined as $\omega t/2$
ρ = ion density
τ_0 = characteristic period of the fundamental ion motion
Φ = electric potential
ω = angular frequency of applied rf field

II. Charged Particle Motion in Quadrupole Fields

A. Electrode Structures and Quadrupole Fields

The potential in a quadrupole field may be expressed in rectangular co-ordinates by the relationship (*13*)

$$\Phi = \Phi_0(\lambda x^2 + \sigma y^2 + \gamma z^2) \tag{1}$$

where λ, σ, and γ are constants characteristic of the particular field. The restoring force on a charged particle depends on the gradient of the potential and is therefore proportional to the first power of the displacement of the particle from the center of the field. This type of restoring force, which increases with larger displacements, is called "strong focusing."

If it is assumed that there is no space charge within the electrode structure, the potential must satisfy Laplace's equation, $\nabla^2\Phi = 0$. Substituting the quadrupole potential into Laplace's equation yields the following relationship between the constants:

$$\lambda + \sigma + \gamma = 0 \tag{2}$$

This condition must be satisfied in all quadrupole field devices.

In the quadrupole mass filter, the constants are chosen so that

$$\lambda = -\sigma = \frac{1}{2r_0^2} \quad \text{and} \quad \gamma = 0 \tag{3}$$

The z axis is then a four-fold axis of symmetry (ignoring the sign of the potentials). The parameter r_0 is the distance from the center of the field to the

                    P. H. DAWSON AND N. R. WHETTEN

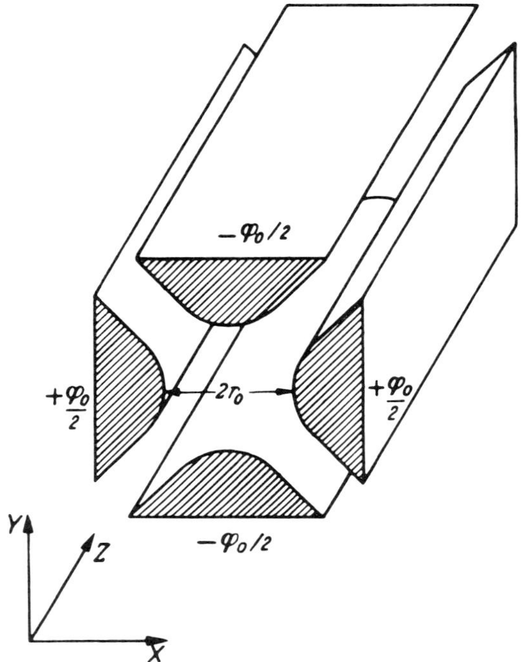

FIG. 1. Quadrupole mass filter geometry. Ions are injected along the z direction.

nearest point of the electrodes. The four electrode structure is shown schematically in Fig. 1. The cross sections of the electrodes in the x–y plane are rectangular hyperbolae.

In this review the voltage $(U - V \cos \omega t)$ is taken as being applied between the opposing sets of electrodes for all three types of instruments, with the center of the field chosen as the zero of the potential. The potential at any point is

$$\Phi = (U - V \cos \omega t)(\lambda x^2 + \sigma y^2 + \gamma z^2) \tag{4}$$

In the monopole, Fig. 2, the constants are chosen so that

$$\lambda = -\sigma = 1/r_0{}^2 \qquad \text{and} \qquad \gamma = 0 \tag{5}$$

The constants differ in magnitude from those of the mass filter by a factor of 2. Some authors have preferred to make the constants equal by defining $(U - V \cos \omega t)$ as one-half the applied potential difference between the two sets of electrodes in the mass filter. This leads to differences by a factor of 2 in some equations as used by different authors.

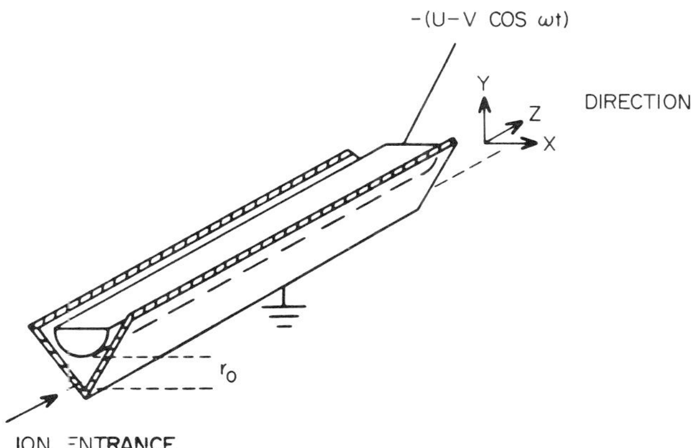

FIG. 2. The monopole mass spectrometer.

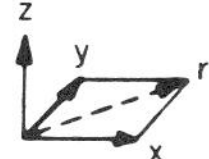

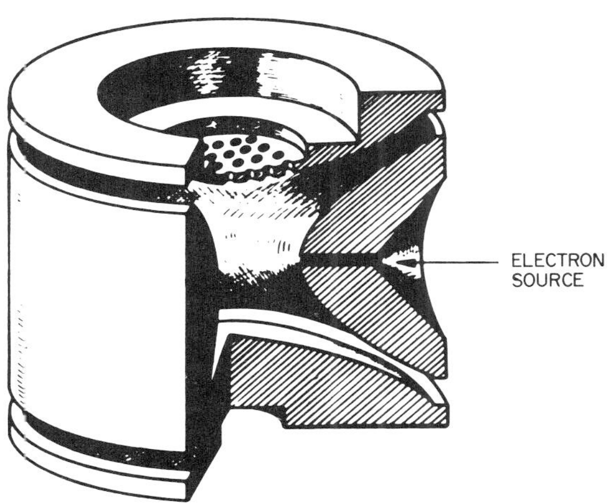

FIG. 3. The three-dimensional quadrupole ion trap. There is rotational symmetry about the z axis.

In the three-dimensional quadrupole ion trap,

$$\lambda = \sigma = 1/2r_0^2 \qquad \text{and} \qquad \gamma = -1/r_0^2 \tag{6}$$

The ion trap is a three-electrode structure, shown schematically in Fig. 3. The x and y directions are equivalent, and are usually denoted by r. The structure is rotationally symmetric about the z axis. The two end caps and the ring electrode have cross sections in any rz plane that are complementary hyperbolae with a ratio of $(2)^{1/2}$ in the semiaxes. Here r_0 is the minimum radius of the ring electrode.

Other structures can be envisioned that would have potentials satisfying Laplace's equation. For example, we could have $\lambda = 2\sigma$ and $\gamma = -3\sigma$. This field has three planes of two-fold symmetry and corresponds to a complex six-electrode structure. Such a device has not been reported in the literature.

B. Mathieu Equation of Motion

The electric fields obtained from the quadrupole potential (Eq. 4) are

$$E_x = -\frac{\partial \Phi}{\partial x} = -(U - V \cos \omega t)2\lambda x$$

$$E_y = -\frac{\partial \Phi}{\partial y} = -(U - V \cos \omega t)2\sigma y \tag{7}$$

$$E_z = -\frac{\partial \Phi}{\partial z} = -(U - V \cos \omega t)2\gamma z$$

The equations of motion of a singly charged ion of mass m are

$$\frac{d^2x}{dt^2} + \frac{e}{m}(U - V \cos \omega t)2\lambda x = 0$$

$$\frac{d^2y}{dt^2} + \frac{e}{m}(U - V \cos \omega t)2\sigma y = 0 \tag{8}$$

$$\frac{d^2z}{dt^2} + \frac{e}{m}(U - V \cos \omega t)2\gamma z = 0$$

Motion in each coordinate direction is therefore independent of motion in the other directions. Consequently, ion motion may be illustrated by considering only one-dimensional motion. Later sections of this article deal with motion in more than one dimension.

The differential equations, Eq. (8), are of the type known as Mathieu equations. They can be put in the canonical form of the Mathieu equation by the substitutions

$$\omega t = 2\xi, \qquad a_x = \frac{8eU}{m\omega^2}\lambda, \qquad q_x = \frac{4eV}{m\omega^2}\lambda \qquad (9)$$

and similar substitutions involving σ and γ. The equations of motion all become

$$\frac{d^2u}{d\xi^2} + (a_u - 2q_u \cos 2\xi)u = 0 \qquad (10)$$

where u represents $x, y,$ or z. The properties of the Mathieu equation are well known (*14, 15*). We are primarily interested in the solutions that result in stable or bounded ion motion.

If $\pm(\pi/2 \pm \xi)$ is substituted for ξ, Eq. (10) is unchanged except for the sign of the $(2q \cos 2\xi)$ term. Some authors use this alternative form. The only difference lies in defining the initial conditions when the equation is integrated to determine ion trajectories. The phase of the rf field when an ion enters the field is taken into account by substituting $(\xi - \xi_0)$ in place of ξ in Eq. (10). The "initial rf phase" is then expressed in this article in terms of $\omega t_0 = 2\xi_0$.

C. Solutions to the Equation of Motion

Solutions to the Mathieu equation can be expressed in the form

$$u = \alpha'e^{\mu\xi} \sum_{s=-\infty}^{\infty} C_{2s} e^{2is\xi} + \alpha''e^{-\mu\xi} \sum_{s=-\infty}^{\infty} C_{2s} e^{-2is\xi} \qquad (11)$$

The solutions are of two types, solutions denoted "stable" where u remains finite as $\xi \to \infty$, and "unstable" solutions where u increases without limit as $\xi \to \infty$. The constant μ determines the type of solution. μ depends on the values of a and q, and is independent of the initial conditions. If μ is real and nonzero, instability will arise from either the $e^{\mu\xi}$ or the $e^{-\mu\xi}$ factor. There are three other possibilities:

1. $\mu = i\beta$ is purely imaginary and β is not a whole number. The solutions are stable.

2. μ is a complex number. The solutions are unstable (assuming that u_0 and $\dot{u}_0$ are not both zero).

3. $\mu = in$ is purely imaginary and n is an integer. The solutions are periodic but unstable.

Since μ depends on the values of a and q, the regions of stable operation can be plotted in a–q space, as shown in Fig. 4. The boundary values can be obtained from tabulations (*15*). The shaded areas represent combinations of a and q giving stable ion motion in this one-dimensional representation. Such a

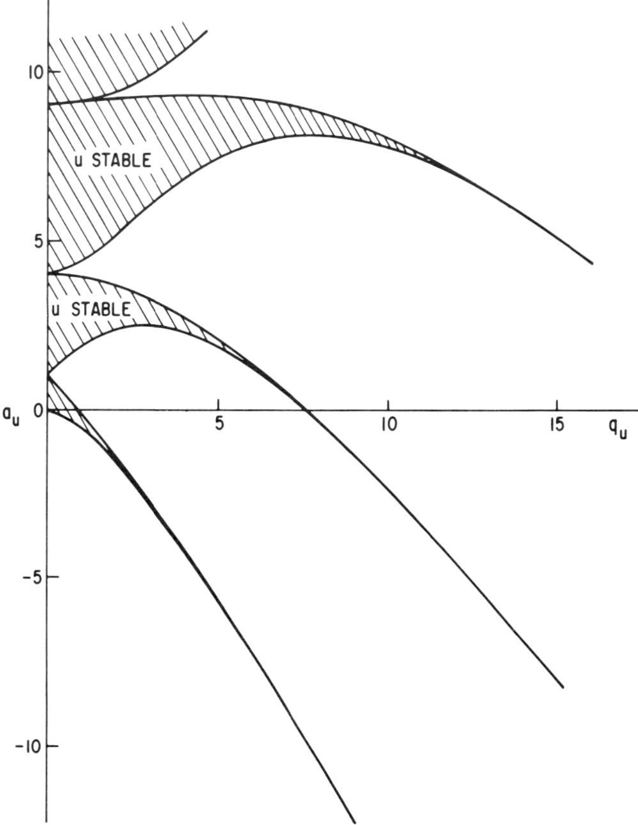

FIG. 4. The Mathieu (a, q) stability diagram for motion in one dimension. The shaded areas correspond to bounded ion trajectories.

figure is referred to as a stability diagram. The stable areas with large values of a and q correspond to ion motion in which the amplitudes of oscillation are large compared with the initial displacement. These amplitudes are generally too large to be useful in practical devices. The stability area near the origin which is of more practical significance is shown in detail in Fig. 5. Points in a–q space having the same value of β are shown as iso-β lines. The stability limits are $\beta = 0$ and $\beta = 1$.

When $\mu = i\beta$, Eq. (11) may be expressed in the form

$$u = \alpha_{\mathrm{I}} \sum_{s=-\infty}^{\infty} C_{2s} \cos(2s + \beta)\xi + \alpha_{\mathrm{II}} \sum_{s=-\infty}^{\infty} C_{2s} \sin(2s + \beta)\xi \qquad (12)$$

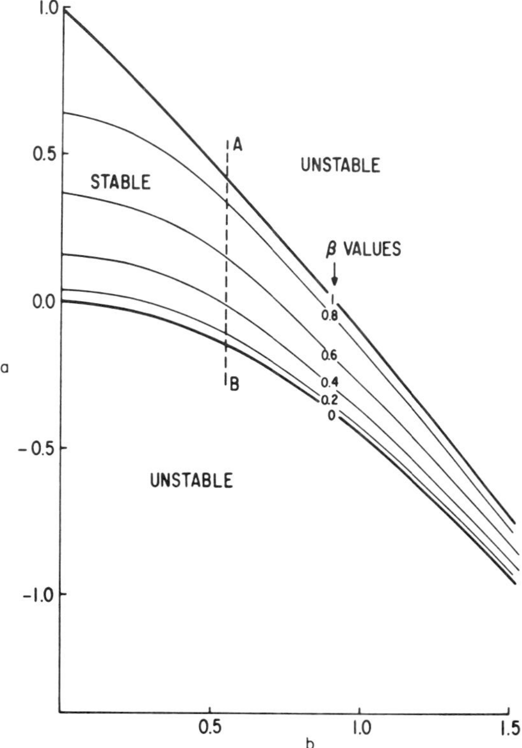

Fig. 5. Detail of the stable area near the origin. The parameter β is related to the frequencies of ion motion.

The integration constants α_I and α_II depend on the initial conditions existing at introduction of the ions, that is, on u_0, $\dot{u}_0$, and the phase of the rf field, ωt_0. The constants C_{2s} and β depend only on a and q. All ions of the same (a, q) value have motion of the same periodicity. β, therefore, is a parameter characteristic of the frequencies of ion motion.

D. Frequency of Ion Motion

The ion motion has a fundamental frequency

$$\omega_0 = \frac{\beta}{2}\,\omega \tag{13}$$

and higher frequencies

$$\omega_1 = \left(1 - \frac{\beta}{2}\right)\omega, \qquad \omega_2 = \left(1 + \frac{\beta}{2}\right)\omega, \qquad \text{etc.} \tag{14}$$

The significance of these frequencies is illustrated by the computed ion trajectory shown in Fig. 6. The conditions were $a = 0$, $q = 0.55$ so that $\beta =$

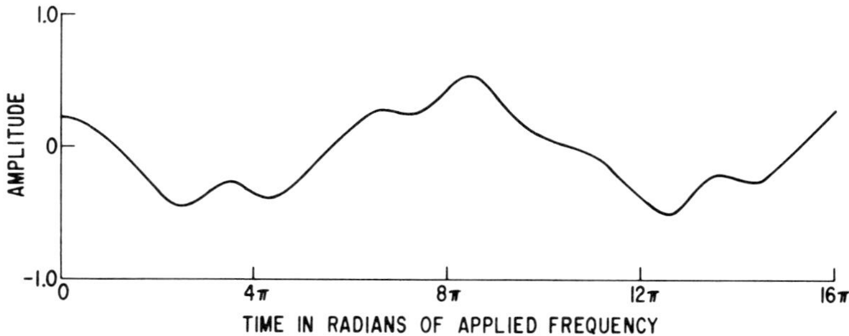

FIG. 6. Ion trajectory for the conditions $a = 0$ and $q = 0.55$. The fundamental frequency of ion motion has a characteristic period of 9.65π. Other periods are 2.44π and 1.66π.

0.414, and

$$\begin{aligned}
\omega_0 &= 0.207; & \tau_0 &= 9.65\pi \text{ rad of applied field} \\
\omega_1 &= 0.793; & \tau_1 &= 2.44\pi \text{ rad of applied field} \\
\omega_2 &= 1.207; & \tau_2 &= 1.66\pi \text{ rad of applied field}
\end{aligned}$$

The characteristic periods τ_0 and τ_1 can readily be seen in the trajectory.

In analytical treatments of quadrupole properties (13), the higher frequencies ω_1 and ω_2 are sometimes neglected in comparison with the fundamental frequency that has the largest amplitude of oscillation. This is equivalent to replacing the Mathieu equation with an equation for simple harmonic motion. That is,

$$\frac{d^2u}{d\xi^2} + \beta^2 u = 0 \tag{15}$$

More rigorous analytical solutions require the calculation of the coefficients C_{2s} with recurrence relations such as

$$\begin{aligned}
\frac{C_{2s}}{C_{2s-2}} = &-\frac{q/(2s + \beta)^2}{1 - a/(2s + \beta)^2} - \frac{q^2/(2s + \beta)^2(2s + \beta + 2)^2}{1 - a/(2s + \beta + 2)^2} \\
&- \frac{q^2/(2s + \beta + 2)^2(2s + \beta + 4)^2}{1 - a/(2s + \beta + 4)^2} - \cdots
\end{aligned} \tag{16a}$$

and

$$\frac{C_{2s}}{C_{2s-2}} = \frac{a - (2s - 2 + \beta)^2}{q} + \frac{q/(2s - 4 + \beta)^2}{1 - a/(2s - 4 + \beta)^2}$$

$$- \frac{q^2/(2s - 4 + \beta)^2(2s - 6 + \beta)^2}{1 - a/(2s - 6 + \beta)^2} - \cdots \qquad (16b)$$

Figure 7 shows some of the coefficients as a function of q and β as calculated by Berkling (9). Only for small values of q and β can all the terms of the series following the first be dropped. Table I presents the coefficients from C_0 to $C_{\pm 6}$ for q values of ± 0.705 and β values of 0.0, 0.02, 0.04, 0.08, and 0.92, 0.96, 0.98, and 1.0 as calculated by Paul *et al.* (*13*). These values are of particular interest in the operation of the quadrupole mass filter. They were also used by von Zahn (*12*) to derive an approximate solution for ion motion

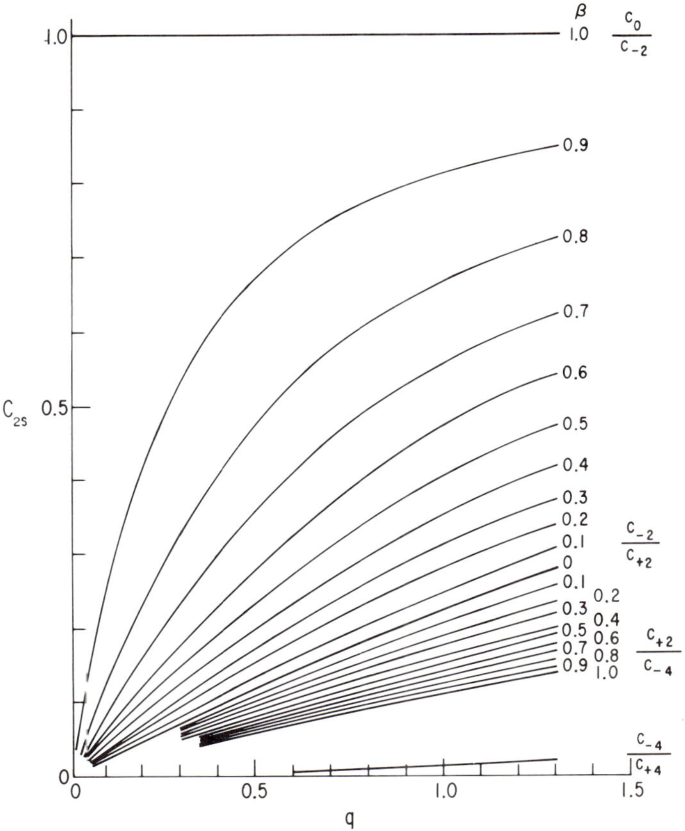

FIG. 7. C_{2s} as a function of β and q, taking C_0 as unity (9). The ordinate gives C_{2s}. The numbers on the right are β values corresponding to the labeled constants.

TABLE I

COEFFICIENTS FROM c_0 TO c_6[a]

q_y	β_y	c_0	c_{-2}	c_2	c_{-4}	c_4	c_{-6}	c_6
0.705	0.00	1.0	-0.16763	-0.16763	0.00729	0.00729	-0.00014	-0.00014
	0.02		-0.17089	-0.16449	0.00750	0.00708	-0.00015	-0.00014
	0.04		-0.17427	-0.16145	0.00773	0.00688	-0.00015	-0.00013
	0.08		-0.18143	-0.15569	0.00821	0.00651	-0.00016	-0.00012
q_x	β_x							
-0.705	0.92	1.0	0.79504	0.08515	0.06065	0.00254	0.00167	0.00004
	0.96		0.89098	0.08288	0.06990	0.00240	0.00196	0.00004
	0.98		0.94383	0.08176	0.07508	0.00235	0.00212	0.00003
	1.00		1.00000	0.08065	0.08065	0.00230	0.00230	0.00003

[a] Paul *et al.* (*13*).

in the monopole. Equations (16a) and (16b) can be used to calculate β, using an iterative procedure, since the two equations are equal for all s.

The iso-β lines in Fig. 5 are taken from detailed calculations by Berkling (9). Figure 8 gives the relationship between β and q (9) along the q axis

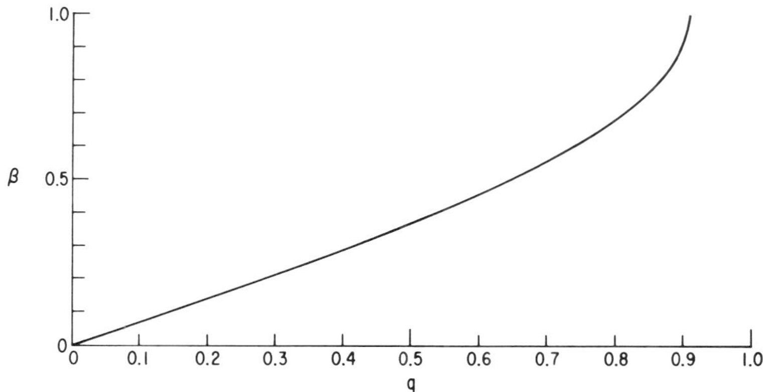

FIG. 8. β as a function of q when $a = 0$ (9).

$(a = 0)$. For small β the relationship can be approximated by

$$q = 2^{1/2}\beta(1 - \tfrac{3}{8}\beta^2) \tag{17}$$

E. Exact Focusing

Values of a and q where $\beta = p/d$ are of special interest for using the focusing properties of quadrupole fields (14, 16). Here p and d denote integers having no common factors. Both functions in Eq. (12) are then periodic, with the period $2\pi d$ in the variable ξ. If p is even (16), the functions have the period πd (addition of πd to ξ leaves all the terms unchanged). If p is odd, the functions are inverted after the period πd. Pairs of (q, a) values satisfying the two conditions $(a, q) = 1/d$ and $(-a, -q) = p/d$ for $2 < d < 13, p \leq d$ are given in Table II, as calculated by Lever (16).

Numerical integration of the differential equations of motion, using digital computers with automatic trajectory plotting, has recently been used to determine the ion motion with different initial conditions (16, 17). This is a much easier approach than attempting analytical or approximate solutions. Figures 9 and 10 illustrate focusing conditions by showing superimposed ion trajectories computed by Lever (16). The initial values of the phase, ξ_0, are varied in steps of $\pi/16$ form 0 to π. The abscissa is $(\xi - \xi_0)$. Figure 9(a) gives trajectories for $u_0 = 0, \dot{u}_0 = 1, a = -0.233982$, and $q = -0.704396$ so that $\beta = \frac{1}{20}$. Figure 9(b) gives trajectories for (a, q) values of the same magnitude but of opposite sign so that $\beta = \frac{19}{20}$. The significance of β is evident, and the focusing properties are excellent in both cases.

TABLE II

Paired Values of a (Upper) and q (Lower) Satisfying the Condition $\beta(a, q) = 1/d$; $\beta(-a, -q) = p/d$, for $2 < d < 13$, $p \leq d$[a]

Values of d	Values of p											
	1	2	3	4	5	6	7	8	9	10	11	12
3	0.000000	0.117076	0.183624									
	0.451105	0.656266	0.752057									
4	0.000000	0.078407	0.166227	0.206116								
	0.344959	0.522848	0.673016	0.732687								
5	0.000000	0.053806	0.126186	0.190656	0.216980							
	0.278436	0.429064	0.575816	0.683276	0.723310							
6	0.000000	0.038706	0.095019	0.156067	0.204406	0.223000						
	0.233150	0.362294	0.495503	0.610395	0.689585	0.718107						
7	0.000000	0.029026	0.073030	0.125161	0.175636	0.212867	0.226671					
	0.200422	0.312945	0.432374	0.542783	0.633358	0.693652	0.714933					
8	0.000000	0.022510	0.057492	0.100984	0.147256	0.189023	0.218429	0.229069				
	0.175699	0.275178	0.382516	0.485382	0.576545	0.649155	0.696401	0.712857				
9	0.000000	0.017941	0.046272	0.082550	0.123236	0.163704	0.198534	0.222274	0.230721			
	0.156378	0.245420	0.342500	0.437523	0.525286	0.601176	0.660405	0.698336	0.711428			
10	0.000000	0.014622	0.037966	0.068448	0.103772	0.140909	0.176176	0.205510	0.225041	0.231906		
	0.140869	0.221402	0.309820	0.397543	0.480593	0.555717	0.619571	0.668668	0.699745	0.710402		
11	0.000000	0.012139	0.031673	0.057528	0.088154	0.121500	0.155027	0.185808	0.210768	0.227097	0.232785	
	0.128150	0.201625	0.282696	0.363872	0.441966	0.514574	0.579307	0.633616	0.674900	0.700802	0.709641	
12	0.000000	0.010235	0.026802	0.048948	0.075590	0.105293	0.136268	0.166407	0.193377	0.214824	0.228667	0.233455
	0.117532	0.185067	0.259856	0.335236	0.408554	0.477963	0.541700	0.597892	0.644554	0.679707	0.701613	0.709061

[a] These are operating points for double focusing in the monopole mass spectrometer (16).

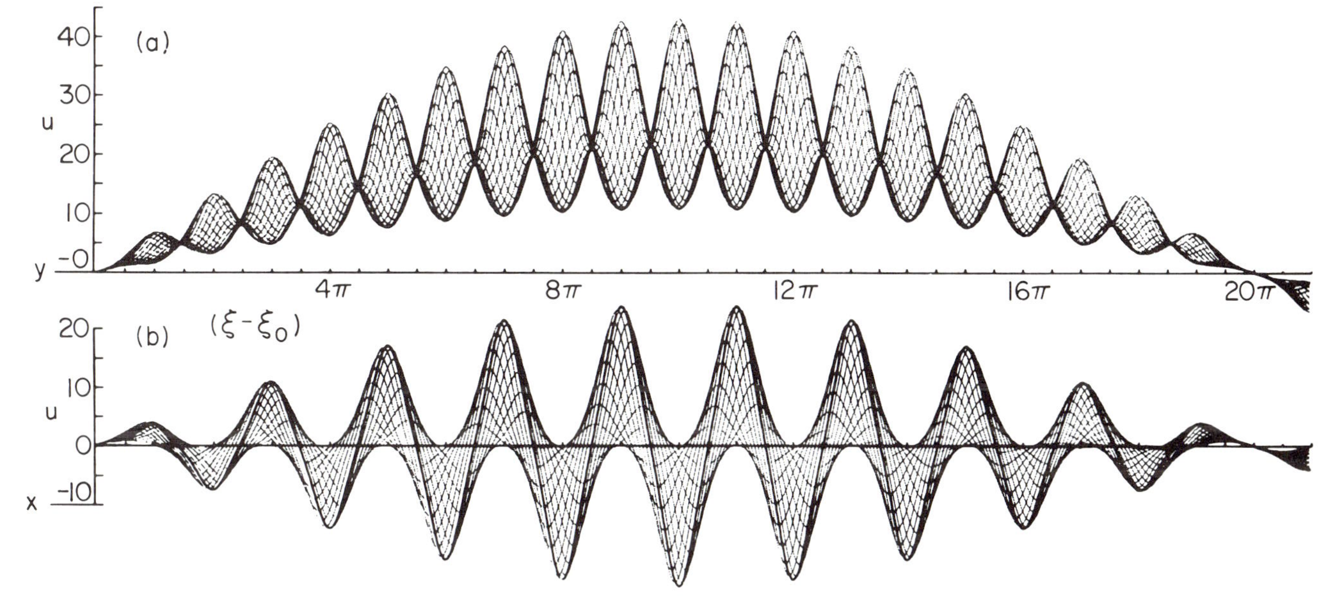

FIG. 9. Trajectories for ions entering at various initial rf phases computed (16) for $u_0 = 0$ and $\dot{u}_0 = 1$ when (a) $a = -0.233982$, $q = -0.704396$ so that $\beta = 1/20$, and (b) $a = +0.233982$, $q = +0.704396$ so that $\beta = 19/20$.

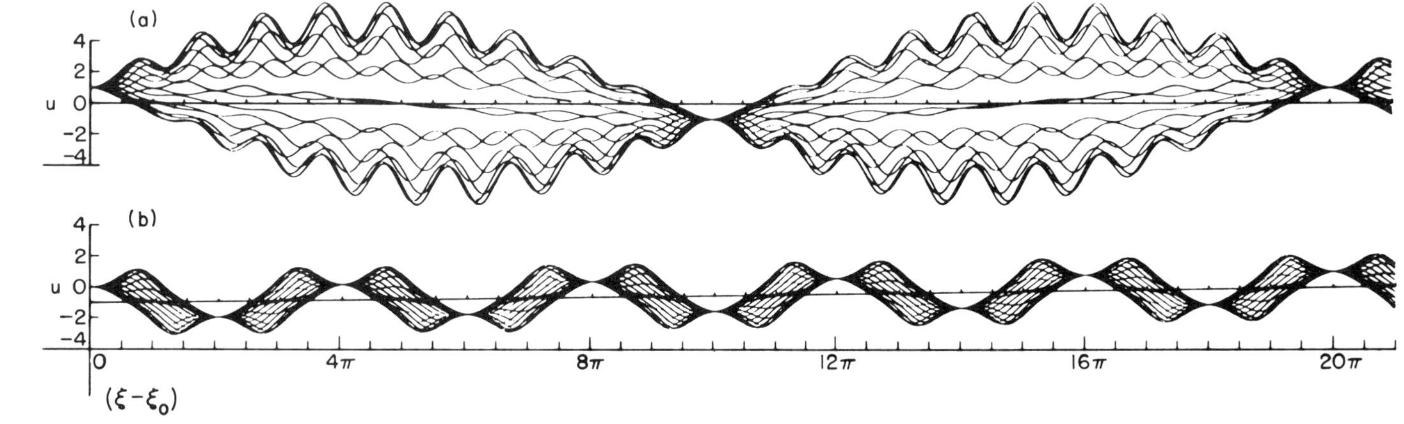

Fig. 10. Trajectories for ions entering at various initial rf phases computed (16) for $u_0 = 1$ and $\dot{u}_0 = 0$ when (a) $a = -0.103772$, $q = -0.480593$ so that $\beta = 1/10$, and (b) $a = +0.103772$, $q = +0.480593$ so that $\beta = \frac{5}{10}$.

Figure 10(a) shows trajectories for $u_0 = 1$, $\dot{u}_0 = 0$, $a = -0.103772$, $q = -0.480593$ so that $\beta = \frac{1}{10}$. Similarly, Fig. 10(b) shows trajectories for $\beta = \frac{5}{10}$. The value of the integer $(2p - 1)$ determines the number of times that ions introduced at a given phase will pass through their initial position before reaching the focus at $2\pi d$. Application of Lever's results to the design of an exact focusing monopole is considered in Section IV.

F. Ion Trajectories

To illustrate the nature of trajectories for different values of β, we consider various (a, q) values on the line AB of Fig. 5. Figures 11(a) and 11(b) show trajectories for $u_0 = 0.001$ and $\dot{u}_0 = 0$ as computed by Dawson and Whetten (17). A single phase of the rf field at ion introduction ($\omega t_0 = 0$) is considered. Figure 11(c) shows trajectories for $u_0 = 0$ and $\dot{u}_0 = 1.44 \times 10^{-4}$ meters/rad of applied field. The type of motion is independent of the initial conditions. At the stability boundaries the frequencies of ion motion are 0 and $\omega/2$. The ordinates in Fig. 11 can be scaled in an arbitrary manner. The influence of the initial phase of the rf field can be judged from Figs. 9 and 10.

G. Maximum Amplitude of Ion Motion

In a practical device, not only must a confined ion have a stable (bounded) trajectory, but also the maximum amplitude of this trajectory must be smaller than the internal dimensions of the device. For a given (a, q) value, the maximum amplitude of motion depends on the phase of the rf field when the ion is introduced.

From Eq. (12), the greatest possible amplitude (13) is

$$|u_M| = (\alpha_I^2 + \alpha_{II}^2)^{1/2} \sum_{-\infty}^{\infty} |C_{2s}| \tag{18}$$

We define $u_I(\xi)$ and $u_{II}(\xi)$ by the equation

$$u(\xi) = \alpha_I u_I(\xi) + \alpha_{II} u_{II}(\xi) \tag{19}$$

This equation, along with its derivative, can be used to solve for α_I and α_{II}. These values can then be substituted back into Eq. (18) to obtain

$$|u_M| = (1/W) \sum_{-\infty}^{\infty} |C_{2s}| \{[u_0 \dot{u}_{II}(\xi_0) - \dot{u}_0 u_{II}(\xi_0)]^2 + [\dot{u}_0 u_I(\xi_0) - u_0 \dot{u}_I(\xi_0)]^2\}^{1/2} \tag{20}$$

where W is the Wronskian determinant and does not depend on ξ_0. For a given value of ξ_0 then, $u_I(\xi_0)$, $u_{II}(\xi_0)$, $\dot{u}_I(\xi_0)$, and $\dot{u}_{II}(\xi_0)$ are all constant, and the quantity under the square root is biquadratic in u_0 and $\dot{u}_0$. For a chosen

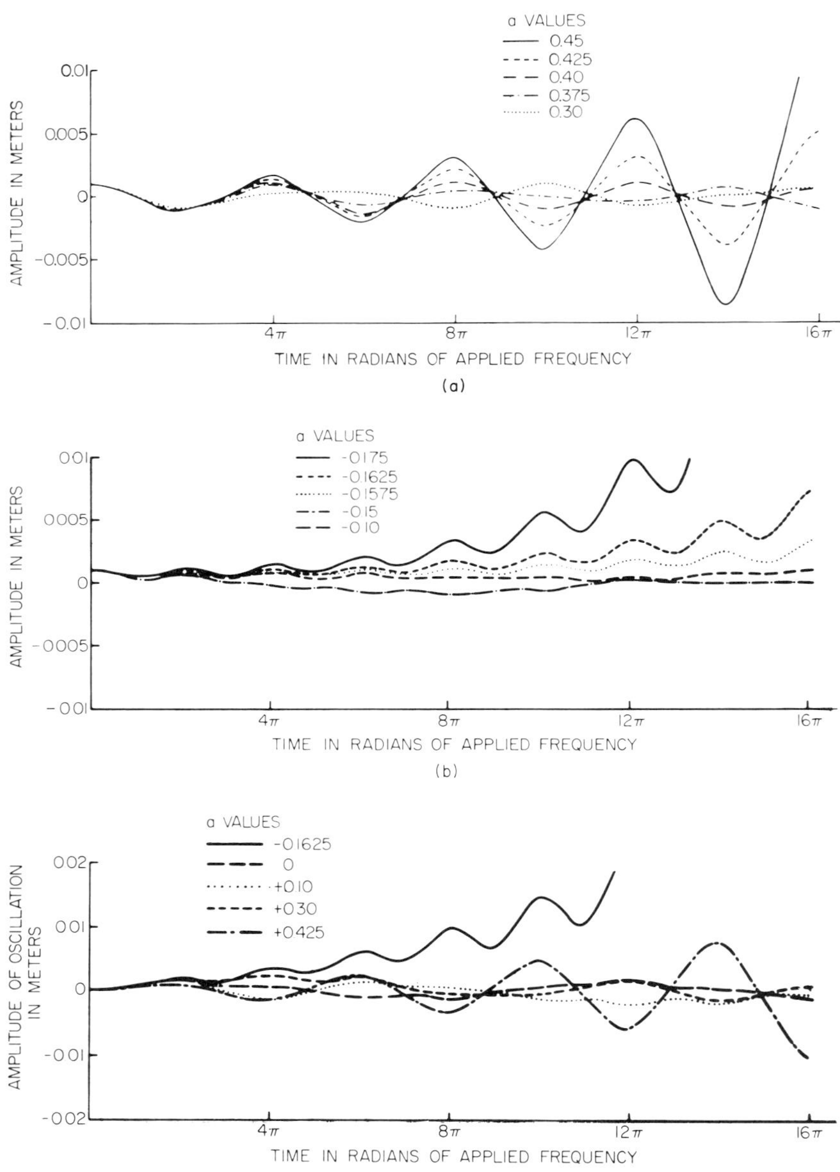

FIG. 11. Illustrations of ion trajectories for various positions on the line AB of Fig. 5. (a) and (b) are for $u_0 = 10^{-3}$ meters and $\dot{u}_0 = 0$, and (c) is for $u_0 = 0$ and $\dot{u}_0 = 1.44 \times 10^{-4}$ meters/rad of applied field. The amplitudes can be arbitrarily scaled (17).

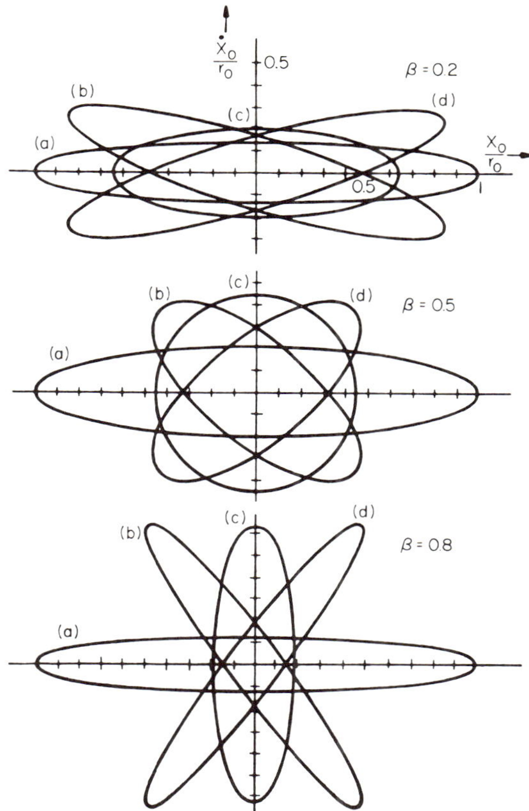

FIG. 12. Families of ellipses for three β values representing the maximum possible initial displacement and/or maximum initial velocity as a fraction of the total allowed amplitude, r_0 (13).The initial phases of the rf field in terms of ωt_0 are (a) 0, (b) $-\pi/2$, (c) π, and (d) $\pi/2$.

value of u_M (for example, r_0), Eq. (20) is an ellipse in the $(u_0, \dot{u}_0)$ plane. Different phases ξ_0 correspond to different ellipses. Figure 12 presents families of ellipses for $a = 0$ and $\beta = 0.2, 0.5,$ and 0.8 as computed by Fischer (10). For $\xi_0 = 0$, the maximum amplitude of oscillation equals the initial displacement. For all other initial phases of the rf field, the maximum amplitude of oscillation exceeds the initial displacement.

Since the maximum amplitude is equal to or greater than the initial displacement for all initial phases of the rf field, the *useful* displacement distance for the introduction of ions is less than r_0. Figure 13 presents ratios of u_M/u_0 for points of interest in mass filter operation as computed by Paul

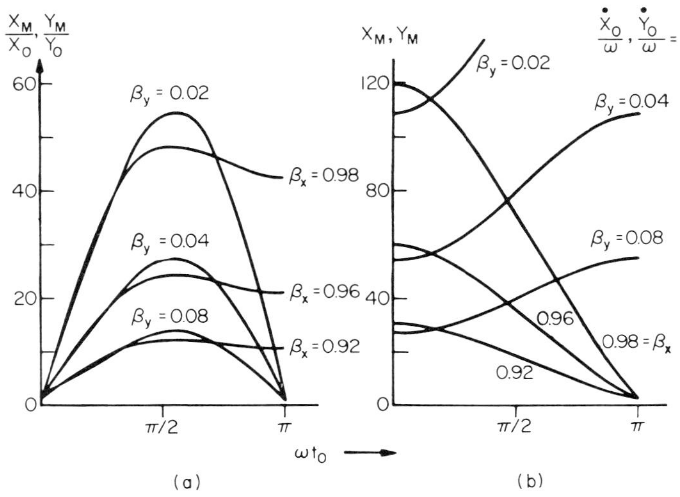

FIG. 13. (a) The maximum amplitude/initial displacement for β values of interest in the mass filter as a function of the initial rf phase. (b) Maximum amplitudes when the initial displacement is zero but the initial radial velocity is not zero (*13*).

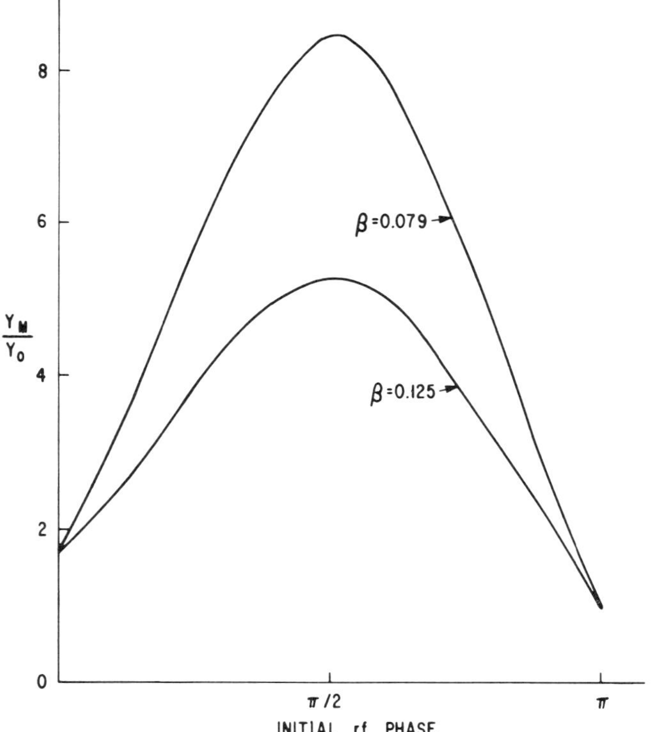

FIG. 14. Maximum amplitude/initial displacement ($\dot{y}_0 = 0$) as a function of the initial rf phase for two points on the scan line $a = 0.2q$. The maximum amplitude is that occurring in the first half-cycle of the fundamental ion motion. This value is important in monopole design.

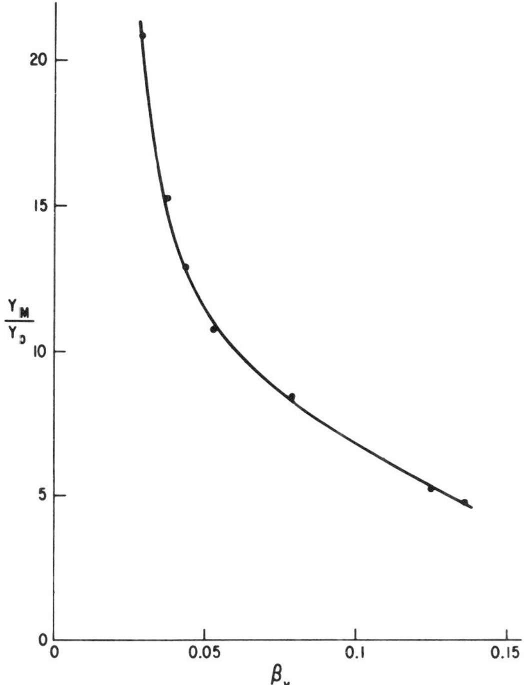

FIG. 15. Maximum amplitude/initial displacement as a function of β for the monopole with an $a = 0.2q$ scan line and $\pi/2$ initial phase.

et al. (*13*). Figure 13(a) corresponds to ions introduced with no initial velocity, and Fig. 13(b) to ions with no initial displacement.

In the moncpole, only the maximum amplitude in the first half-beat length is important. Figure 14 shows values for points on the scan line $a = 0.2q$, obtained by inspection of ion trajectories that were computed by Dawson and Whetten (*17*). Figure 15 gives the maximum amplitudes at the $\pi/2$ initial phase as a function of β for the same mass scan line. These amplitude calculations are important in instrument design, as will be illustrated in later sections.

III. THE QUADRUPOLE MASS FILTER

A. *Introduction*

The principle of operation of the mass filter was first described by Paul and Steinwedel (*4*) in 1953. This was followed by a description of an operating instrument by Paul and Raether (*18*) in 1955. The history of the device is presented in more detail in Section I. The mass filter, as the monopole and the ion trap, operates with no magnet and is small and light compared with magnetic instruments. It is called a mass filter rather than a spectrometer because ions are sorted by using path stability instead of focusing properties. For some applications a major advantage of the mass filter over the monopole and ion trap is its insensitivity to the component of ion velocity along its axis. This feature has made it attractive for the analysis of charged particles in space and the sampling of ions from gas discharges.

B. Principle of Operation

1. Electrode Configuration

The quadrupole field in the mass filter is obtained with four parallel rods, as shown schematically in Fig. 1. A dc voltage, U, and an rf voltage, $V \cos \omega t$, are impressed between the opposite pairs of rods. The z direction is defined as the axis of the instrument, parallel to the rods. In this review, the x direction is that through the centers of the pair of rods for which the dc component is positive with respect to the second pair.

The rods should be hyperbolic in cross section, located so as to correspond to equipotential lines of the quadrupole field. For convenient fabrication, most mass filters have been constructed with rods that are circular in cross section. The best approximation to the hyperbolic field using cylindrical rods is to space the rods so that $r_{\mathrm{rod}}/r_0 = 1.16$ (*19*), where r_{rod} is the rod radius and r_0 is the minimum distance from the axis of the instrument to the edge of a rod.

2. Equations of Motion

For the mass filter, the constants in Eq. (2) are $\gamma = 0$ and $\lambda = -\sigma = 1/2r_0^2$. The equations of ion motion, Eq. (8), are

$$d^2x/dt^2 + (e/mr_0^2)(U - V \cos \omega t)x = 0$$
$$d^2y/dt^2 - (e/mr_0^2)(U - V \cos \omega t)y = 0 \qquad (21)$$
$$d^2z/dt^2 = 0$$

The third equation expresses the fact that there is no acceleration in the z direction. The equations in x and y are identical except for the sign in the second term. If we define $a \equiv a_x = -a_y$ and $q \equiv q_x = -q_y$, then we can write both equations in the canonical form of the Mathieu equation:

$$d^2u/d\xi^2 + (a - 2q \cos 2\xi)u = 0 \qquad (22)$$

where u represents x or y (see Section II-B for definitions of a, q, and ξ).

As described in Section II-C, solutions to the Mathieu equation give bounded or "stable" trajectories for certain values of the constants a and q. The stability area in (a, q) space was shown in Fig. 4 for the one-dimensional case. For an ion to traverse the mass filter, it must have a stable trajectory in both the x and y directions. We can superimpose stability diagrams for the x and y directions to determine the stability region for the mass filter (Fig. 16). The diagram of Fig. 4 is symmetrical about the a axis, so the x and y regions differ only by a reflection through the q axis. The stable region of operation used in the quadrupole mass filter is that closest to the origin, which is shown enlarged in Fig. 17.

There was an early suggestion by Post (7) to use the second stable region

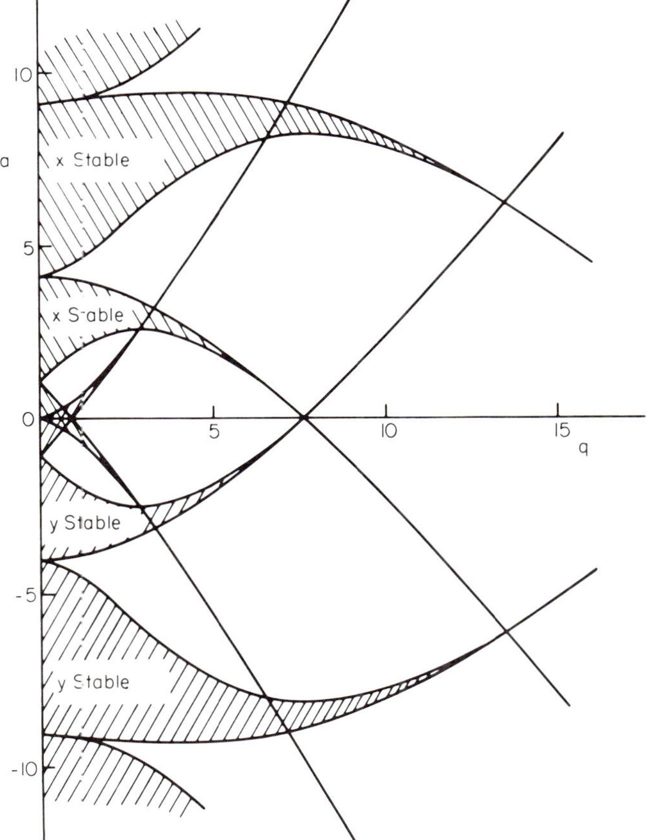

FIG. 16. Mathieu (a, q) stability diagram for the mass filter. In this two-dimensional case, stability diagrams for the x and y directions, which differ by the factor -1, are superimposed.

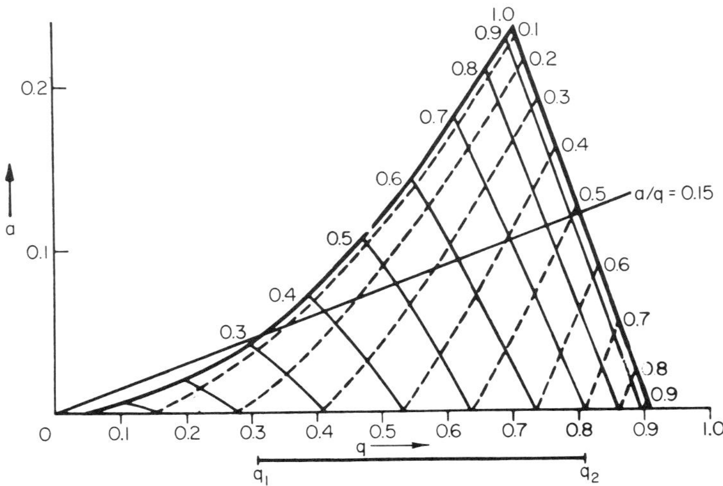

FIG. 17. Detail of mass filter stability diagram near the origin. The dashed lines are iso-β_y contours, the solid lines are iso-β_x contours. A mass scan line of fixed a/q ratio is shown (*13*).

on the $a = 0$ line which would offer a resolution of about 100. His calculations showed, however, that large amplitudes of oscillation occur and no details of the experimental use of this region have been published.

3. Mass Scan Line

The constants a and q, Eq. (9), are given by $a = 4eU/mr_0^2\omega^2$, $q = 2eV/mr_0^2\omega^2$. The ratio a/q is therefore $2U/V$. If the ratio of dc to rf voltage is fixed, this defines a line on the stability diagram of slope a/q, which is called the mass scan line. The apex of the stability triangle lies at $q_{\text{limit}} = 0.70600$ and $a_{\text{limit}} = 0.23699$. Consider a straight line through the origin and intersecting the stability triangle, as shown in Fig. 17. For any chosen value of U (and therefore V), ions of differing e/m will be spread out along the line according to their (a, q) value, with ions of larger e/m farthest from the origin. By varying the magnitude of U and V, keeping constant the ratio U/V, each ionic species may be brought in turn into the area for stability in both x and y directions, and can be transmitted through the device. Ions with e/m greater than the species passing through the analyzer have unstable trajectories in the x direction and will strike the x electrodes or pass out of the quadrupole field. Ions with e/m less than the species with stable trajectories are unstable in the y direction.

4. Ion Trajectories—Qualitative Description

As discussed in Section II-D, the type of motion undergone by an ion is determined by the values of β_x and β_y. Iso-β lines are shown on Fig. 17. The apex of the stability triangle is at $\beta_x = 1$ and $\beta_y = 0$. Trajectories for the point $a = 0.233982$ and $q = 0.704396$ are illustrated in Fig. 9. Figure 9(a) is the trajectory in the yz plane and Fig. 9(b) that in the xz plane. The type of motion differs greatly because of the difference in β values in the two directions.

The difference in the xz and yz trajectories may be better understood by the following qualitative description, after the treatment of Brubaker (20). The difference is caused by the dc voltage applied to the x electrodes with respect to the y electrodes. The rf voltage differs in phase only by 180 deg between the x and y electrodes. The positive potential on the x electrodes tends to make ions stable in the x direction even in the absence of an rf voltage (as at $q = 0$). If we eliminate the rf term for Eq. (22), we have

$$d^2x/d\xi^2 + ax = 0 \tag{23}$$

which is the equation for simple harmonic motion, with the ion oscillating through the center of the instrument. With rf voltage applied ($q \neq 0$), we must use the Mathieu equation of motion, but the trajectories in the x direction retain characteristics of the oscillatory motion through the center of the field as in Fig. 9(b). With increasingly large rf voltages, the motion remains oscillatory, but has a growing amplitude so that the ion becomes unstable in the x direction.

Ion trajectories in the y direction are unstable if no rf voltage is applied, since the dc field is defocusing in the y direction. Sufficient rf voltage must be applied to overcome this dc defocusing effect, and the trajectory in the y direction becomes oscillatory but returns through the center of the device with a much lower frequency.

5. Peak Shape

The maximum amplitudes x_M/x_0 and y_M/y_0 were shown in Section II-G to depend on β_x and β_y (Fig. 13). The iso-β lines may be replaced by lines of constant x_M/x_0. It is assumed here that ion entry is parallel to the axis ($\dot{x}_0, \dot{y}_0 = 0$). x_M and y_M can be put equal to r_0, since ions with greater excursions may strike the electrodes. The iso-β lines then correspond to lines of initial displacement, x_0^M and y_0^M, which are the maximum possible initial displacements for ions entering at all phases of the rf field to pass through the instrument without striking the electrodes (Fig. 18). Values of

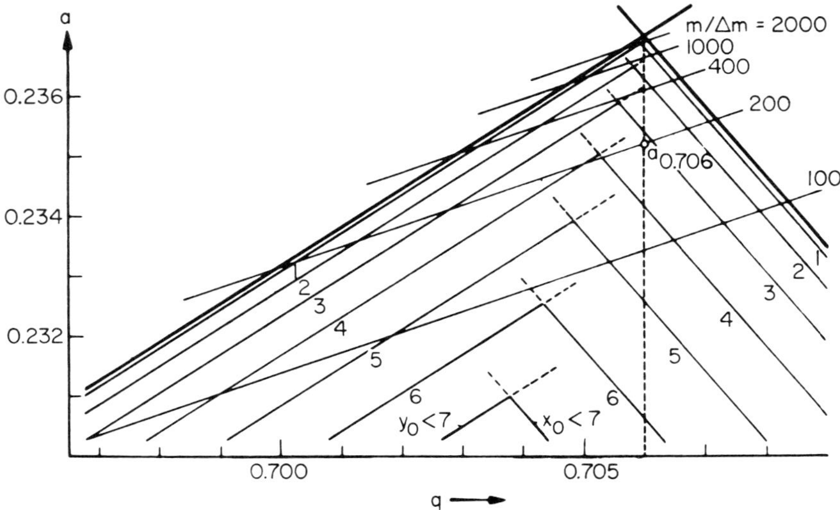

FIG. 18. The tip of the stability triangle with contours representing the maximum initial displacements $x_0{}^M$ and $y_0{}^M$ for 100% ion transmission in the mass filter when $r_0 = 100$ and $\dot{x}_0 = \dot{y}_0 = 0$ (13).

a and q within a pair of x and y lines result in 100% transmission for ions with initial displacements less than the corresponding lines. The initial displacement may be restricted (as by an entrance aperture) so as to define a triangle corresponding to 100% transmission. For a uniform current distribution at the aperture and a mass scan line intersecting the 100% transmission triangle, consider increasing the voltages so that a given e/m species passes through the stability triangle. Just inside the $\beta_y = 0$ boundary, some of the ions will be transmitted through the instrument. The transmitted ion current increases until the (a, q) value corresponding to the $y_0{}^M$ line is reached. Beyond this the ion current is constant until the $x_0{}^M$ line is reached. The ion current then decreases to zero at the $\beta_x = 1$ boundary of the stable region. The ion peak is flat-topped (trapezoidal), but the sides are not symmetrical because the mass scan line intersects the iso-β_x and β_y lines at different angles.

If the mass scan line intersects the stability diagram beyond the 100% transmission triangle, the peaks will be triangular rather than trapezoidal. As before, the sides of the peak will not be symmetrical. Since ions entering at some initial phases of the rf field are not being transmitted, the sensitivity is correspondingly smaller. The resolution, however, is improved, since the peak becomes narrower when the mass scan line intersects a narrower region of the stable area. One can define two operating ranges: (1) Region I: low resolution. The peaks are flat-topped (trapezoidal) with 100% ion transmission, and the maximum intensity is independent of the resolution. (2) Region II: high resolution. The peaks are triangular with less than 100% ion transmission, and the intensity is inversely proportional to the resolution (*13*).

6. Resolution

In this review, unless otherwise stated, we shall define the resolution as $(M/\Delta M)$, where ΔM is the width of the peak at half-height, expressed in mass units. At low resolution, where the peak has a broad flat top, the half-width may be nearly as great as the full width. The resolution is then the ratio of the length of the mass scan line, from the origin to the center of the stable area, divided by the length of the line between the boundaries of the stable region.

Using accurate measurements of the stable region (and assuming that the mass scan lines are parallel and of equal length to the center of the mass peak), Paul *et al.* (*13*) obtained the following relationship for resolution in the trapezoidal region:

$$(M/\Delta M) = 0.178/(0.23699 - a_{0.706}) \tag{24}$$

Here, $a_{0.706}$ is the value of a at the point where the mass scan line intersects $q = 0.706$.

In the triangular peak region, the half-width is approximately half of the full width, giving the expression

$$(M/\Delta M) = 0.357/(0.23699 - a_{0.706}) \tag{25}$$

At $q = 0.706$ in the region near the apex of the stable area, β_x and β_y can be expressed in terms of a by

$$\left.\begin{aligned} 1 - \beta_x &= (0.23699 - a)/1.9375 \\ \beta_y{}^2 &= (0.23699 - a)/0.79375 \end{aligned}\right\} q = 0.706 \tag{26}$$

Consequently the resolution can be related to β and therefore to x_M and y_M (Section II-G and Fig. 13). For injection parallel to the axis, the maximum amplitude ratio is [Paul *et al.* (*13*)]

$$x_M/x_0, \, y_M/y_0 < 1.8(M/\Delta M)^{1/2} \tag{27}$$

The maximum amplitude increases as the square root of the resolution. We can now relate the resolution to the optimum entrance orifice for 100% transmission of the ions. Setting x_M (or y_M) equal to r_0, and letting the diameter of the entrance hole be $D = 2x_0$ (or $2y_0$), we have

$$D \approx r_0/(M/\Delta M)^{1/2} \tag{28}$$

for the injection parallel to the axis.

When the ions are injected on the axis but with a radial velocity $\dot{x}_0$ or $\dot{y}_0$, the maximum permitted radial velocity for 100% transmission may be obtained from Fig. 13(b) and Eqs. 25 and 26, giving (13)

$$\dot{x}_M, \dot{y}_M < 0.16 r_0\,\omega(\Delta M/M)^{1/2} \tag{29}$$

C. Design Criteria

1. Voltage, Frequency, Power, Ion Velocity

For a singly charged ion of atomic weight M at the apex of the stable region, we have

$$V = 14.438 M f^2 r_0{}^2 \quad \text{and} \quad U = 2.424 M f^2 r_0{}^2 \tag{30}$$

where f is the frequency in megahertz and r_0 is in centimeters. Note that V is the zero-to-peak rf voltage applied between the pairs of rods (Section II-A).

The rf power is given by

$$P = 6.5 \times 10^{-4} C M^2 f^2 r_0{}^4/Q \tag{31}$$

in watts, where C in micromicrofarads is the capacitance of the system, and Q is the figure of merit of the power supply.

From Section III-B for 100% ion transmission, the diameter of the input aperture is

$$D \approx r_0/(M/\Delta M)^{1/2} \tag{32}$$

A considerable number of rf cycles may be required in order to remove an ion that is unstable but whose (a, q) value is close to the stable area. This number of cycles, n, depends on the required resolution, and it puts an upper limit to the input axial velocity of the ion. It was found experimentally (21) that for a resolution near 100,

$$n \approx 3.5(M/\Delta M)^{1/2} \tag{33}$$

The maximum accelerating voltage $V_{\text{acc}}^{\text{max}}$ is then given by

$$V_{\text{acc}}^{\text{max}} \approx 4.2 \times 10^2 f^2 L^2 M/(M/\Delta M) \tag{34}$$

where $V_{\text{acc}}^{\text{max}}$ is in volts and L is the length of the field in meters.

Ions with radial *energy* U_r, entering along the field axis, will be 100% transmitted when from Eqs. (29) and (30),

$$U_r < V/30(M/\Delta M) \tag{35}$$

The angle of entry becomes critical when the optimum performance is required; that is, when the maximum resolution is required for a given accelerating voltage and 100% ion transmission. Combining Eqs. (30), (34), and (35) gives for the maximum angle of injection, ψ_M,

$$\tan \psi_M = 1.44 \times 10^{-3} r_0^2 / L^2 \tag{36}$$

The maximum angle of injection under the conditions of optimum performance for a given accelerating voltage depends only on the geometry of the device.

2. Accuracy and Uniformity

Any variation in the power supplies should cause a maximum fluctuation in the position of a peak that is small compared with the required resolution. That is, a and q (and therefore U and V) should be stabilized to at least $\frac{1}{2}(M/\Delta M)$ of their value. Since frequency ω and field radius r_0 appear in a and q to the second power, they should be stabilized to at least $\frac{1}{4}(M/\Delta M)$ of their value. Local variations in the field radius r_0 are not so important as errors throughout the field region, since errors occurring over many field cycles may have a cumulative effect. Errors in the field, and their effect on peak shape and transmission, are discussed in detail in Section VI.

When sweeping the voltages U and V at a fixed frequency, the ratio U/V must remain constant if operating in region II. Otherwise the sensitivity will vary during the scan and the measured peak intensities will not correspond to true relative abundances. For masses M_2 and M_1 with measured intensities I_2 and I_1 and true relative abundances I_2° and I_1°, Eq. (25) shows that

$$\frac{I_2^\circ/I_1^\circ - I_2/I_1}{I_2^\circ/I_1^\circ} = \frac{(U/V)_{M_2} - (U/V)_{M_1}}{U_g/V_g - (U/V)_{M_1}} \tag{37}$$

where U_g and V_g are the values of U and V at the apex of the stability triangle.

If we now assume that U and V are not strictly proportional and let $U = cV^{1+\alpha}$, where $\alpha > 0$, then if $V = kM$,

$$\frac{I_2^\circ/I_1^\circ - I_2/I_1}{I_2^\circ/I_1^\circ} = \frac{M_2^\alpha - M_1^\alpha}{(0.1678/cK^\alpha) - M_1^\alpha} \tag{38}$$

The anomalous change in relative intensity with mass derived from this equation is shown in Fig. 19 for two values of α. The four sets of curves are for the intensity of mass 12 normalized at resolutions of 100, 200, 400, and 1000. Even for small values of α there is a considerable departure from true relative abundances. Marchand and Marmet (22) suggested using a scan line such as $a = 0.34q - (0.33/M)$ to obtain a resolution always approximately proportional to the ion mass. The resulting percentage transmission and ion injection voltages can then be kept to a maximum for each species.

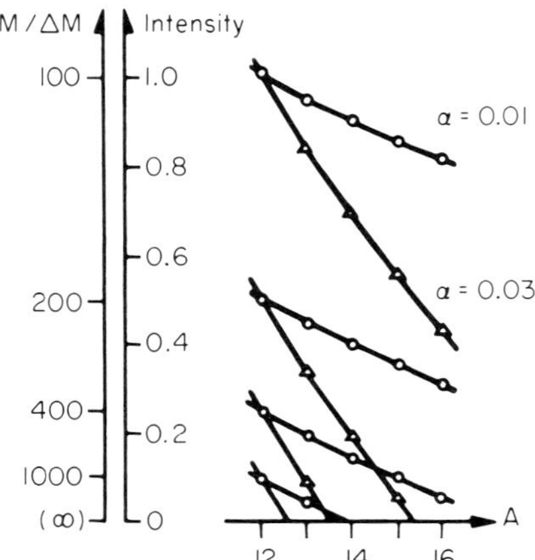

FIG. 19. The variation in the intensity from its true relative value when the U/V ratio is not constant during a mass scan. Mass filter operation is in the triangular peak-shape region II and $U \propto V^{1+\alpha}$. Values normalized at mass 12 are shown for resolutions of 100, 200, 400, and 1000 (*13*).

Some manufacturers of mass filters have also chosen to vary the U/V ratio during the scan. This is used either to obtain an approximately constant peak width for all mass peaks (analogous to the monopole output) or sometimes, in a more complex way, to compensate for mass discrimination caused by ionization cross section and electron multiplier sensitivity differences. Several mass ranges are sometimes obtained by simultaneously changing not only the rf frequency but also the mass scan line. In these cases, corrections are required to correlate the apparent ion intensities.

D. Experimental

1. Design Example

The compromises that must be made in designing a quadrupole mass filter (based on the criteria of Sections III-B-6 and III-C-2) are illustrated in this section. The example described here is an instrument constructed by Paul *et al.* (*13*). The design is for a resolution of 1500, and a mass range up to $M = 16$. The field radius r_0 must be held to an accuracy less than $\frac{1}{4}(M/\Delta M)$. Allowing machining tolerances of 0.01 mm for the rod diameter d, and taking $\Delta r_0 = 0.2 \Delta D$ (*13*), the minimum field radius is 1.2 cm. A field radius of 1.5 cm was chosen. The number of rf cycles the ions must spend in the quadrupole field is given by $n = 3.5(1500)^{1/2} = 135$ cycles, Eq. (33). An acceleration voltage of 75 volts was chosen.

There is a necessary compromise between the rf frequency and the length of the field, Eq. (34). To reduce the power requirements, the frequency should be low, since power varies as the fifth power of the frequency, Eq. (31). Paul *et al.* (*13*) chose a length of 1 meter, and a frequency of 4 MHz, requiring a power of 300 watts at mass 16 and an rf voltage of 8 kV. The maximum entrance aperture was 0.4 mm diameter, Eq. (28). The ion radial energy on entry must be less than 0.15 eV, Eq. (29). Rods with circular cross section rather than hyperbolic were used to facilitate fabrication, in accordance with the relation $r_{\rm rod}/r_0 = 1.16$ for the best approximation (*19*) to a hyperbolic field.

The mass filter was tested by its ability to separate clearly the methane–oxygen mass 16 doublet (16.0364 : 16). The mass difference was measured as $36.7 \pm 0.5 \times 10^{-3}$ amu, comparing favorably with a range of 36.370 to 36.419 in the best values measured with magnetic spectrometers (*23–25*).

Figure 20 presents the methane spectrum taken at low resolution in the trapezoidal peak-shape region. Figure 21 gives the normalized peak heights

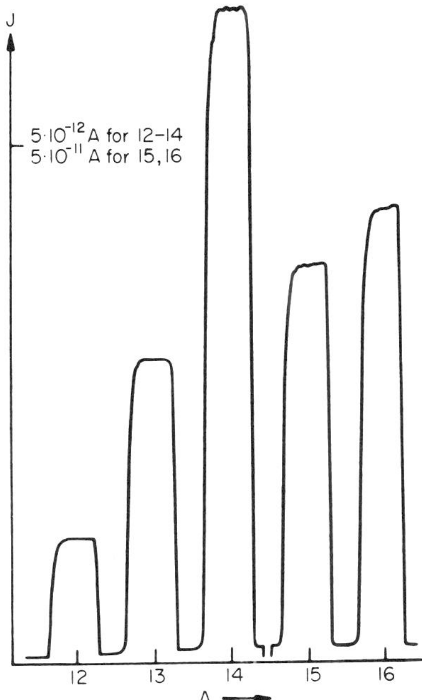

FIG. 20. Methane spectrum (*13*) in the trapezoidal peak-shape region I (*13*).

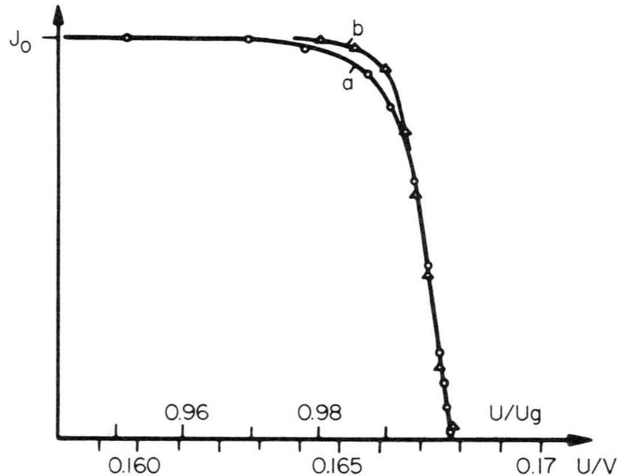

FIG. 21. Ion current as a function of the U/V ratio for (a) CH^+ and (b) CH_3^+ showing the trapezoidal peak-shape region I of constant current and the triangular peak-shape region II where sensitivity decreases with resolution (*13*).

for CH^+ and CH_3^+ as the U/V ratio was changed. It shows the division into the trapezoidal region of constant intensity and into the triangular region where the intensity is inversely proportional to the resolution. The relative intensities of the two peaks should remain constant. The slight deviation was

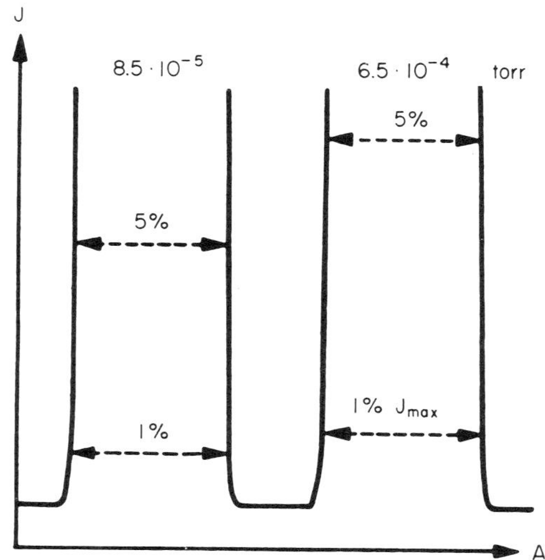

FIG. 22. The base of the mass 15 peak at high pressures for $(M/\Delta M) = 40$. There is little pressure broadening (*13*).

attributed to a nonzero α caused by curvature in the characteristic of the diode used to obtain the dc voltage.

An advantage of the quadrupole mass filter over magnetic instruments is that it can operate at higher total pressures. Small-angle collisions with background gas molecules (excluding charge exchange) do not affect the stability of the ions to a first approximation. This is in contrast to focusing instruments, where the ion energy is critical. Pressure broadening of the peaks should be smaller with the mass filter. Figure 22 shows the base of the peak of mass 15 (CH_3^+) for methane pressures of 8.5×10^{-5} torr and 6.5×10^{-4} torr. The broadening with increased pressure is small.

2. Angular Entry of Ions

The analysis of the transmission of ions through a mass filter as a function of the angle of entry of the ions has given considerable insight into the effect of fringing fields. As a result, Brubaker (26, 27) has been able to modify the fringing fields so as to reduce their undesirable effects. Figure 23 presents the transmission of 140-eV K^+ ions through a quadrupole mass filter for various angles of entry in the xz plane, with the ions entering on axis (26). Figure 24 presents the corresponding case for ions incident at angles in the yz plane. The conditions were such that Paul's criterion, Eq. (36), would indicate an angle of up to 16 deg for 100% transmission.

The much greater dependence of ion transmission on the injection angle in the yz plane results from the fringing fields before the entrance to the analyzer. If it is assumed that the dc and rf voltages are in the same proportion, but decrease in magnitude with distance from the entrance aperture, an ion that is stable in the mass filter will be unstable in the y direction but stable in the x direction as it passes through the fringing field (see Fig. 17). In Fig. 24, the transmission is greatest on the high mass side, where the y stability is greatest.

Figures 25 and 26 present more data on the anomalous effects produced by fringing fields with angular injection of ions. Such effects are important in trying to optimize performance for satellite applications. For 30 deg entry in the xz and yz planes, respectively, these figures show the maximum transmitted ion current as a function of the entrance *position* at different rf frequencies (constant scan line). The energy of the K^+ ions was 15 eV. There are two surprising effects. For the xz plane the transmission is smallest at the highest frequency. For the yz plane the transmission is poor except at the highest frequency and is greatest at large initial displacements. These results are not fully understood (26).

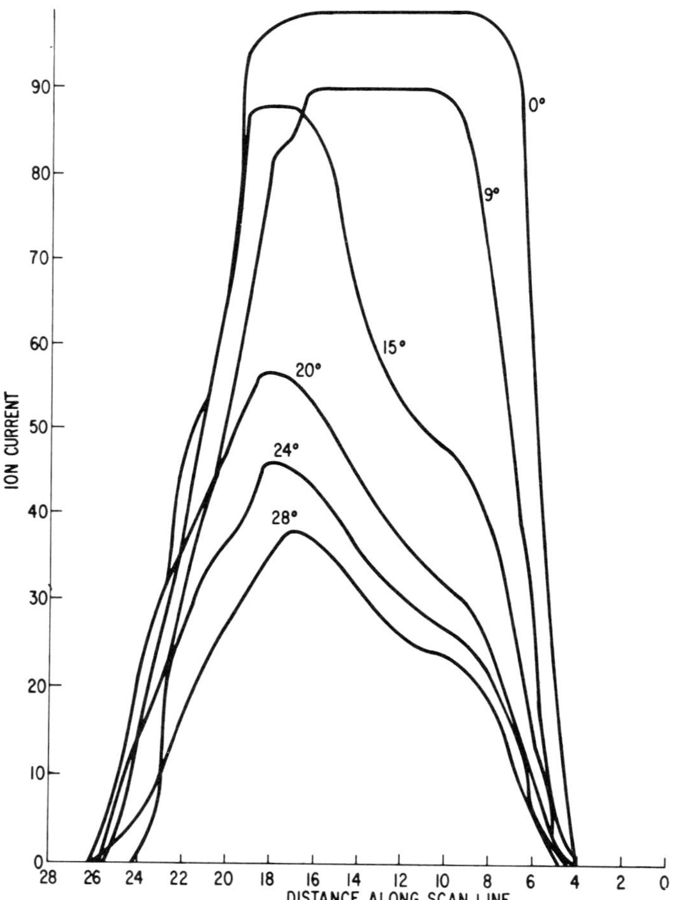

FIG. 23. Ion transmission for various angles of ion entry in the xz plane and on the z axis (20). The abscissa is the corresponding point on the scan line, measured from a fixed point.

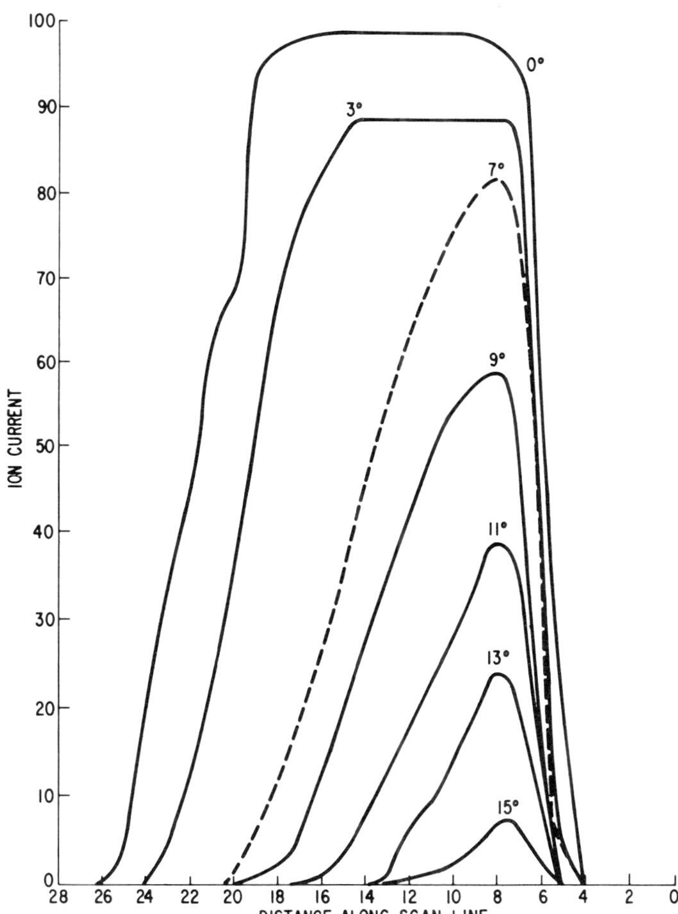

FIG. 24. Ion transmission for various angles of ion entry in the yz plane and on the z axis (*20*). The abscissa is the corresponding point on the scan line, measured from a fixed point.

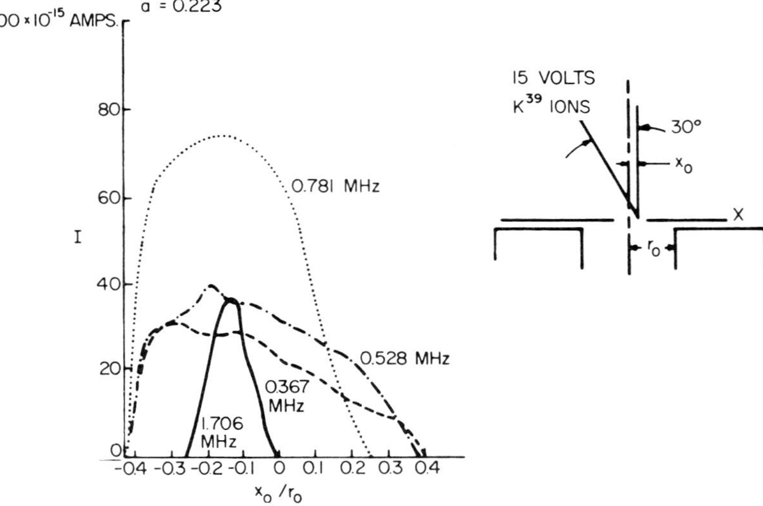

FIG. 25. Dependence of ion transmission on the position of ion injection in the x direction, for injection at a 30 deg angle (*20*).

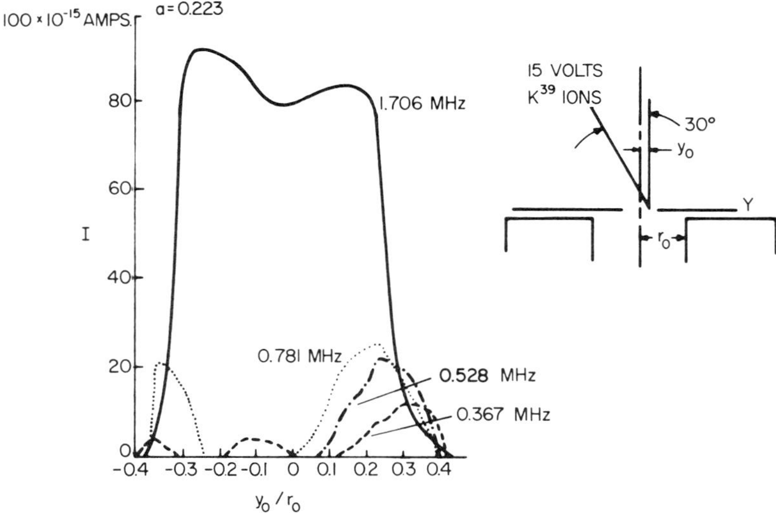

FIG. 26. Dependence of ion transmission on the position of ion injection in the y direction, for injection at a 30 deg angle (*20*).

3. The Delayed dc Ramp Technique

a. Computer simulation. A useful technique for reducing the loss of ions due to fringing fields has been proposed and demonstrated by Brubaker (*27*). A set of auxiliary rods (Fig. 27), insulated electrically from the main electrodes, is located at the entrance to the mass filter. The potentials on these rods are adjusted so that the ion remains in a stable area of the (a, q) diagram, even when the fields are of less than full magnitude (Fig. 28). This is done by

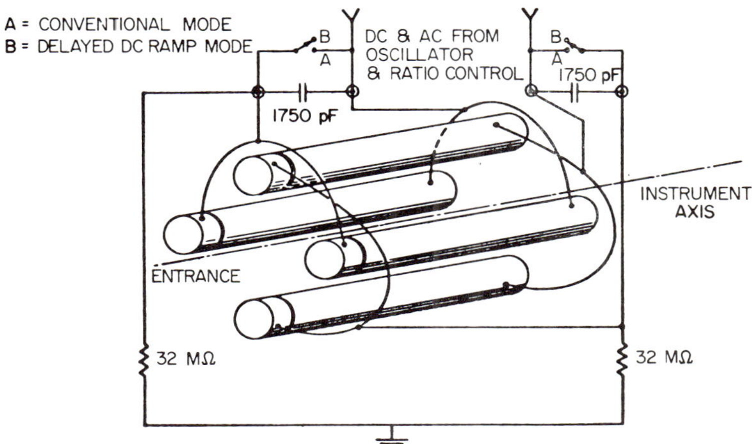

FIG. 27. Delayed dc ramp technique for improved quadrupole performance. The short preliminary sections of the rods have only the rf voltage applied to them (*27*).

applying the full rf voltage to the auxiliary rods, but no dc voltage. Because the dc fields are delayed for the ions entering the instrument, this has been called the "delayed dc ramp" mode of operation.

A summary of computer calculations of trajectory amplitudes showing the advantages of the delayed dc ramp is given in Fig. 29. The ratio of the maximum amplitude to the initial displacement is given for ions with no initial radial velocity. Amplitudes are shown for trajectories in the xz and yz planes for four initial phases of the rf field. The first case is for no fringing fields; therefore the ions were assumed to start within the uniform field region. The second and third cases are for ions entering with the rf to dc ratio fixed, but with the magnitude of the voltages increasing linearly from zero to full value over 2 and 10 cycles of the rf field, respectively. For the 10-cycle ramp, the amplitude of oscillation is greater than 1000 times the initial displacement for trajectories in the yz plane. The xz trajectories are very much smaller in amplitude. A reduction in amplitude was found by using the delayed dc ramp

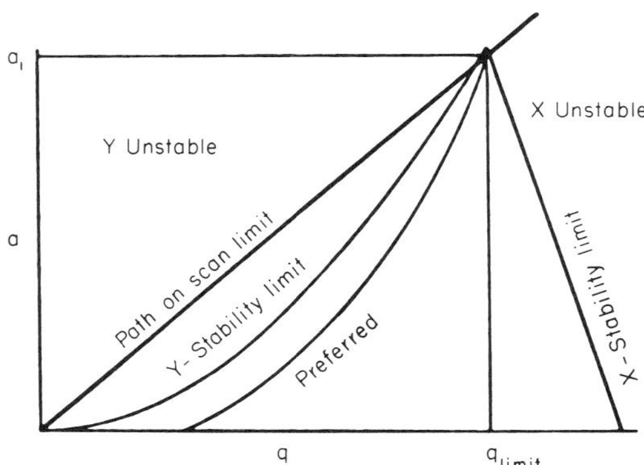

FIG. 28. Stability diagram showing the usual movement of ions along the scan line as they approach the entrance to the mass filter, assuming a linear variation of the fringing fields. The ions pass through a region of y instability. Using the delayed dc ramp, a preferred (a, q) path, which is always in the stable region, can be achieved (27).

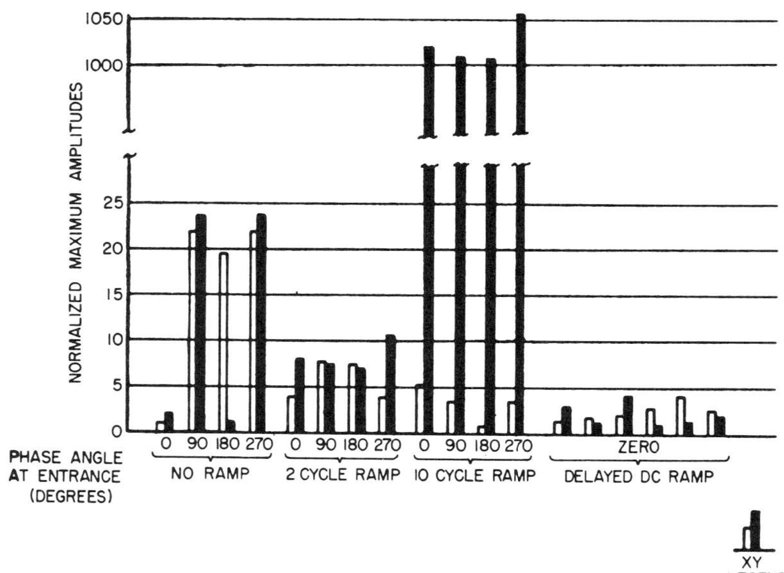

FIG. 29. Results of trajectory computations that demonstrate the effect of various voltage ramps as the ion approaches the quadrupole field. The 10-cycle linear ramp is highly defocusing in the y direction. The delayed dc ramp is beneficial (27).

method. A number of delayed ramps were tried. The steepest increased the rf field uniformly to full value during 12 cycles, with the dc being zero until the seventh cycle and then increasing to full value at the twelfth cycle. For all the delayed dc ramps, the amplitude remained less than five times the initial displacement.

b. Experimental. Figure 30 presents experimental data on the effectiveness of the delayed dc ramp technique (*27*). Rods of 1.5-cm diameter, 48-cm length, and 1.5-cm length end segments were used, with an rf frequency of 1.6 MHz.

Measurements of resolution were based on the width of the ^{84}Kr peak at 10% of its height. The ions with 4-eV energy remain in the field about 185 rf cycles; therefore high resolution should be possible. Without the delayed dc ramp the sensitivity is low. At a resolution of 200 the delayed dc ramp increases the sensitivity by a factor of 500. With ions of 15-eV energy, the resolution is limited by the ion transit time. Although the ions spent less than 3 cycles in the fringing field, there is still some improvement at low resolutions.

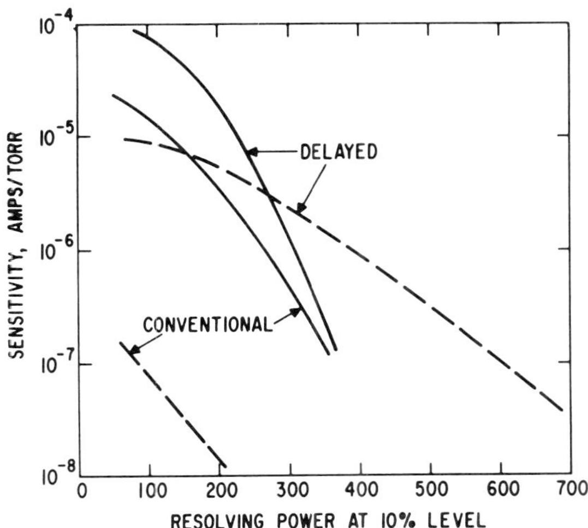

FIG. 30. Sensitivity versus resolution for conventional mass filter operation and that with the delayed dc ramp. The solid lines are for 15-eV ions and the broken lines for 4-eV ions (*27*).

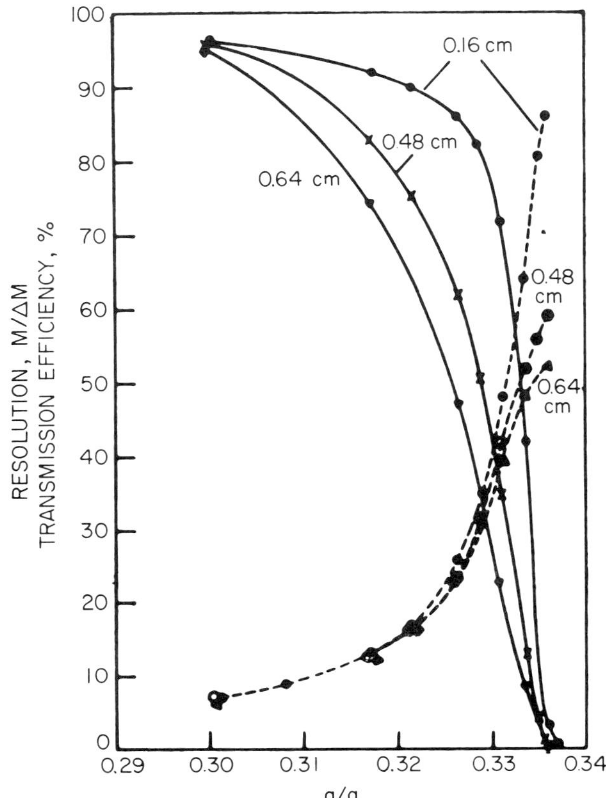

FIG. 31. Transmission efficiency (solid curves) and resolution (dashed curves) for mass 28 as functions of a/q, using a 1.66 MHz power supply. Ion injection energy was 9.5 eV. The numbers on the curves are entrance aperture diameters (28).

4. Optimization of Aperture Size, Applied Frequency, and Scan Line

For fixed ion energies, as might be found in satellite applications, the required resolution may be obtained with several combinations of aperture size, frequency, and a/q ratio. The optimum choice will depend on the resolution. Figure 31 shows both the ion transmission efficiency and resolution (at 10% of peak height) as a function of the entrance aperture diameter and the a/q scan line (28). The rf frequency was 1.66 MHz, and the ions were mass 28 with 9.5-eV energy in a device with rods 25 cm long and 1.5 cm diameter. The smallest aperture permits greater resolution and greater transmission at high resolution (high a/q). The peak intensity is the product of transmission efficiency and aperture area if the aperture is properly used. The peak intensity is shown in Fig. 32 as a function of resolution, taken from data of Brubaker (28). The largest aperture provides greater sensitivity only at low resolutions. It is interesting to note that Paul's formula for aperture size at 100% transmission, Eq. (32), predicts quite well the break in the 0.16 cm diameter curve in Fig. 31.

The dependence on rf frequency is illustrated in Fig. 33 for the 0.16-cm aperture. Higher frequencies permit marked increases in performance at high a/q. The improvement in resolution results from the increased number of cycles the ion spends in the field, but the increased frequency and associated increased voltages also improve the transmission.

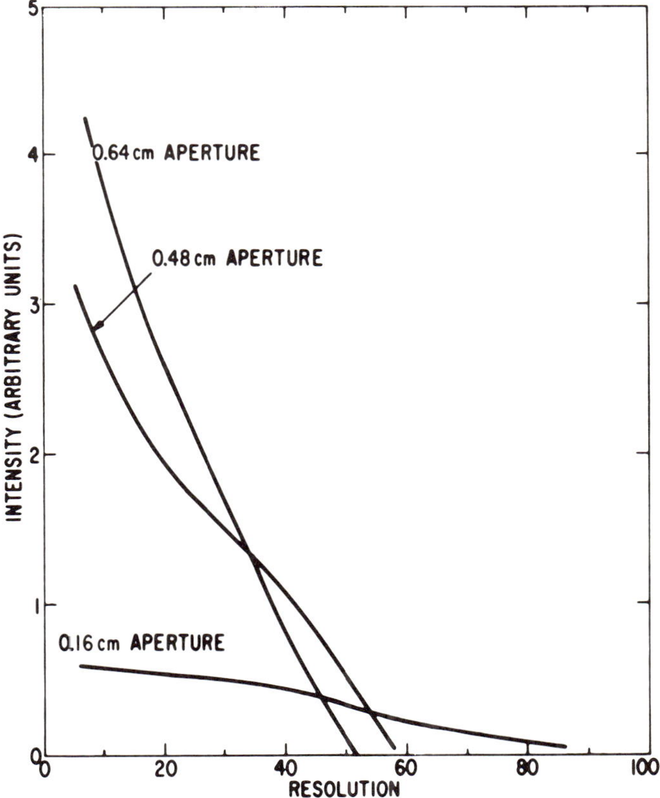

FIG. 32. Ion intensity (transmission efficiency × aperture size) as a function of resolution for 1.66 MHz applied frequency. (Derived from data in Ref. 28.)

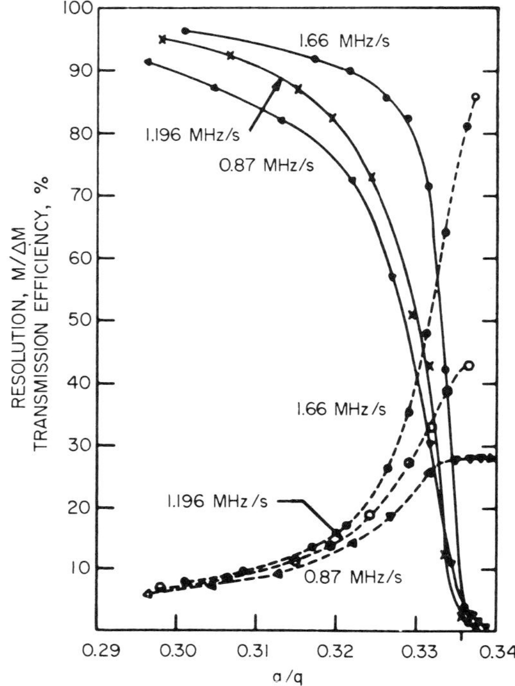

FIG. 33. Transmission efficiency (solid curves) and resolution (dashed curves) for mass 28 as functions of a/q for various applied frequencies and a 0.16-cm entrance aperture. Ion injection energy was 9.5 eV (*28*).

5. Special Sources and Detectors

The relative insensitivity of the mass filter to ion energy along its axis has led to special applications, such as the sampling of ions directly from the upper atmosphere (*29*) and from gas discharges (*30*), and also to the development of special ion sources.

Böhm and Günther (*31, 32*) and Günther and Hänlein (*33*) have used a Penning type of discharge as a cold cathode source. With this technique, partial pressures of 10^{-9} torr were detected with a mass resolution of 50 to 100. The Penning source (Fig. 34) had a total pressure range from 10^{-3} to 5×10^{-7} torr. Figure 35 shows a spectrum taken with the Penning ion source (*33*).

Blum and Torney (*34*) used a modification of Redhead's cold cathode magnetron gauge (*35*) as the ion source for a quadrupole mass filter. This source, shown in Fig. 36, is useful down to pressures below 10^{-13} torr, and has a high sensitivity. A spectrum taken with a mixture of rare gases at 2.3×10^{-6} torr is presented in Fig. 37. By retarding the ions before they enter

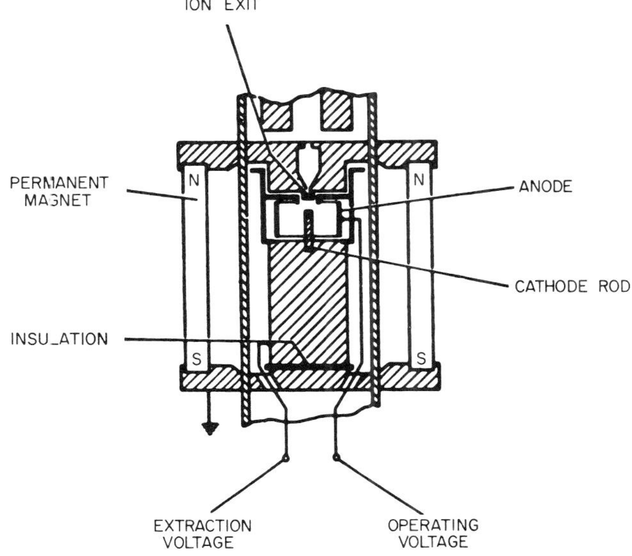

FIG. 34. Penning type of ion source used by Günther and Hänlein (33).

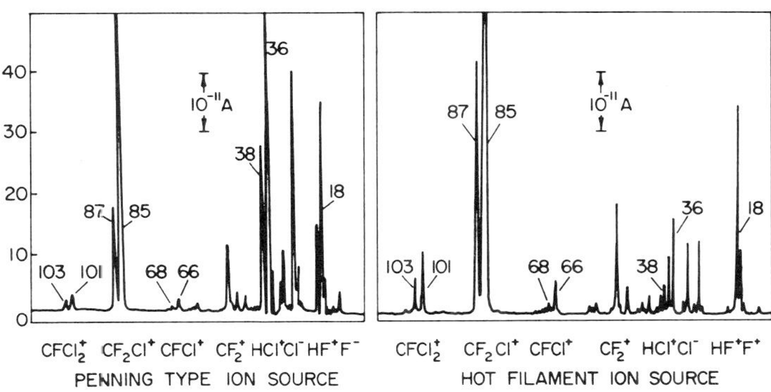

FIG. 35. Mass filter spectra comparing the performance of a Penning ion source with a conventional hot filament source (33).

the analyzer, the resolution could be made equal to that with the hot filament source. It was limited by the analyzer rather than by the cold cathode ion source. A sensitivity of 0.82 mA/torr for mass 28 at 10^{-7} torr was achieved, compared with 0.25 mA/torr, using a hot filament ion source at 3-mA emission current. No electron multiplier was used in determining these sensitivities.

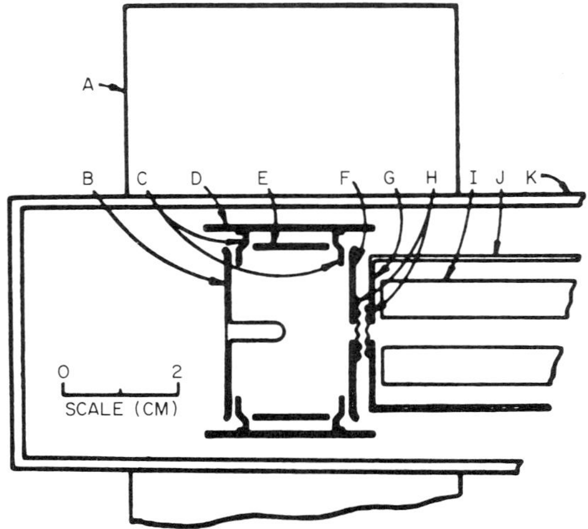

FIG. 36. A cold cathode source designed by Blum and Torney (*34*). A, magnet; B cathode K_1; C, auxiliary cathodes; D, source housing; E, anode; F, cathode K_2; G, quadrupole entrance aperture; H, tungsten screens; I, quadrupole rods.

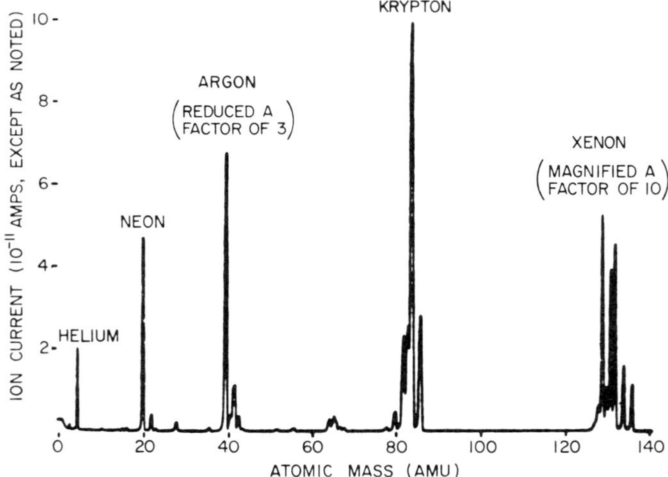

FIG. 37. Mass spectrum of a mixture of rare gases obtained using the cold cathode ion source of Blum and Torney (*34*).

A wide variety of ion detectors have been used with the mass filter. Frequently an electron multiplier is used to enhance the signal. The exit aperture of the filter may be large and the ion beam may be divergent. The electron multiplier therefore must have a large aperture, and ion focusing on the collector may be necessary for a high level of collection efficiency. The design and construction of an oversized 14 stage Ag–Mg electron multiplier was described by Marchand *et al.* (*36*) for use with a 1.1-meter mass filter with a 5-cm exit aperture. The reported ion detection efficiency was near 100%.

6. Mounting the Rods

Correct spacing of the rods is essential if high resolution is to be attained. Figure 38 shows a simple mounting technique for the four rods (*37*). The

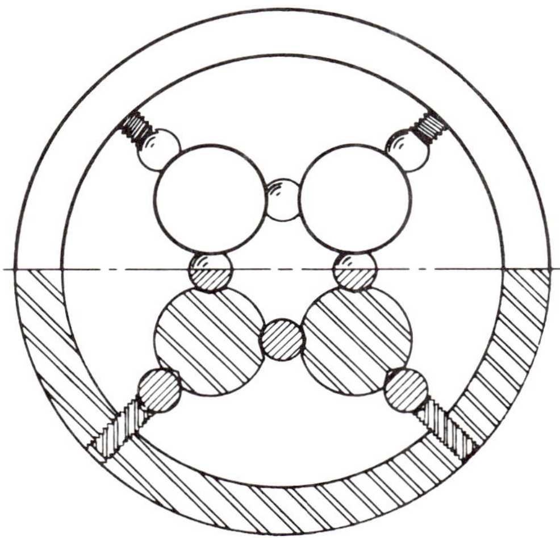

FIG. 38. End-on and cross-sectional view of a simplified method for mounting quadrupole rods, by Munro (*37*).

precision ground stainless steel rods (commercially available, 0.791-cm diameter) were dimpled with a drill point while held in a dividing head. Sapphire ball spacers, 0.3175 ± 0.00013-cm diameter, were placed in the dimples between adjacent rods. Stainless steel set screws were used to hold the assembly and to adjust the diagonal measurements to an accuracy of 0.0003 cm with the aid of a telescoping feeler gauge. The rod plus ball measured 1.074 ± 0.0013 cm. The quadrupole was 26 cm long and had a resolution of about 200.

An instrument with matched thermal expansion so that the mass filter can be operated at high temperature was constructed by Story (*38*). At resolutions near 1000, he found by experiment that the nonuniformity of rod spacing must be less than 2.5 μm for a mass filter with a length of 14 cm and field radius of 0.25 cm. Figure 39 shows a typical mounting arrangement for

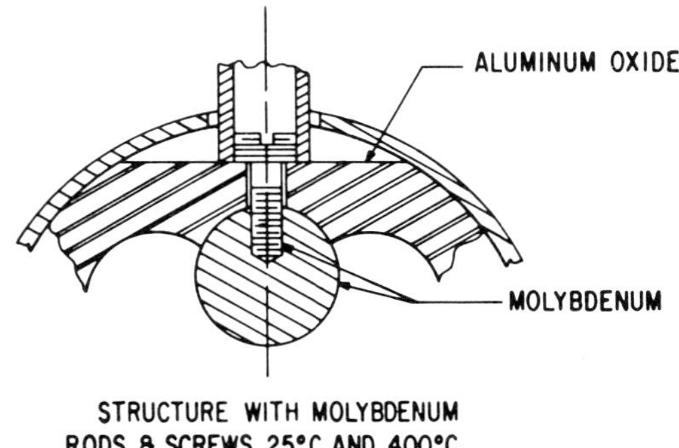

FIG. 39. Method of mounting quadrupole rods in a structure designed to operate at high temperatures, by Story (*38*).

a quarter-section of the quadrupole structure. An aluminium oxide ceramic was used to position the four molybdenum rods, and the tolerances on the seat dimensions were 2.5 μm. The rods were held in place by screws that also make the electrical connections. The ceramic insulator was 2.5 cm OD, and the rods were 0.64 cm diameter. At room temperature, stainless steel rods could be used, but at 400°C the expansion of the stainless steel relative to the alumina ceramic would leave a gap of 10 μm between the rods and the ceramic seat. This greatly exceeds the permissible tolerance. The molybdenum expansion is actually slightly less than that of the ceramic. No measurable loss in resolution was found between room temperature operation and operation at 325°C.

Platinum-coated quartz rods were used by Fairbairn (*38a*) to fit directly into circular holes in stainless steel end plates. The resolution exceeded 160, with rods of 0.95-cm diameter and 20-cm length. The frequency was 3 MHz, and the initial ion energy was 10 eV.

7. *Hyperbolic versus Circular Rods*

Most workers have used rods of circular cross section, but a few mass filters have been constructed with hyperbolic rods. A mass filter, 5.82 m long with a field radius of 3.5 cm, was constructed by von Zahn (*39, 40*) for precision mass measurements. His hyperbolic surface consisted of molybdenum wires stretched to 80% of their breaking load to ensure axial uniformity. Each electrode was made up of 61 wires, each 2.9×10^{-4} meters in diameter, held in place by hyperbolic forms at the ends of the instrument. The use of wires avoids the difficulty of machining long rods to close tolerances. The rf frequency was 471 kHz, and the maximum resolution was

16,000 (*40*). Mosharrafa and Oskam (*41*) have also constructed a hyperbolic mass filter using wires. The analyzing field was 1 meter in length with a field radius of 2 cm. The instrument was used for sampling plasmas at gas pressures above 0.1 torr. No deterioration of the performance of the analyzing field was observed with pressures as high as 10^{-3} torr in the analyzing region. The maximum resolution was not reported.

A careful comparison of quadrupole mass filters with round and hyperbolic rods was made recently by Brubaker (*42, 43*). Rods of 304 stainless steel, 25-cm long, were ground to a hyperbolic cross section. The resolution was about a factor of 2 greater when hyperbolic rods were used. Figure 40 A

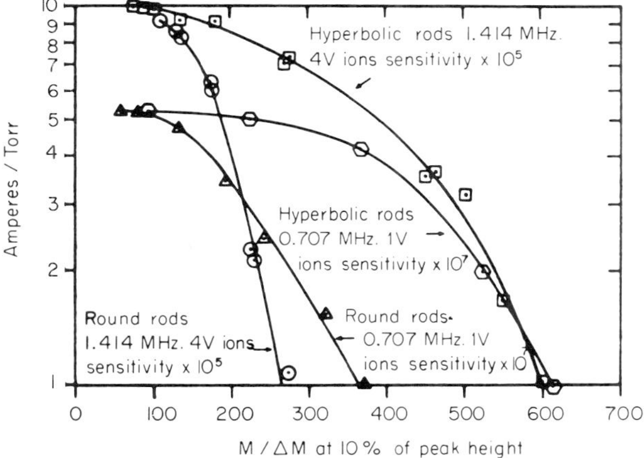

FIG. 40 A. Comparison of mass filter performance with round and hyperbolic rods (*42, 43*).

shows some of the results. The resolution here is $(M/\Delta M)$, where ΔM is the width at 10% of the peak height. Two different rf frequencies are shown, 1.414 MHz and 0.717 MHz, with incident ion energies of 4 eV and 1 eV, respectively. Both instruments were operated in the " delayed dc ramp mode " (Section II-D-2). Tests on mass 84 (krypton) showed that the flat-topped, 100% transmission region, which corresponds to the trapezoidal region of the stability diagram (Section III-B-5), extended out to a resolution of 400. With 1-eV ions, the transmission efficiency was down to only 20% of its peak value for a resolution of 1000, Fig. 40 B.

Brubaker also made computer studies of ion trajectories for the case of round and hyperbolic electrodes. The analysis is complicated in the case of round rods because the motion in the x and y directions is not independent. However, under some simplifying conditions it was found that ions with excursions exceeding 70% of the field radius become unstable and are lost when using round rods. Brubaker reasoned that one could achieve comparable

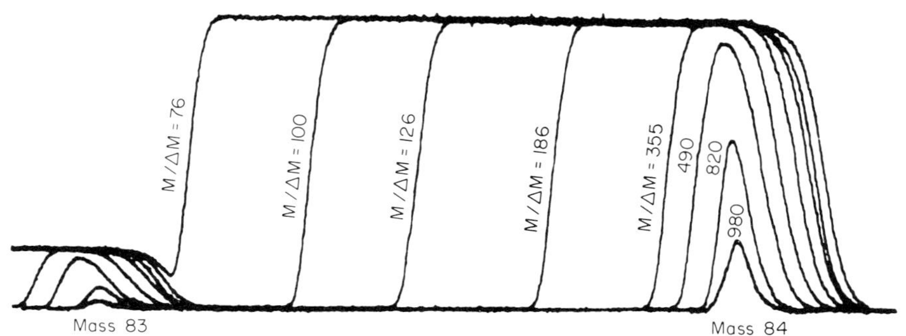

FIG. 40 B·　　Multiple scans of mass 84 with different resolutions (*42, 43*).

sensitivity using hyperbolic rods placed at 70% of r_0, where r_0 is the field radius when round rods are used. Consequently the rf power requirement could be reduced to 25% of that for round rods, since the power varies as r_0^4, (Eq. 31). This is important for experiments in the upper atmosphere and outer space, where minimizing the power is a major consideration.

8. *Electronics*

An electronic circuit used by Paul *et al.* (*13*) to operate a quadrupole mass filter is shown in Fig. 41. The 4-MHz rf voltage was produced by a

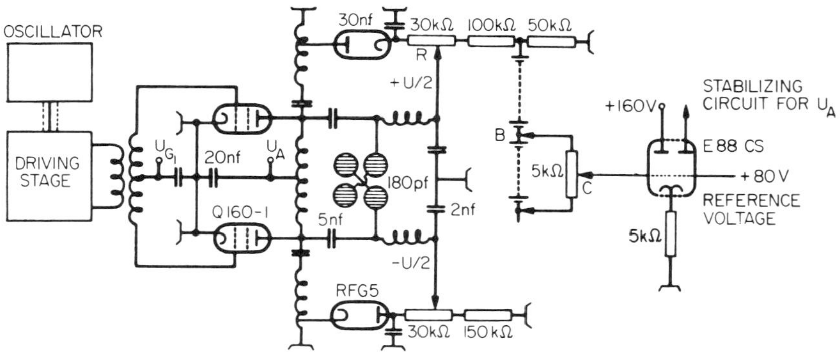

FIG. 41. Circuit used by Paul *et al.* (*13*) to supply the combination of rf and proportionate dc voltages to the mass filter.

separately excited push-pull stage, with the quadrupole system itself being the tank capacitance. The dc voltage, U, was obtained directly from the rf voltage by means of an inductive voltage divider, thereby maintaining a constant ratio between the rf and dc voltages. Symmetrization of the field was obtained with helipots, one for $+U/2$ and another for $-U/2$. One helipot division corresponded to a change of 10^{-4} in $U/2$. The high frequency amplitude was stabilized to $\pm 1.5 \times 10^{-4}$ V by regulation of the plate voltage. Considerable difficulty was experienced in symmetrizing the field. This is necessary because the input conditions are modified if the quadrupole axis is not at ground potential. Both U and V must be symmetrical with respect to the axis. To adjust the symmetry of the rf voltage, a wire 30 cm in length was positioned along the axis, and the high frequency voltage was adjusted to minimize the induced voltage on the wire. A more recent circuit showing the electronic components in detail is presented in Fig. 42 (*44*). A simple rf generator using a commercial signal generator with frequency scanning has recently been described by Burt (*44a*).

E. Isotope Separation

1. Theory

The quadrupole mass filter has been used as an isotope separator by Paul *et al.* (*13*) and von Busch and Paul (*45*). When operating the mass filter in the normal manner, it is difficult to obtain a high ion density or large cross-sectional area of the beam. For example, ions entering the field parallel to the axis are completely transmitted only when the entrance aperture is less than $D = r_0/(M/\Delta M)^{1/2}$, Eq. (28). At high ion currents, the divergence in the trajectories caused by space charge will tend to make an even smaller fraction of the field available for ion injection. A technique suitable for high currents was described by Paul *et al.* (*13*) and the mathematical description given below closely follows their treatment. The mass filter is operated so that the isotopes are near the center of the stability diagram, such as near $a = 0$, $q = 0.6$. There is no mass separation with this field alone, but x_M/x_0 and y_M/y_0 are small; therefore the entrance aperture can be large and a large space charge distortion of the fields (high current density) can be tolerated.

To obtain a mass filter effect, an auxiliary rf resonance field is applied, tuned to the fundamental frequency of ion motion of one of the isotopes. That isotope then has a constantly increasing amplitude of oscillation and will strike the sides of the device. (See Section II-D for a discussion of the fundamental frequency of ion motion and its relation to β.) The fundamental frequency, and therefore the auxiliary rf frequency, depends on the isotopic

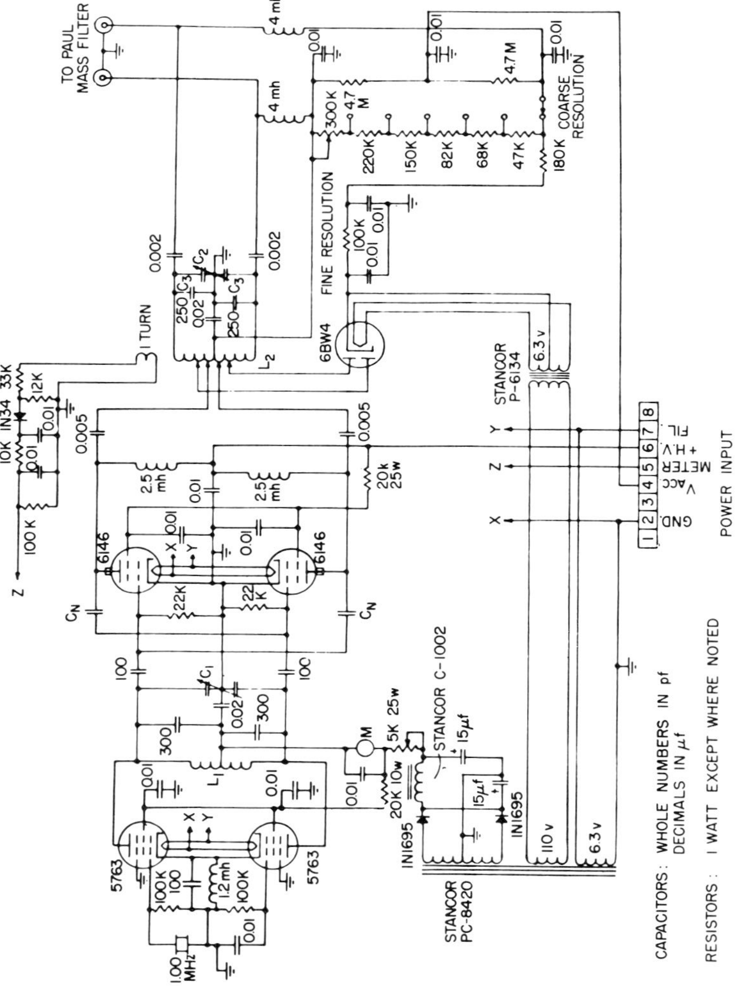

Fig. 42. Mass filter power supply used by Brink (44).

mass. The trajectories of ions of nearby masses are modulated to produce beats, but they are stable and will be transmitted if the beat amplitude is not too great.

The auxiliary resonant field may be a quadrupole field or a field applied in only one direction. Assume a uniform field in the x direction and no change in the y and z directions. The equation of motion in the x direction becomes the inhomogeneous Mathieu equation:

$$m \frac{d^2x}{dt^2} + 2e(U - V \cos \omega t) \frac{x}{r_0^2} = eE'e^{i\omega't} \tag{39}$$

where E' is the resonance field in the x direction with frequency ω'. This can be changed to the conventional form

$$\frac{d^2x}{d\xi^2} + (a - 2q \cos 2\xi)x = \rho e^{i(\omega'/\omega)2\xi} \tag{40}$$

where $\rho = 4eE'/m\omega^2$, and ξ, a, and q are as defined in Eq. (9).

The general solution of this equation combines the solution of the homogeneous equation and a particular solution of the inhomogeneous equation. The latter represents the resonance perturbation of the original motion and is the factor of interest here.

A particular solution of the inhomogeneous equation is

$$x_p = \frac{\rho}{W}\left(x_{\mathrm{II}} \sum_{s=-\infty}^{\infty} C_{2s} \frac{\exp[2i(s + \beta/2 + \omega'/\omega)\xi]}{2i(s + \beta/2 + \omega'/\omega)} \right.$$
$$\left. + x_{\mathrm{I}} \sum_{s=-\infty}^{\infty} C_{2s} \frac{\exp[-2i(s + \beta/2 - \omega'/\omega)\xi]}{2i(s + \beta/2 - \omega'/\omega)} \right) \tag{41}$$

where x_{I} and x_{II} represent a fundamental solution to the homogeneous equation and W is the Wronskian determinant.

If we choose the resonance frequency ω' to correspond to a fundamental frequency of an ion, such as $\omega' = (s' + \beta/2)\omega$, we obtain the solution for an ion exactly at resonance:

$$x_p = \frac{\rho}{W}\left(x_{\mathrm{II}} \sum_{-\infty}^{\infty} C_{2s} \frac{\exp[2i(s + \beta/2 + \omega'/\omega)\xi]}{2i(s + \beta/2 + \omega'/\omega)} \right.$$
$$\left. + x_{\mathrm{I}} \sum_{-\infty}^{\infty} C_{2s} \frac{\exp[2i(s + \beta/2 - \omega'/\omega)\xi]}{2i(s + \beta/2 - \omega'/\omega)} + x_{\mathrm{I}} C_{2s'}\xi \right) \tag{42}$$

When ξ is large, the last term dominates, with

$$x_p \approx \frac{\rho C_{2s'}}{W} \xi x_{\mathrm{I}} \tag{43}$$

Substituting for ρ and ξ,

$$x_p \approx 2eE'C_{2s'}\, tx_{\mathrm{I}}/m\omega W \tag{44}$$

This describes an oscillation x_{I} with a linearly increasing amplitude. The increase in amplitude is proportional to the magnitude of the resonance field and is greatest when the resonance frequency equals the fundamental frequency ($s' = 0$), since the C_{2s} decrease rapidly for increasing s.

When the resonance frequency is close, but not equal to the unperturbed frequency of ion motion, we may use Eq. (41). The second term with the smaller denominator will be dominant. Letting $\Delta\omega' \equiv (s' + \beta/2)\omega - \omega'$

$$x_p \approx \frac{\rho C_{2s'}}{W}\, \frac{\exp(-2i(\Delta\omega'/\omega)\xi)}{2i\, \Delta\omega'/\omega}\, x_{\mathrm{I}}$$

or

$$x_p \approx \frac{2eE'C_{2s'}}{im\omega W\, \Delta\omega'}\, \exp(-i\, \Delta\omega't)x_{\mathrm{I}} \tag{45}$$

The original motion has superimposed on it a series of beats with the beat frequency $\Delta\omega'$. The amplitude of the beats is proportional to the magnitude of the resonant field, and it increases as the resonance frequency is approached.

When separating one isotope from another, then, the resonance frequency may be chosen equal to the fundamental frequency of the isotope to be eliminated. The maximum amplitude of the transmitted isotope ions must be smaller than the field radius r_0. If the relative mass difference between the two isotopes is small, the beat amplitude of the transmitted isotope will be relatively large. However, the beat amplitude can be reduced by using a small resonance field.

It should be pointed out that the amplitude of the resonant ions increases linearly with time rather than exponentially as with an ion in an unstable region of the stability diagram. The minimum transit time must be nearly one beat period, $1/\Delta\omega'$, in order to filter the isotopes fully. This places a limitation on the maximum permissible accelerating voltage, for a given value of frequency and length of field. A high accelerating voltage is, however, necessary in order to use high beam currents.

2. Experiment

a. Design. An experimental design requires a compromise between a large ion accelerating voltage to minimize space charge in the beam, the device length, and the applied frequency, since the transit time must be at least one beat period. The power increases with f^5 but only linearly with the length, so a long device is preferable. Paul *et al.* (*13, 46*) built a separator for magnesium isotopes with a length of 3 meters, a field radius of 1.5 cm, and an applied frequency of 2.56 MHz. The device had a capacitance of 450 $\mu\mu f$, and required 0.3 kW of high frequency power for magnesium, with a circuit figure of merit Q of 400.

Figure 43 shows the circuit used by Paul for isotope separation. The dc

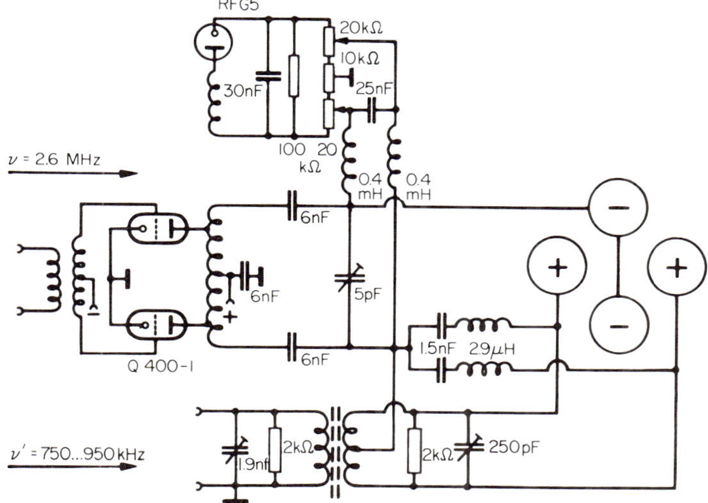

FIG. 43. Power supply used by Paul *et al.* (*13*) with the isotope separator.

voltage was obtained by rectifying part of the rf voltage so that the ratio U/V remained constant. The rf voltage was 4 to 6 kV, with a dc voltage of 200 to 600 volts for magnesium ions near the middle of the stable region.

The resonance voltage V' was applied to the two electrodes in the x direction. The resonance field therefore was only an approximation to the field used in developing the theory (Section III-E-1). The field was not homogeneous, and rods of circular cross section were used. Nevertheless, the predictions using the theory were in satisfactory agreement with experiment. The resonance frequency could be varied between 75 and 950 kHz, with an amplitude of 40 to 100 volts.

A low voltage, magnetically stabilized arc was used as an ion source, since the energy spread of the ions is unimportant. Experimental results showed that a pressure as high as 10^{-4} torr could be used, since the strong focusing of the quadrupole field allows considerable small angle scattering without significant loss in resolution.

To test the transmission, the device was first operated without isotope separation, using the conditions $a = 0.08$, $q = 0.6$. The maximum allowable accelerating voltage for mass 25 was 2 kV. With this accelerating voltage the collector current was 2×10^{-3} amperes which, for magnesium, corresponds to 2.8×10^{-3} g/hr. It was estimated that this was about 50% of the current from the ion source, with the loss being evenly divided between the drawout lens and the mass filter.

b. Space charge. Space charge has several effects. It increases the cross section of the beam and contributes to the spread in the angle of entry of the incident ions. In addition it distorts the hyperbolic field. A crude approximation of the latter effect may be obtained by assuming that a cylinder of radius r_0 is filled with a uniform density space charge. Its effects can be represented by an additional dc field on the ion which is proportional to the space charge density and to the distance of the ion from the field axis, and which is always defocusing. Expressed in terms of the Mathieu constant a, we have an additional increment a' whose value is given by (*13*)

$$a' = \frac{1.8I}{f^2 (M V_{\mathrm{Acc}})^{1/2}} \tag{46}$$

where I is the ion current, M the atomic weight of the ions, and V_{Acc} the ion accelerating voltage. Consequently,

$$I_{\max} = 0.56 a'_{\max} f^2 (M V_{\mathrm{Acc}})^{1/2} \tag{47}$$

In the most favorable case, with no imposed dc field, $a = 0$; therefore, with $q = 0.66$, the maximum allowable $a'_{\max}$ is 0.2 (Fig. 17). For $V_{\mathrm{Acc}} = 2.4$ kV, $f = 2.56$ MHz, and $M = 25$, this corresponds to a maximum ion current of 6 mA, which is only a factor of 2 greater than that observed experimentally (*13*). Consequently, a better ion source might not greatly increase the ion current. The space charge limitation on current could be reduced by increasing the length (and V_{Acc}) or the frequency.

Paul *et al.* (*13*) suggested the use of four parallel mass filters, with a total of nine electrode rods, as shown in Fig. 44.

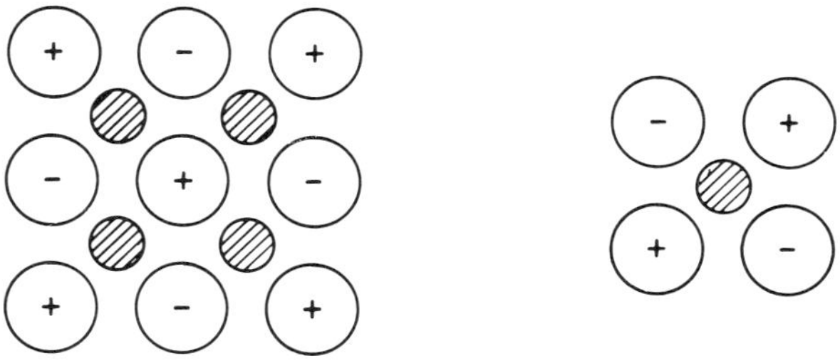

FIG. 44. Single and multiple channel mass filters (*13*).

The transmission would be greater by a factor of 4, but the power consumption would increase by only a factor of 2.6. With nine parallel filters (16 rods), the power would increase by a factor of 4.5.

The effect of space charge was studied by determining the shift in the resonance frequency of ^{23}Na as a function of ion current, other variables being held constant. The conditions were: $V = 4.8$ kV, $U = 340$ volts, a constant magnitude of the resonance field sufficient to reduce the ion current to zero at resonance, and $V_{Acc} = 2$ kV. The fundamental frequency of ion motion should be 875 kHz under conditions of no space charge. Figure 45

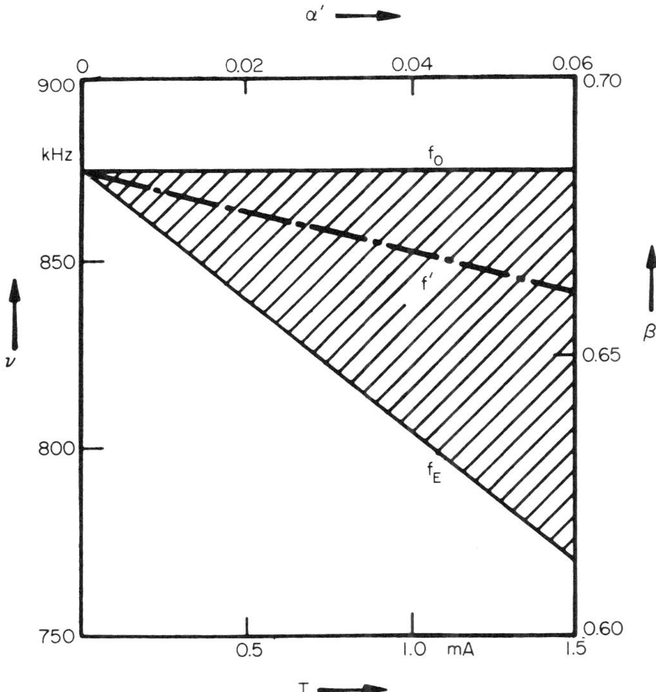

FIG. 45. Dependence of fundamental frequency of ion motion (or resonance frequency) on space charge. f_0 is the frequency with no space charge; f_E is the frequency calculated from the current entering the mass filter; f' is the experimental resonance frequency (13).

shows the resonance frequency f' necessary to reduce the ion current to zero as a function of the ion current. The required frequency decreases with increasing ion current. This is consistent with the view that space charge shifts the operating point and therefore the resonance frequency of the ions. The calculated change in a value due to the ion space charge, Eq. (46), is also shown on the abscissa, and f_E is the resonance frequency calculated using the incident ion current. At resonance, the ion current varies along the device, and the experimental resonance frequency f' lies only about one-third the distance from f_0 (the value with no space charge) to f_E.

It might be possible partially to neutralize the space charge, as is done in magnetic isotope separators. A very high frequency field could be used to stabilize electrons (13) without seriously affecting the much heavier ions. However, the power requirements would be greater. A second alternative might be to use a magnetic field along the axis of the mass filter (13) to contain the electrons in tight spirals without much distortion of the ion trajectories. Neither of these techniques appears to have been attempted.

c. Isotope separation. Figure 46 shows the transmission of rubidium,

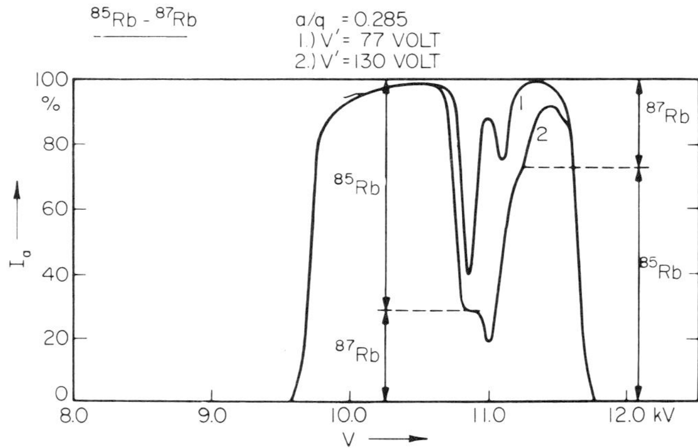

FIG. 46. Absorption curve for rubidium in the isotope separator, showing the effect of two resonance fields of differing strengths and the isotope separation that can be achieved (13).

which has isotopes with atomic weights 85 and 87, as a test of the use of the auxiliary field in removing selected isotopes. The mass scan line was chosen at $a/q = 0.285$. The operating point is in the stable region for rubidium between about 9.6 and 11.8 kV. Ions of ^{85}Rb are in resonance at 10.84 kV, where the iso-β line, $\beta = 0.79$, is crossed. Similarly, resonance for ^{87}Rb ions takes place at 11.08 kV. Curve 1 is for an auxiliary rf amplitude of 77 volts. This amplitude is insufficient to remove the isotope completely. A resonance voltage of 130 volts (about 1% of the focusing field) was found to be necessary for complete removal. The ion current was about 10^{-6} amperes.

Sodium was used at higher currents, and since it has only one isotope, ^{23}Na, its resonance curve could be easily determined (Fig. 47). The half-

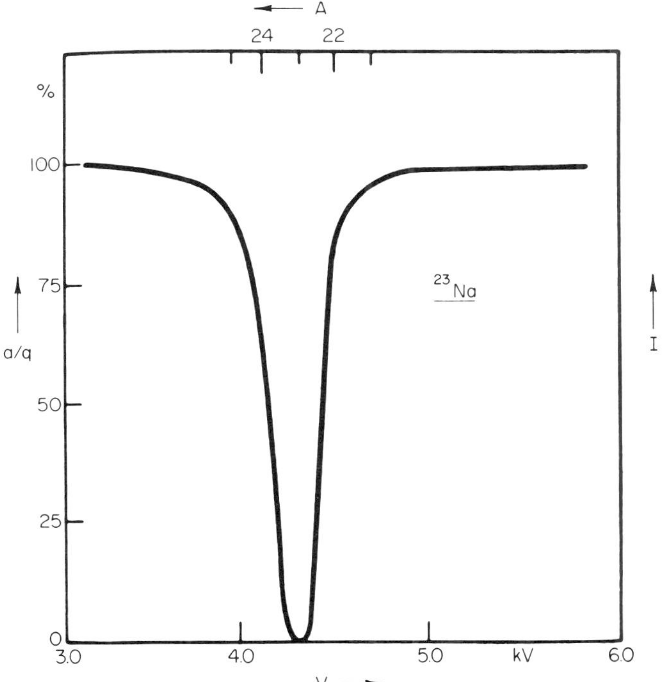

FIG. 47. Absorption curve for sodium. The half-width of the absorption region was about 1.6 mass units (*13*).

width of the absorption region is 1.6 mass units, so that for complete removal of mass 23, some ions of masses 22 and 24 are partially removed.

In cases where more than two isotopes are present, several resonance fields may be used simultaneously, each corresponding to a fundamental frequency of one of the isotopes to be removed. The trajectory of the transmitted isotope will be modulated in a complicated manner, but its maximum excursion will remain within the electrode structure if the resonance fields are not too large. In suitable cases, one resonance field large enough to remove several neighboring isotopes might be used.

The operation of the isotope separator was monitored by Paul *et al.* (*13*), using a second mass filter to determine the composition of the transmitted ion current. The results using magnesium isotopes are shown in Fig. 48. The frequency of the resonance field was changed to bring masses 26, 25, and 24 into resonance in turn. If these are compared with the spectrum with no resonance field, it is apparent that isotope separation can be quite effective.

In separating the rarer isotopes ^{25}Mg and ^{26}Mg, von Busch *et al.* (*45*) were able to obtain a yield of 30 μg/hr with a purity near 50% for ^{25}Mg. The yield

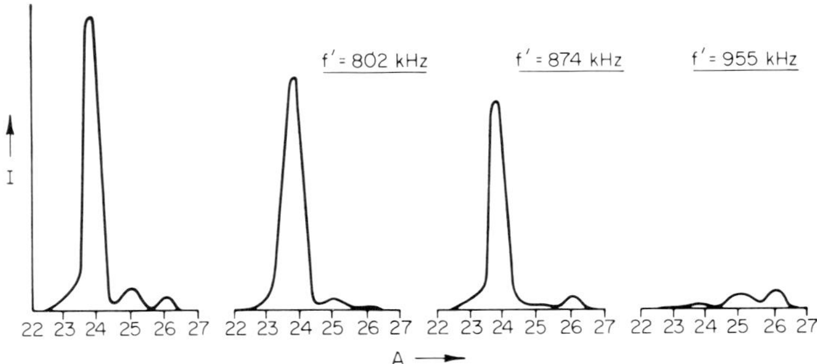

FIG. 48. Analysis of the ion current from the isotope separator, with a second mass filter showing the separation of magnesium isotopes (*13*).

of ^{26}Mg was the same, but the purity was 84%. The total power used by the ion source and electronics was 2.0 kW. A serious problem in this work was the difficulty of keeping the voltages sufficiently constant for long periods of time, even though attempts were made to monitor the separation with a second mass filter mounted behind a pinhole in the collector.

Isotope separation using resonance fields with the quadrupole mass filter is an interesting technique with considerable potential use, and a subject that has not been thoroughly explored.

F. Other Uses of the Quadrupole Mass Filter

This section is intended to be only a brief introduction to some of the varied applications of the quadrupole mass filter. Quadrupole mass filters are commercially available from a number of suppliers (see Appendix), and their use in determining the partial pressure of gases in vacuum systems (*47*) is routine. They are used to monitor evaporation processes and surface reactions as well as in a wide variety of vacuum experimentation.

A major advantage of the quadrupole mass filter over most mass spectrometers is its relative insensitivity to the initial velocity of the ion along the axis. Consequently the mass filter is useful in many experiments where there is a spread in the initial ion energy. For example, Oskam and his co-workers at the University of Minnesota (*48–51*) have carried out studies of collision processes in gaseous plasmas. Figure 49 shows a spectrum of ions taken from a dc discharge in neon. The neon pressure in the discharge was 27.6 torr, and the pressure in the mass-analyzing field region was 10^{-3} torr. This illustrates another useful property: The mass filter performs well at relatively high pressures. The peaks near mass 42 are diatomic molecular neon ions with different isotopic compositions.

The mass filter was used as a molecular beam detector by Bennewitz and

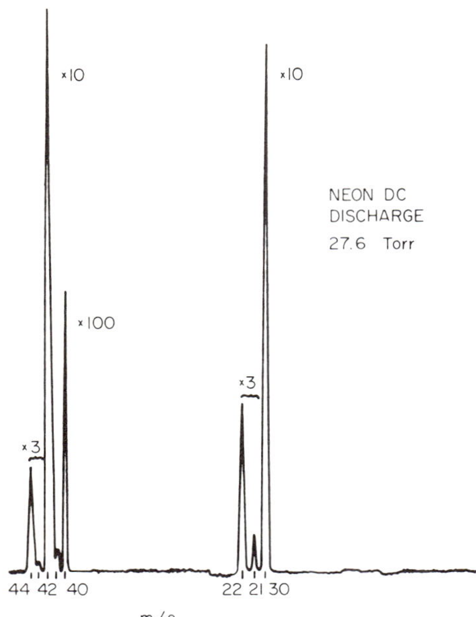

FIG. 49. Mass spectrum of ions from a dc discharge in neon obtained with a hyperbolic mass filter by Mosharrafa and Oskam (41).

Wedemeyer (52). In this application the cylindrical geometry is advantageous. The molecular beam was ionized by electron impact and analyzed in the mass filter. The smallest beam detected had a density of 1.8×10^6 molecules/sec mm^2 in a background pressure of 8×10^{-7} torr. Beam chopping and phase-sensitive detection were used to improve the signal-to-noise ratio.

Other uses have included the study of electron-impact ionization cross sections of metals by C. K. Crawford (53), charge transfer reactions at very low energies by Golden et al. (54) and rapid reactions in shock waves by Gutman et al. (55, 56). Plasma plumes generated by laser surface interactions were investigated by Gilmour and Giori (57).

The lifetime of metastable molecule ions was studied by von Zahn and Tatarczyk (58, 59) in an unusual application of a quadrupole field. An electron impact source and a small double focusing magnetic mass spectrometer produced a beam of ions with a given mass-to-charge ratio (Fig. 50). The ions then traveled along a drift tube 3 meters in length consisting of a quadrupole mass filter with no dc voltage applied. Consequently both parent and fragment ions from the metastable decay were transmitted and entered a second quadrupole mass filter where they were mass-analyzed. An appreciable number of ions were found with metastable states having lifetimes longer than 10 μsec.

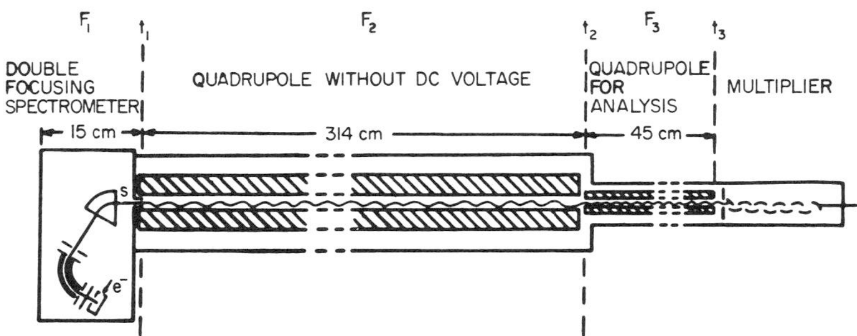

FIG. 50. An apparatus used by von Zahn and Tatarczykn (*59*) to determine lifetimes of long-lived metastable ions. The central quadrupole structure, which has no dc field, acts as a drift tube.

Charged macroscopic particles of salt powder were suspended in a quadrupole field by Gradewald (*60*), and the damping of their motion due to gas pressure was observed. This work is similar to that of Wuerker *et al.* (*11*) using the three-dimensional quadrupole, and discussed in Section V-D.

Mosharrafa *et al.* (*60a*) have described a miniaturized quadrupole mass filter for medical applications.

A considerable body of work exists on upper atmosphere research using quadrupole mass spectrometers in rocket, satellite, and balloon applications. Figure 51 shows a schematic view of a quadrupole mass filter mounted in a rocket nose cone. Mauersberger *et al.* (*61*) used this to measure neutral constituents of the upper atmosphere. The rocket nose cone and ceramic cover of the ion source were ejected after the rocket was aloft so that the mass analyzer was exposed directly to the atmosphere. The field length was 29 cm and its radius 0.7 cm. The sensitivity of the spectrometer was 0.28 mA/torr for nitrogen.

IV. THE MONOPOLE MASS SPECTROMETER

A. Introduction

The monopole was first described by von Zahn (*12*) in 1963, ten years after the conception of the quadrupole mass filter. The later introduction is reflected in the paucity of literature on the monopole. There was rapid development of it as a commercially available partial pressure analyzer (*62*) even though no detailed theoretical analysis of the mode of operation had then been published. Hudson (*63*) has discussed qualitatively the operation of the monopole, and recently computer simulations have been carried out

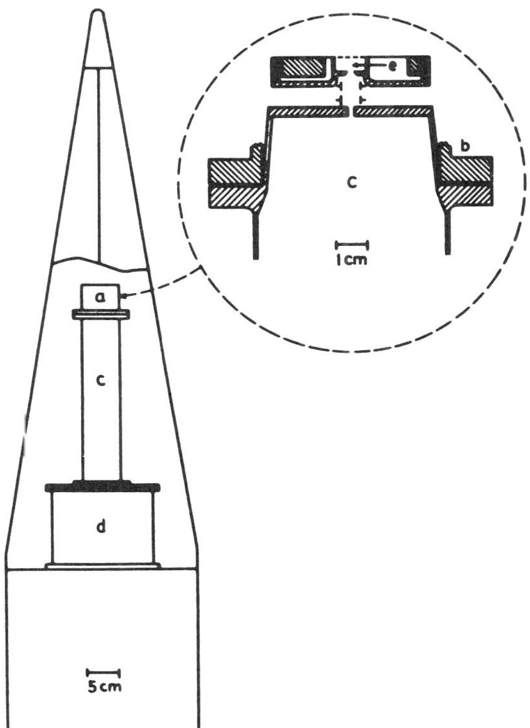

FIG. 51. Schematic view of a rocket-borne spectrometer, the split nose cone, and ion source (*61*). *a*, ceramic cover; *b*, broken cap after ejection of upper part; *c*, analyzing field; *d*, electronics; *e*, electron beam.

(*64, 65*). A different type of monopole spectrometer, utilizing exact focusing simultaneously in two directions, was proposed and analyzed by Lever (*16*) in 1966. This promised very high performance, but no such instrument has been constructed.

The monopole field is equivalent to one quadrant of the field used in the mass filter (Fig. 2), but the monopole uses partial focusing properties in the y direction in addition to the mass filtering action of the rf quadrupole field. It is therefore called a "mass spectrometer" rather than a mass filter. The focusing requirement limits the area in the stability diagram for ion transmission to a band parallel to the $\beta_y = 0$ line, such as that shown shaded in Fig. 52. Ions with (a, q) values that are in the stable area but not in the shaded band have stable trajectories, but they intersect the vee block before the exit slit is reached. The monopole may be operated with a mass scan line intersecting the stable area anywhere along the $\beta_y = 0$ line rather than just

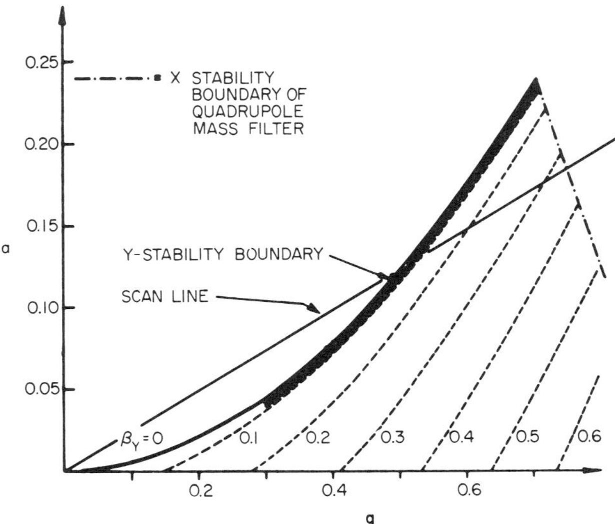

FIG. 52. Mathieu (a, q) stability diagram for the monopole. The shaded area is the region where ions taking 10 or more rf cycles to traverse the device are transmitted through it. The iso-β lines represent positions of similar motion in the y direction. *(12)*.

near the apex of the stability triangle. This leads to properties of the monopole that are different from those of the mass filter.

In this section we first present the theory of the monopole, applying the approaches developed in Section II, and give a qualitative description of its operation. The computer simulations and predicted properties for perfect fields are considered next. The experimental achievements, which in some ways surpass the computed properties, are described, followed by an account of Lever's proposed instrument. Finally, some advantages specific to the monopole are mentioned. The properties resulting from imperfect quadrupole fields, caused, for example, by the use of a rod with a circular instead of hyperbolic cross section, are considered in Section VI.

B. *Mode of Operation*

1. *Theory*

For the monopole, the constants in the expression for the potential, Eq. (1), are

$$\lambda = -\sigma = \frac{1}{r_0^2} \quad \text{and} \quad \gamma = 0$$

so that

$$\Phi = \frac{x^2 - y^2}{r_0{}^2}(U - V \cos \omega t) \qquad (48)$$

with $y \geq |x|$. The right-angled electrode (vee block) is held at ground potential, and the dc and rf voltages, $-U$ and $V \cos \omega t$, are applied to the rod. The equations cf ion motion are

$$\frac{d^2u}{d\xi^2} + (a_u - 2q_u \cos 2\xi)u = 0 \qquad (49)$$

and $d^2z/d\xi^2 = 0$, where u represents either x or y. From Eq. (9), a and q for the monopole are defined as $a_x = -a_y = 8eU/m\omega^2 r_0{}^2$, $q_x = -q_y = 4eV/m\omega^2 r_0{}^2$.

von Zahn (12) analyzed the motion in the yz plane given by Eq. (49). The stable solution is

$$y = \alpha_I y_I + \alpha_{II} y_{II}$$

$$= \alpha_I \sum_{s=-\infty}^{\infty} C_{2s} \cos \left(s + \frac{\beta}{2}\right)\omega t + \alpha_{II} \sum_{s=-\infty}^{\infty} C_{2s} \sin \left(s + \frac{\beta}{2}\right)\omega t \qquad (50)$$

One can introduce new coefficients

$$a_0 = C_0, \qquad a_s = C_{2s} + C_{-2s}, \qquad b_s = C_{2s} - C_{-2s} \qquad (51)$$

and tabulate values derived from the earlier calculations of C_{2s} (13) (Table I). α_I and α_{II} can be eliminated from Eq. (50), as in Eqs. (18–20). Assuming $\dot{y}_0 = 0$, y can be given in terms of y_I, y_{II}, and the initial position y_0 as

$$y = [y_0/W][\dot{y}_{II}(t_0)y_I(t_0 + t) - \dot{y}_I(t_0)y_{II}(t_0 + t)] \qquad (52)$$

By substituting y_I and y_{II} and the new coefficients, this equation gives a sum of eight products, each involving two infinite series. However, for high resolving power, the β_y value at maximum ion transmission is small (Fig. 52) and all the terms can be neglected except one, so that

$$y \approx gy_0\beta^{-1} \sin \frac{\beta\omega t}{2} \sum_{s=0}^{s=\infty} sa_s \sin (s\omega t_0) \sum_{s=0}^{s=\infty} a_s \cos (s\omega[t_0 + t]) \qquad (53)$$

where $g = 2.38$ for $q = 0.705$ from the tabulated coefficients. The equation is valid only for

$$\omega t_0 \neq n\pi \qquad \text{and} \qquad \omega t \neq n2\pi\beta^{-1}, \qquad n = 0, 1, 2 \cdots \qquad (54)$$

The term $\sin \beta\omega t/2$ gives an oscillation that depends on β but not on the initial conditions when the ion enters the field. The period of this oscillation, τ_0, is $4\pi/\beta\omega$ or $2/\beta$ rf periods.

The sum $|a_1| + |a_2| + \cdots$ is smaller than $|a_0|$, so the second summation term represents a high frequency oscillation within the fundamental motion and does not cause the ion to cross the instrument z axis between the times $\tau_0/2$. In other words, the ion motion is similar to that shown in Fig. 10(a), as opposed to the x direction of motion, which is similar to Fig. 10(b).

Provided the geometrical condition, $y > |x|$, is satisfied, an ion may pass through the monopole if the time it takes to pass through the rf field is less than $\tau_0/2$ and if its amplitude is within the bounds of the instrument. The ion will strike the vee block at $\tau_0/2$ if it remains long enough in the field.

The first summation in Eq. (2) has an alternating sign, and for small t, $(y - y_0)$ will be positive only when $(2m - 1)\pi < \omega t_0 < 2m\pi$, $m = 1, 2, 3 \ldots$. Therefore, for about half of each rf cycle the ions will be driven immediately toward the vee block and will be lost. This does not apply when $\dot{y}_0 > 0$, as is evident in Fig. 9.

2. Qualitative Description

The monopole can be mass-scanned by frequency variation while keeping U and V constant, or by varying the magnitudes of U and V (amplitude scanning) with a fixed frequency and a fixed U/V ratio. With frequency scanning, which was used by von Zahn, ions of different masses stay the same number of rf cycles within the analyzing field, for a fixed energy in the z direction. The spectrum has the peak positions dependent on $1/(\text{frequency})^2$. Recent use of the monopole has generally been at a fixed frequency with voltage amplitude variation (62, 66). This facilitates the design of circuitry for a wide mass range and fast sweep capability. With amplitude sweeping and a fixed ion energy, the length of the device in terms of rf cycles depends on $(M)^{1/2}$. The peaks are approximately equally spaced.

In both methods of scanning, the ions pass through the device when the half-beat length of their fundamental motion in the y direction is longer than the time (number of rf cycles) they spend in the field. The ions must be injected with a narrow energy spread. Ions of lower masses (higher q) than those just passing through the device will be spread out in a spectrum intersecting the vee block until the q value is such that the x motion becomes unstable for the particular U/V scan line being used. Ions of higher mass or lower (a, q) value will to a first approximation pass through the device unless their (a, q) value is such that they are y unstable. In this way one might expect two sharp boundaries to the mass spectral peak in the (a, q) diagram, the lower one being the y stability boundary and the higher one the appropriate $\beta_f = 1/n$ line. However, for the low (a, q) side of the peak, the implied assumption is that the allowable amplitude of oscillation is infinite. The finite dimensions of the instrument result in a gradual edge on the low (a, q) side of the peak, as the computer simulation results demonstrate (Section IV-C). For the infinite amplitude approximation, the resolution can be estimated for any scan line if the relationship between β and q (or a) is known, since

$$(M/\Delta M) = (q_{\beta_f} + q_{\beta=0})/2(q_{\beta_f} - q_{\beta=0}) \tag{55}$$

From Fig. 52 it is evident that the resolution for a given n would be expected to decrease as the slope of the scan line U/V is decreased owing to the lower q of the boundary and the greater width between the iso-β lines. This presumably provides a limit to lowering the U/V ratio so as to operate at lower voltages (see Section IV-D-3).

The relationship given by Paul *et al.* (*13*) for the apex of the stability diagram is

$$\beta_y^2 = (a_{\beta=0} - a_\beta)/0.79375 \tag{56}$$

$a_{\beta=0}$ is 0.23699 at the apex so that, from Eq. (55),

$$(M/\Delta M) \approx 1/3.34\beta_f^2 = n^2/3.34$$

where $(M/\Delta M)$ is the resolution at the base of the peak.

Near the apex of the stability area, von Zahn (*12*) predicted (without giving its source) the relationship

$$(M/\Delta M) = n^2/2.25 \tag{57}$$

where $(M/\Delta M)$ is the full width at half-height.

Lever (*16*) computed the approximate empirical relation

$$q_\beta n^2 = 2/(q_\beta - q_0) \tag{58}$$

When β is small, $q_\beta \approx q_0$ and the resolution is approximately $(M/\Delta M) = n^2 q_0^2/2$. That is, at the apex, $(M/\Delta M) = n^2/4$. For the $a/q = 0.2$ scan line, one would predict $(M/\Delta M) = n^2/10$ at the base of the peak.

The relationship between resolution and n is explored further in both the computer simulation and in the experimental data, Sections IV-C and D.

When $n \propto M^{1/2}$ (amplitude sweeping), and if a formula of the type $(M/\Delta M) = n^2/n$ is valid, then $(M/\Delta M) \propto M$. One would expect to a first approximation a mass spectrum of equally spaced peaks of constant half-width. This is a useful property when simplicity in the output signal is advantageous.

C. Computer Simulation

Computer simulation of the monopole operated in the conventional fashion (all ions having the same energy in the z direction and with partial focusing in the y direction only) has been reported by Dawson and Whetten

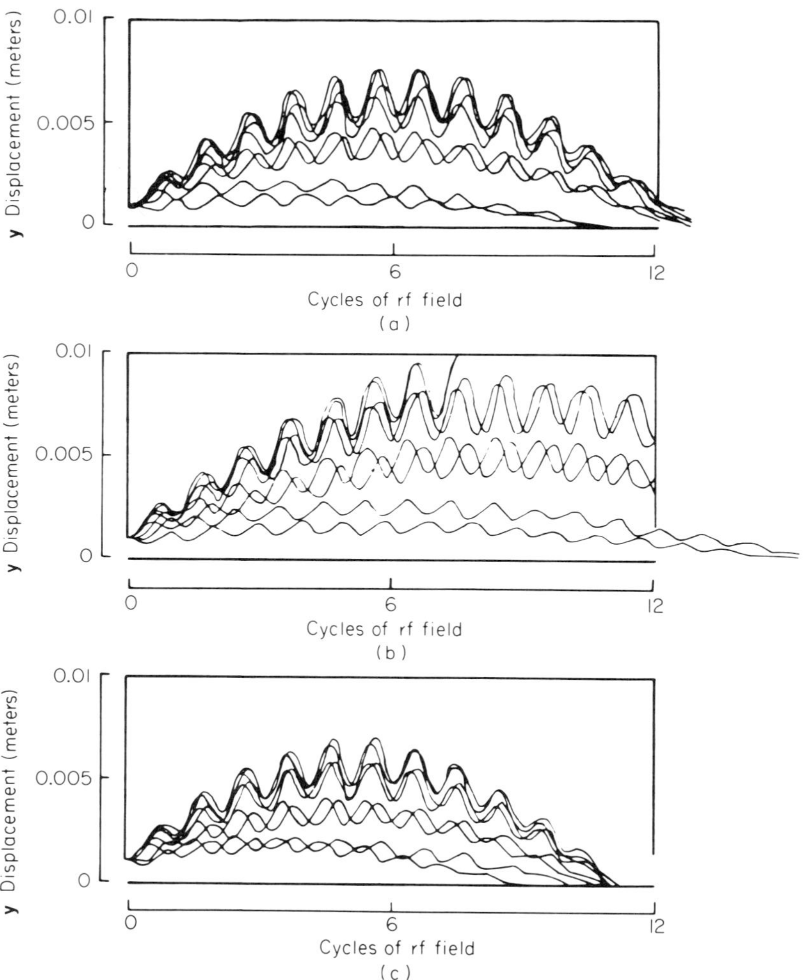

FIG. 53. (a) y trajectories for $a = 0.085$, $a = 0.425$. The ions have an initial displacement of 10^{-3} meters. Trajectories are shown for ions that enter only at phases between $\pi/32$ and $15\pi/32$ in ξ. The trajectories are given for intervals of $\pi/16$ in ξ. Ions entering at other phases would strike the vee block. The dimensions of the simulated monopole are superimposed on the trajectories. The exit slit is 1.5×10^{-3} meters wide and ions spend 12 cycles within the field (64). (b) y trajectories for an ion of higher mass when voltages are set for the transmission of ions of $M = 72$ illustrated in (a). Here, $a = 0.827$ and $q = 0.4135$ so that $M = 74$ for the trajectories shown. (c) y trajectories under the same conditions for $M = 71$.

(64, 65). The simulation was carried out by numerical integration of the equations of ion motion with automatic trajectory plotting on an X–Y recorder. The scan line $a/q = 2U/V = 0.2$ was considered. For ion entrance to the analyzer with an initial displacement from the axis, but no radial velocity, peak shapes were computed for ions taking 5.8, 12, and 20 rf cycles to pass through the device. For the 20-cycle case, ions entering with a radial velocity were also considered. To make the simulation specific in terms of ion mass, it was assumed that an ion of 72 amu spent 12 rf cycles within the field. For a monopole 0.1 meter long operated at 1 MHz, this corresponds to an ion energy of 26.2 eV in the z direction.

An example of the data obtained for the y trajectories is shown in Fig. 53. The ions are considered to enter with $\dot{y}_0 = 0$ and an initial y displacement of 10^{-3} meter. The figures can be scaled arbitrarily in the y direction. The illustrated exit slit is 1.5×10^{-3} meter wide. The trajectories are for phases of the rf field at ion entry between $\pi/32$ and $15\pi/32$ in ξ, with intervals of $\pi/16$. Ions entering at other phases quickly strike the vee block. The ion illustrated has a z velocity such that it takes 12 rf cycles to pass through the device. In Fig. 53(a) the (a, q) values are 0.085 and 0.425, and ion transmission is near the maximum In Fig. 53(b), $a = 0.0827$ and $q = 0.4135$, so that the illustration is for the same ion species as in (a) and a lower scanning voltage or, alternatively, for the same scanning voltage as in (a) and an ion of higher mass. If Fig. 53(a) is for an ion of mass 72, Fig. 53(b) is for $M = 74$ under the same conditions, and Fig. 53(c) is for $M = 71$. By determining the percentage of the rf cycle during which ions are transmitted through the exit aperture as the scanning voltage is varied, the peak shapes can be predicted

Figure 54 illustrates trajectories in the xz plane for those phases of the rf field at ion entry which are likely to result in ion transmission in the yz plane. An initial displacement but no initial velocity has been assumed. The x trajectories do not vary much in nature until the x stability boundary is approached (that is, for very low masses). However, because of the condition $|x| < y$, the x trajectory is important, particularly when the y displacement approaches zero near the exit from the monopole.

The shift in the (a, q) position for maximum ion transmission with the ion transmission time (that is, with M) that is evident in Fig. 53 is shown more fully in Fig. 55. Here the voltage required for maximum transmission, given as a multiple of that corresponding to the y stability boundary, is plotted as a function of the number of cycles an ion spends in the field, and is also plotted against m/e for the simulated monopole. Peak separation is almost constant except at low masses. The lower m/e limit for any device is set by the approach to the x stability boundary, which focuses at 2.4 cycles in the y direction for the $a = 0.2q$ scan line. Higher mass ions can always be focused by a closer approach to the $\beta_y = 0$ boundary, but even if other con-

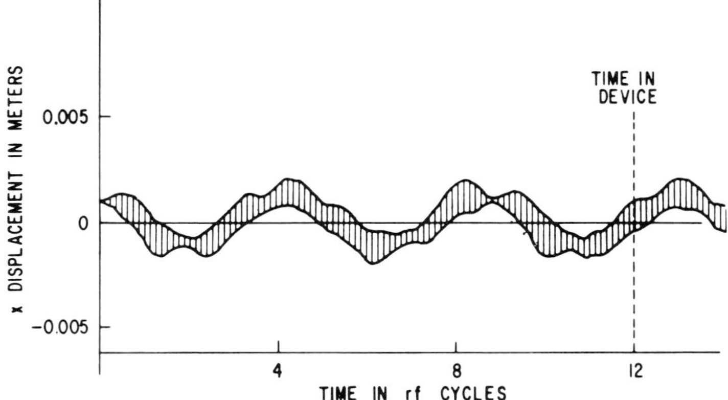

FIG. 54. Envelope of the trajectories in the x direction for initial phases between 0 and $\pi/2$ in ξ, and an initial displacement of 10^{-3} meters. The trajectories are for $a = 0.085$, $q = 0.425$, where ions are focused in the y direction after 12 rf cycles (*64*).

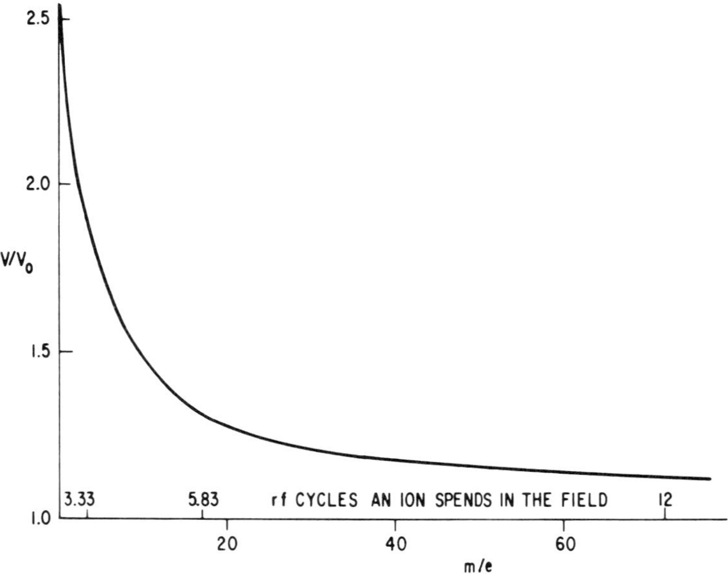

FIG. 55. Variation of peak spacing with the number of cycles an ion spends in the quadrupole field. V is the voltage at which the ion peak occurs and V_0 is the voltage corresponding to the $\beta_y = 0$ stability boundary. The abscissa corresponds to both the m/e ratio of the ions for this simulated instrument and to the number of rf cycles that any ion spends in the field before collection. The latter is of general validity for the $a = 0.2q$ scan line (*64*).

siderations do not interfere (e.g., the voltage required), a limit may be set by the finite rod/vee block separation, which restricts the maximum allowable amplitude. Figure 15 is useful in monopole design, since the maximum amplitude is given as a function of β_y.

Figure 56 shows the peak shapes for the 12 cycle case ($M = 72$), when there is no initial y velocity and the exit slit width is 1.5, 2, and 6 times the initial displacement of the ions. Only the yz plane was considered. It should

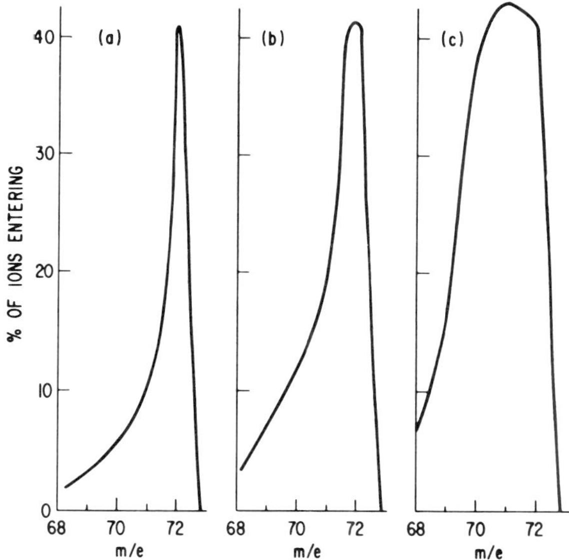

FIG. 56. Computed peak shapes for $M = 72$ ions spending 12 cycles in the field, and entering at 10^{-3} meters displacement in the y direction (64). $a = 0.2q$. (a) Exit slit = 1.5×10^{-3} meters. Half-height resolution = 97. (b) Exit slit = 2×10^{-3} meters. Half-height resolution = 58. (c) Exit slit = 6×10^{-3} meters. Half-height resolution = 24.

be emphasized that in a particular instrument the rod/vee block separation may limit the amplitude of oscillation and provide an "effective" exit slit width. The device has its own natural aperture. Figure 57 shows a comparison of peak shapes for the 5.8-, 12-, and 20-cycle cases (corresponding to masses 17, 72, and 200).

Information relating the resolution to the number of cycles an ion spends in the field is summarized in Table III. It is evident that the formula ($M/\Delta M$) $= n^2/h$, given in Eq. (57), is approximately correct; here h is constant for a given exit slit width.

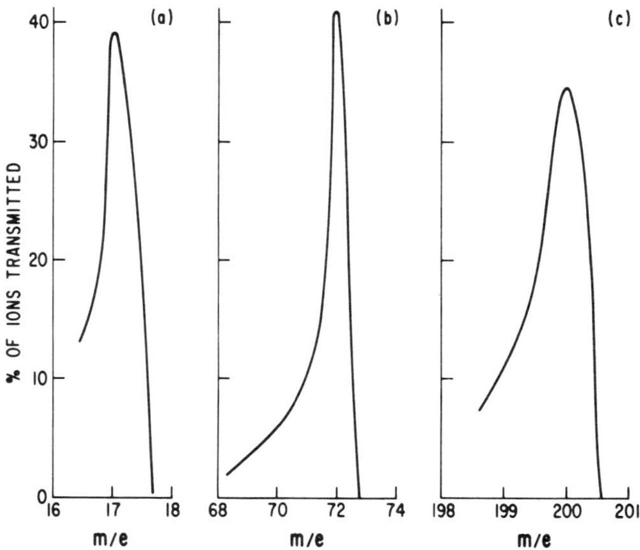

FIG. 57. Computed peak shapes for $a = 0.2q$. (a) Ions spending 5.83 cycles in the field. Exit slit $= 1.5 \times 10^{-3}$ meters. (b) Ions spending 12 cycles in the field. Exit slit $= 1.5 \times 10^{-3}$ meters. (c) Ions spending 20 cycles in the field. Exit slit $= 2 \times 10^{-3}$ meters. All ions were assumed to enter with 10^{-3} meters displacement and parallel to the axis (65).

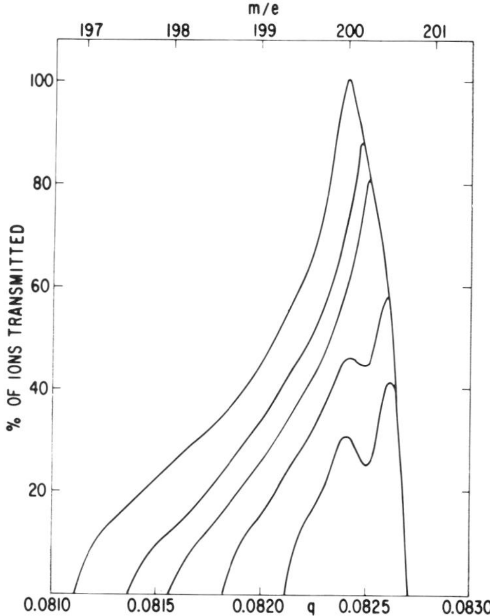

FIG. 58. Peak shapes for ions spending 20 cycles in the field with ion injection at $y = 10^{-3}$ meters with an initial velocity in the y direction of 3×10^{-4} meters/rad of the applied field (65). Assumed exit slit widths were 1, 2, 3, 4, and 6×10^{-3} meters.

Peak shapes when ions are injected at an angle to the z axis are illustrated in Fig. 58 for 20 cycles spent in the field. Ion injection was considered at $y = 10^{-3}$ meter with an initial velocity in the y direction of 3×10^{-4} meter/rad

TABLE III

COMPUTED MONOPOLE RESOLUTION[a]
(*All Ions Enter at $10^{-3}m$ Displacement*)

	No. of cycles in field (*n*)	Exit slit ($\times 10^{-3}$ meters)	Resolution ($M/\Delta M$) (half-height)	h where $(M/\Delta M) = n^2/h$
		Scan Line $a = 0.2q$		
Parallel injection	5.8	1.5	24	1.44
Parallel injection	12	1.5	97	1.49
		3	58	2.46
		6	24	6.0
Parallel injection	20	2	210	1.92
		3	160	2.48
		4	140	2.85
		6	100	4.0
Injection with y velocity of 3×10^{-4} m/rad of applied field	20	1	230	1.75
		2	200	2.0
		3	200	2.0
		4	180	2.2
		5	170	2.33
		Scan Line $a = 0.08q$		
Parallel injection	12	1	15	9.6
		1.5	12.7	11.3
		2	9.6	15

[a] P. H. Dawson and N. R. Whetten (*65*).

of the applied field. For a monopole of 0.1-meter length this is equivalent to injection at a 20 deg angle. If the exit slit is too small, the peaks exhibit a central dip. Maximum possible ion transmission is higher with angular injection and there is less broadening of the peak on the low mass side. In practice, experimental entrance conditions for the monopole may be difficult to control or keep constant throughout the complete rf cycle because the applied rf can be very large compared with the ion energy. This may partially account for the differences between the simulated and experimental peak shapes.

D. Experimental

1. Ion Source Design

Low ion energies are usually required in the monopole, and ion sources without magnets are used to avoid disturbance of ion trajectories. An electron impact source described by von Zahn (*12*) is reproduced in Fig. 59. The electrons were focused by two electrostatic lenses. Electrodes J_1 and J_2 focused

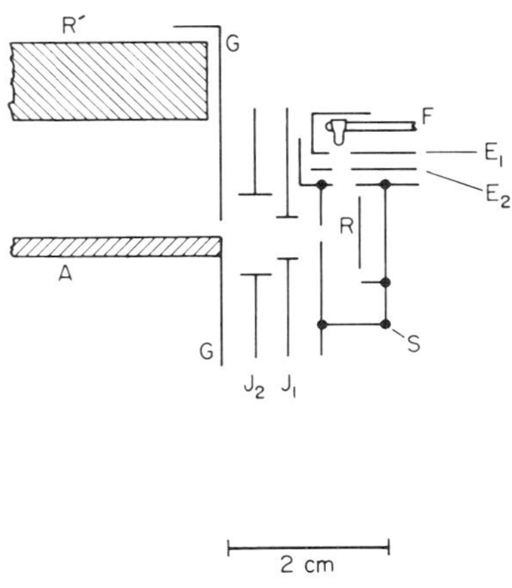

Fig. 59. Schematic drawing of von Zahn's ion source. Hole diameters for E_1, E_2, S, J_1, and J_2 were 2, 2, 2, 4, and 10-mm diameter, respectively. Typical voltages were: $F = -70$, $E_1 = -74$, $E_2 = +380$, $S = R = +90$, $J_1 = +83$, $J_2 = +60$, $G = 0$. R′ is the rod electrode and A is the vee block (*12*).

the ions on a V-shaped entrance slit of 2 mm^2 area. The field radius was 1.51×10^{-2} meter. The triangular exit hole was 36 mm^2. These dimensions were chosen to achieve the best compromise between sensitivity and peak shape—presumably indicating more severe peak shape problems with larger entrance and exit holes. However, monopoles are often constructed without exit and entrance apertures, the rod/vee block spacing defining the fraction of the available field that can be effectively used at the entrance and exit.

2. Electronic Circuitry

von Zahn (*12*) used a self-excited oscillator of about 1 kV maximum amplitude and scanned by frequency variation between 1.11 and 2.5 MHz. No other details are available. The advantages and disadvantages of the frequency or amplitude sweeping have been discussed by Hudson and Watters (*62*). They also discuss the trade-off between such parameters as rf voltage, rf frequency, rod length, and ion energy to obtain a given performance. Hudson and Watters used frequencies near 1.8 MHz and a peak amplitude of 3000 volts peak-to-peak, which was sufficient to reach 300 amu with $a/q = 0.20$. The rf circuit consisted of a quartz crystal oscillator driving a modulated amplifier and a sweep generator, connected in a feedback circuit. The resonant high voltage circuit ("tank" circuit) with an associated ac/dc ratio circuit was located at the device itself. This permitted the use of a simple receiving tube as the rf amplifier. The dc output was subtracted from the output of the function generator, using feedback to keep the difference small. Loop gain was greater than 10^4. Scan rates as high as 10^{-3} sec/amu were achieved.

A solid-state field generator for the monopole has been described in detail by Lins and Paul (*66*). Their circuit is reproduced in Fig. 60. Voltage amplitude sweeping was used. The frequency could be switched manually from 1.0 to 0.5 MHz to cover mass ranges of 30 to 0 and 130 to 0 in their instrument. The maximum peak-to-peak rf amplitude was 1 kV. The final amplifier is shown in detail. The monopole is a capacitive component of the tuned output and is a low dissipation load so that the output has a high Q and little harmonic distortion. The dc voltage was obtained in the conventional way by detecting and filtering part of the rf signal. The ramp for the amplitude sweep was taken from an oscilloscope. The minimum sweep time was 5×10^{-2} sec.

3. Scan Line and Mass Range

Figure 61 presents several scan lines on the (a, q) stability diagram, and the distribution of masses in (a, q) space when mass 200 is being transmitted through the monopole (*67*). A change from the line AC to line AD decreases the voltage for focusing a given mass by more than a factor of 2. It is therefore a useful change to make if the performance does not suffer. A change of scan line from AC to AD has been reported (*67*) with, surprisingly, no loss of resolving power and with only slight changes in the ion transmission efficiency as a function of mass. A further reduction in the a/q ratio would seem attractive, although a limitation may be that the field required for the lowest mass becomes smaller than is practical. Of course the a/q ratio could be changed for different mass ranges. One would expect that the resolution would decrease as the a/q ratio is reduced, due to the lower q and the greater width between iso-β lines. Operation at lower voltages should have little effect on the high mass side of the peak. However, the tailing on the low mass side should become more severe at lower voltages when the amplitudes of the trajectories tend to remain small. This is illustrated by the computed peak shapes of Fig. 62. The

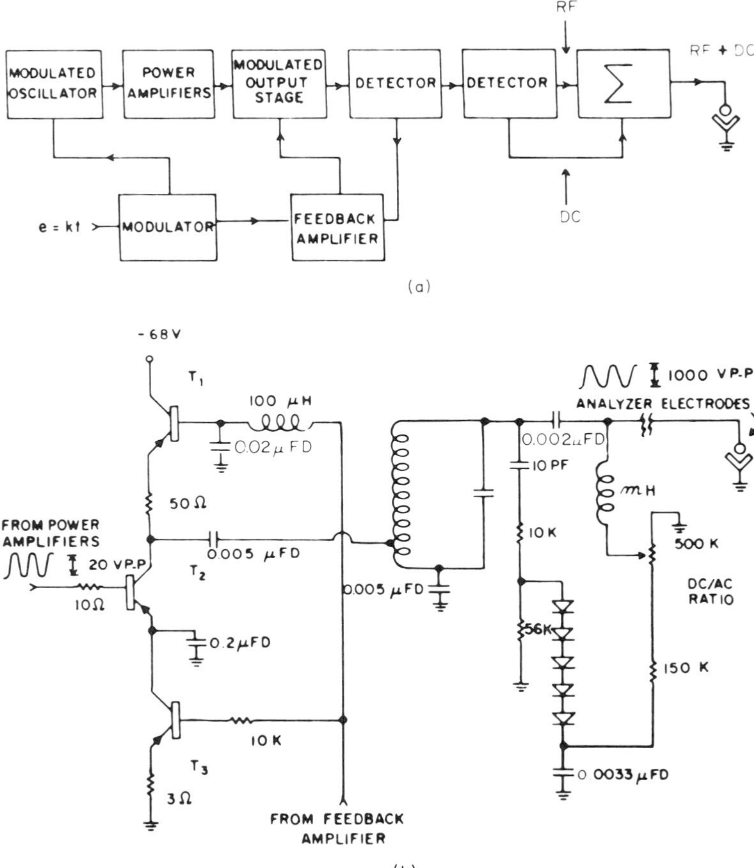

FIG. 60. Solid-state field generator for the monopole: (a) schematic; (b) output stage; from Lins and Paul (*66*).

ions in this figure spend 12 cycles in the field, but the scan line is $a = 0.08q$ (compare with Fig. 56 and see Table III). The performance has deteriorated as expected, but there is no *experimental* evidence to indicate at what voltages such effects become important.

Other methods of reducing the rf power (or alternatively extending the mass range) are to decrease the frequency and to decrease the field radius r_0. For a given tube length, the resolution decreases with the square of the frequency, since $(M/\Delta M) = n^2/h$. Decreasing the field radius may decrease the relative mechanical precision in the electrode structure and increase the field faults (Section VI). It also reduces the natural aperture of the monopole, which is particularly serious for high masses (small β) and eventually leads to a loss in both sensitivity and resolution.

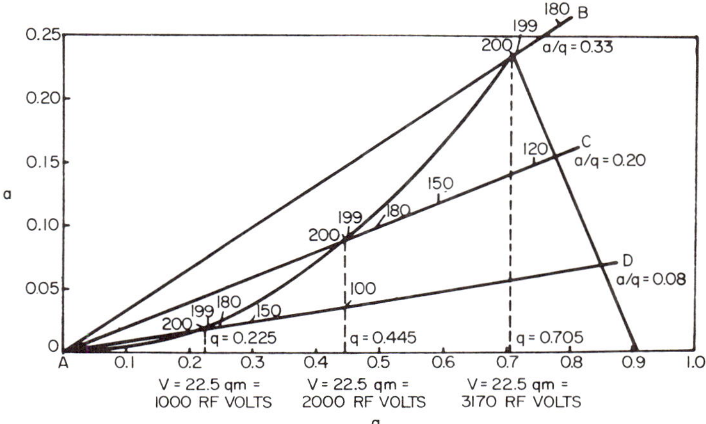

FIG. 61. Stability diagram for the monopole, showing several scan lines and the distribution of masses in (a, q) space when mass 200 is being transmitted. The voltages are those required to transmit mass 200 (*67*).

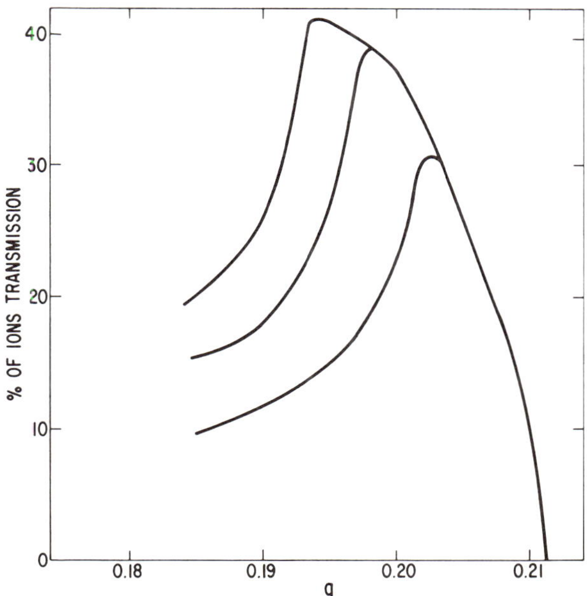

FIG. 62. Computed peak shapes for ions spending 12 cycles in the field for the scan line $a = 0.08q$. The ion entrance is at 10^{-3} meters, parallel to the axis. The exit slit widths are 1, 1.5, and 2×10^{-3} meters.

4. Performance

von Zahn's original monopole (*12*) had an analyzing field 27 cm in length. With an ion energy of 92 eV, the sensitivity was 3×10^{-4} amperes/torr at a resolution of 190. Figure 63 shows a spectrum taken at a pressure

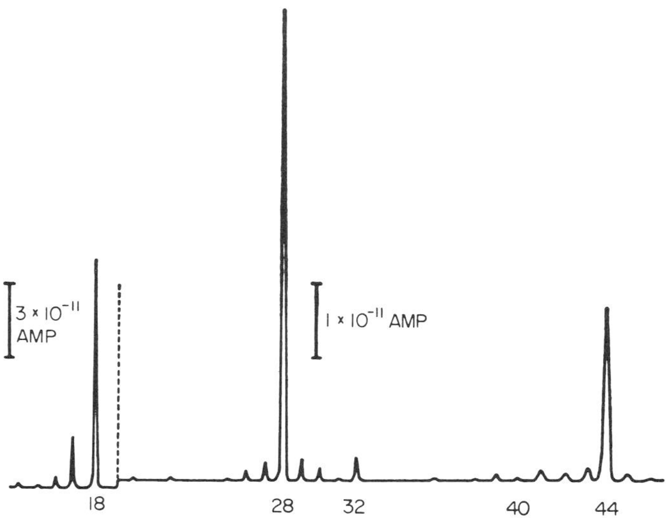

FIG. 63. Monopole spectrum at 10^{-6} torr. Ion accelerating voltage was 92 volts and ions spent 16.5 rf cycles in the field. Frequency scanning through a point near the stability apex was used (*12*).

of 10^{-6} torr. For $M = 28$, the frequency was 1.54 MHz, so that n was 16.5 cycles. Lower ion energies gave a higher resolution, the highest reported being 470 with an accelerating voltage of 35 volts and $n = 30$. Figure 64 shows a semilogarithmic plot at a higher pressure and reveals details of the peak shape. These spectra were taken near the apex of the stability diagram. von Zahn found experimentally that $h \ [=n^2/(M/\Delta M)]$ was between 1.45 and 2.25, in good agreement with his theoretical prediction. When the scan line was changed to lower a/q ratios, there was a deterioration in peak shape with a tail developing on the high mass side. von Zahn ascribed this to mechanical imperfections in the analyzer. Figure 65 shows an example of the effect. The value of h was about 3.8. These results were obtained by using frequency scanning, so the resolution $(M/\Delta M)$ was constant.

With a voltage amplitude scanning monopole, Hudson and Watters (*62*) reported a resolving power of 40 with an ion energy of 70 eV, and 600 with an ion energy of 20 eV. The mass M at which the resolutions apply was not

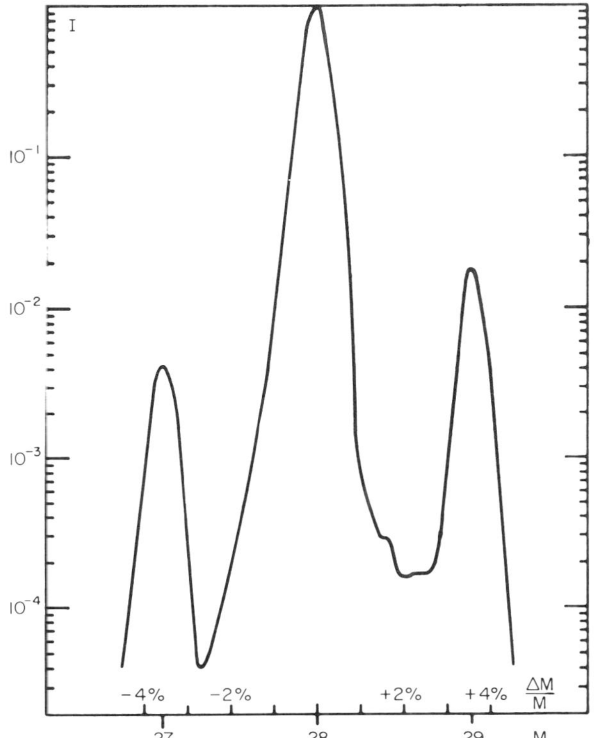

FIG. 64. Monopole peaks taken with a slightly higher pressure than in Fig. 63. The semilogarithmic scale emphasizes the peak shapes (*12*).

reported. Their device was about 20 cm long and operated at 1.8 MHz. If the quoted figures were for $M = 300$, the values of h would be very high—17.5 and 75, respectively. The scan line was $a/q = 0.2$. The sensitivity was about 5×10^{-6} amperes/torr at the multiplier input.

In a later device, with an $a/q = 0.08$ scan line, Grande *et al.* (*67*)obtained a resolution of about 2000 at mass 580, as shown by the germanium tetraiodide spectrum reproduced in Fig. 66. Table IV shows the relative sensitivity for various gases. The absence of a larger variation was attributed to two opposing effects; the tendency for an increase in ionization cross section with the size of the particle, and a decrease in multiplier gain with the square root of the ion mass.

Monopoles are available with sufficient sensitivity to observe partial pressures of 10^{-12} torr routinely, and 10^{-14} to 10^{-15} torr with greater care, using electron multipliers. For trace analysis, claims have generally been

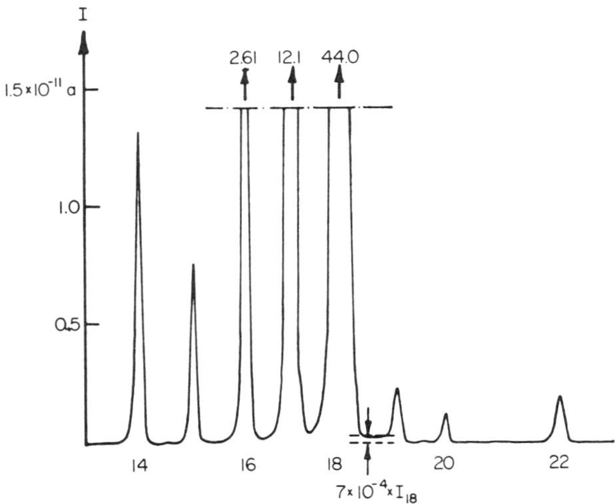

FIG. 65. Monopole spectrum taken by von Zahn with a scan line $a \approx 0.25q$ and an ion energy of 70 volts. This shows the problem of "tailing" on the high mass side (*12*).

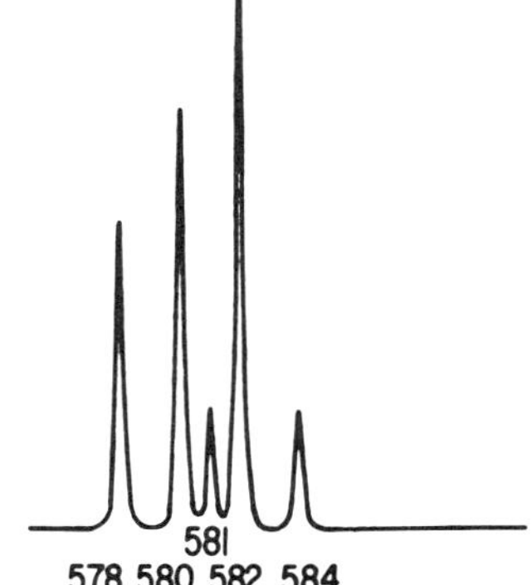

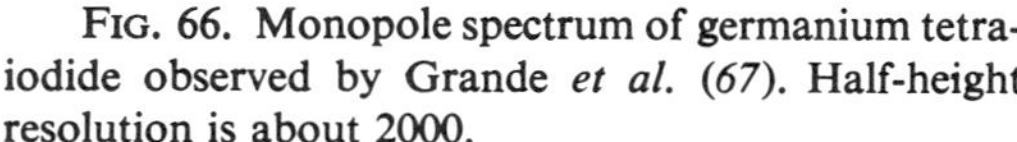

FIG. 66. Monopole spectrum of germanium tetra-iodide observed by Grande *et al.* (*67*). Half-height resolution is about 2000.

limited to about one part in 10^6, largely due to a pressure-dependent background noise.

The superiority of the experimental peak shapes over those predicted by the computer simulation is very noticeable. One explanation may be that the ions are accidentally injected at an angle to the axis. However, since the low mass "tailing" results from ions that always have a very small y displacement, a more likely reason is that these ions strike the vee block on their way down the tube. They may also be discriminated against by end effects at the entrance to the field. For improved peak shapes, the computer simulation results suggest the use of a stop placed halfway along the tube. In practice it is apparently not necessary.

TABLE IV

RELATIVE SENSITIVITY OF MONOPOLE
FOR VARIOUS GASES (67)[a]

Gas	Relative sensitivity (S_{gas}/S_{N2})
H_2	0.3
He	0.05
N_2	1.0
Ar	1.5
Xe	0.6
Hg	0.4

[a] Conditions: Pressure Range:
4×10^{-3} to 1×10^{-5} torr
Ionizing Current: 1.0 ma
Ionizing Voltage: 50 volts

E. The Exact Focusing Monopole

von Zahn, in his original paper (12), mentioned the possibility of utilizing focusing properties in both x and y directions. Lever (16) developed this idea in a computer study. His tabulations of double focusing points in the (a, q) diagram and examples of the trajectory calculations (Fig. 9 and 10) were given in Section II. Since the image after $\tau_0/2$ is inverted in the y direction and negative y values are not allowed in the monopole, Lever proposed that ion entry should be with $y_0 = 0$ and $\dot{y} \neq 0$ (Fig. 10). The image is then formed on the axis. The ions may be injected through an aperture in the vee-block electrode so that end effects are minimized. The angle of injection should be as large as possible without the ions striking the rod electrode. An approximate guide is that the undeflected beam would strike the rod electrode about three-quarters of the way down the tube. The initial x velocity, $\dot{x}_0$, should also be nonzero so as to use x focusing. The ions of a particular mass should be monoenergetic in the z direction and all masses must spend the same number of cycles within the field. This requires using frequency scanning, which presents the difficult problem of achieving wide mass range or of using voltage amplitude scanning and altering the initial ion energy (in all directions) in proportion to the voltage.

The attainable resolution depends on the variation in focus with varying (a, q). Figures 67 and 68 show parts of the y and x trajectories for times near

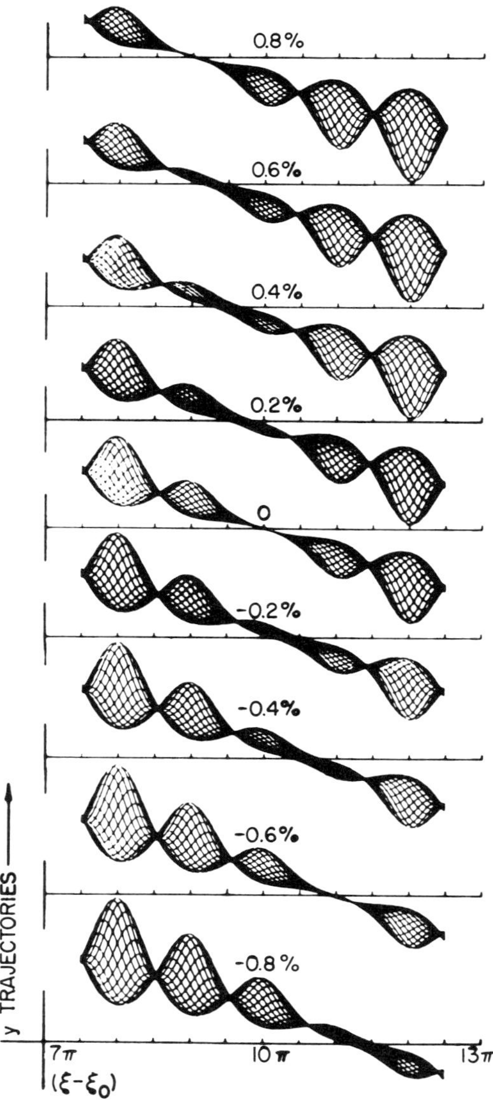

FIG. 67. Ion trajectories in Lever's proposed exact focusing monopole, showing the effect of varying q at constant a/q on the quality and position of focus in the yz plane. Initial conditions were $y = 0$, $\dot{y} = 1.0$, $0 < \xi_0 < \pi$ at intervals of $\pi/16$. The operating point for the curve for 0% variation in q was $(-0.225041, -0.699745)$ so that $\beta = 1/10$ (16).

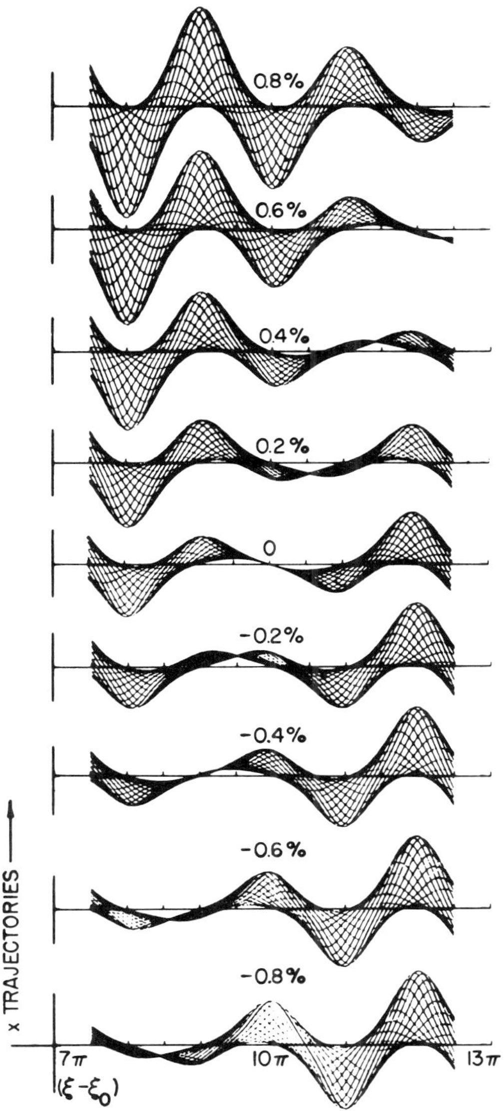

FIG. 68. Ion trajectories as in Fig. 67, but for the corresponding x trajectories. $(a, q) =$ +0.225041, +0.699745, and $\beta = 9/10$ (*16*).

10 cycles in the field and for varying q with a contant a/q. The chosen (a, q) operating point was $(\pm 0.225041, \pm 0.699745)$, where $d = 10$ and $p = 9$. As one moves away from the operating point, both the position and the quality of the focus changes. The discussion of resolution will be limited to the y direction. The spatial length of the focusing distance z_f depends on the number of cycles n and on (energy)$^{1/2}$ for a particular e/m and applied frequency. As a consequence of the method of scanning, the energy is dependent on the q value. Hence,

$$z_f \propto nq^{1/2} \tag{59}$$

However, an empirical relation was found relating $nq^{1/2}$ to q, viz.,

$$n^2q = 2/(q - q_0) \tag{60}$$

By substituting and differentiating, one can obtain

$$\begin{aligned} \Delta z_f/z_f &\approx -\Delta q/2(q - q_0) \\ &= -(qn/2)^2(\Delta q/q) \\ &= (qn/2)^2(\Delta M/M) \\ &= (qd/2)^2(\Delta M/M) \end{aligned} \tag{61}$$

Thus, for example, if d is chosen as 20 and p is 19 so that q is about 0.7, and it is assumed that $z_f/\Delta z_f = 100$ is a practical value, then $(M/\Delta M) = 10^4$. The focusing monopole therefore offers the possibility of very high resolution. Achievement of this would, of course, require accurate control of the variables U, V, r_0, and ω and a low spread in the z velocity. If the z velocity is well defined but is different from that for which the instrument is designed, the exact focusing will not occur at the exit slit, and when the ions are transmitted, d will be nonintegral and the peaks broadened. However, if qd is large, an error in the z velocity, v, would give only a small change in the apparent mass, since

$$\Delta v/v = (qd/2)^2(\Delta M/M) \tag{62}$$

F. Uses and Advantages

The electronics and the field configuration in the monopole are both simpler than in the quadrupole. Electronically, the simplicity results from the single-ended rf output and the fact that the constancy of the dc/rf voltage ratio is not critical. With voltage amplitude sweeping, the spectrum has a simple and useful form consisting of a sequence of almost equally spaced peaks of equal half-width. Changes in ion transmission as a function of mass have been found to be slight. The monopole also has a convenient geometry for utilizing the exact focusing properties of quadrupole fields, although this has not yet been accomplished. In contrast to the quadrupole, the monopole requires ions of a well-defined energy. This may make it inappropriate for some applications. On the other hand, the energy dependence might be utilized to provide information about ion energies. Since the monopole operates with a mass scan line that does not pass near the apex of the stability triangle, nonlinear resonances due to errors in the field (Section VI) may be less important than with the mass filter. The ability to operate at low dc/rf ratios allows the rf voltage (and therefore the power) to be smaller than with the mass filter. The use of only one-quarter of the quadrupole field permits a further reduction in power. This has made monopoles attractive for space and upper atmosphere studies. The first rocket-borne monopole was recently reported (*68, 69*).

V. The Three-Dimensional Quadrupole Ion Trap

A. Introduction

The quadrupole ion trap generates a quadrupole field in all three coordinate directions (Fig. 3) and ions may be trapped or stored within the field. The trapping can be extremely effective, and ions have been stored for days at low pressures (*70*). The three-dimensional trapping contrasts with the sorting action of the two-dimensional quadrupole field of the mass filter and the monopole, where ion motion in the z direction is unaffected by the field. The ion trap may be operated so as to store a wide range of ionic species or to trap a single mass species.

The three-dimensional quadrupole field was described by Paul and Steinwedel in their original 1953 patent (*6*). Berkling (*9*) and Fischer (*10*), colleagues of Paul, demonstrated the ion storage principle. Their results suggested that the device had considerable potential. At about the same time as Fischer's publication, Wuerker *et al.* (*11*) reported trapping macroscopic charged particles in a three-dimensional field. Recently Rettinghaus (*71*) improved on the earlier work of Fischer by using a more sophisticated ion detection technique. However, he encountered problems in the interpretation of the mass spectra which were apparently inherent to the device in the way in which it was being operated. Berkling, Fischer, and Rettinghaus stored ions of a wide mass range and detected individual ionic species by a resonant absorption (Purcell) (*71a*) technique. A problem encountered was that the presence of cne ionic species affected the storage of other species.

The ion trap was operated by Dawson and Whetten (*17, 70, 72, 73*) in a different manner, avoiding ion interactions by trapping only one species at a time. The stored ions were detected by pulsing them out through holes in one of the electrodes into an electron multiplier.

The ion trap has been used in studies in atomic physics, particularly in the determination of rf hyperfine structure spectra by Dehmelt and others (74–78). Burnham and Kleppner (79) studied the loss mechanism of ions from the trap. These types of applications of the ion trap seem likely to increase in the future. An analysis of the ion trap operation for storage experiments has been published by Dehmelt (80).

The ion trap is not now available as a commercial partial pressure measuring device. Whether it will become an accepted general purpose tool is not yet clear, although its unique properties seem certain to find application in special cases and in the study of the properties of ions.

In this section, we first apply the one-dimensional theory developed in Section II to the three-dimensional case, and describe the ways in which the ion trap can be operated. We then discuss the theoretically expected performance for the two ways of using the trap. Experimental methods and achievements in mass spectrometric use are described next, followed by a separate account of the processes by which ions are lost from the trap. The non-mass-spectrometric uses of the ion trap are briefly described. Finally the advantages and possible uses of three-dimensional quadrupole devices are summarized.

B. Theory of Operation

1. The Stability Diagram

The theory developed in Section II presents the solution to the Mathieu differential equation of motion for a one-dimensional field. The ion trap uses a three-dimensional quadrupole field, with the geometry shown in Fig. 3. The device is rotationally symmetric about the z axis and there are three electrodes that are hyperboloids of revolution—two identical cap electrodes and a ring electrode. The two sets of hyperbolae seen in any cross section of the structure containing the z axis have common asymptotes and are related, since $r_0{}^2 = 2z_0{}^2$. For the ion trap, the constants in Eq. (2) are $\lambda = \sigma = \frac{1}{2}(r_0{}^2)$, and $\gamma = -1/r_0{}^2$. The potential with reference to the center is

$$\Phi = (x^2 + y^2 - 2z^2)(U - V \cos \omega t)/4z_0{}^2 \qquad (63)$$

U and $V \cos \omega t$ are dc and rf voltages applied to the ring electrode. The cap electrodes are grounded. r_0 is the minimum radius of the ring electrode and z_0 is half the minimum distance between the end caps.

It is convenient to simplify the consideration of ion motion by using coordinates r, z, and θ, since there is rotational symmetry. No forces act in the θ direction. The equations of motion in both the r and z directions are of the form

$$(d^2u/d\xi^2) + (a_u - 2q_u \cos 2\xi)u = 0 \tag{64}$$

where

$$\begin{aligned}
\xi &= \omega t/2 \\
a_z &= -2a_r = -4eU/mz_0^2\omega^2 \\
q_z &= -2q_r = -2eV/mz_0^2\omega^2
\end{aligned} \tag{65}$$

To determine the conditions for stable trajectories in both the r and z directions simultaneously, two Mathieu diagrams differing by the factor -2 in both a and q are superimposed as in Fig. 69. This shows several areas of

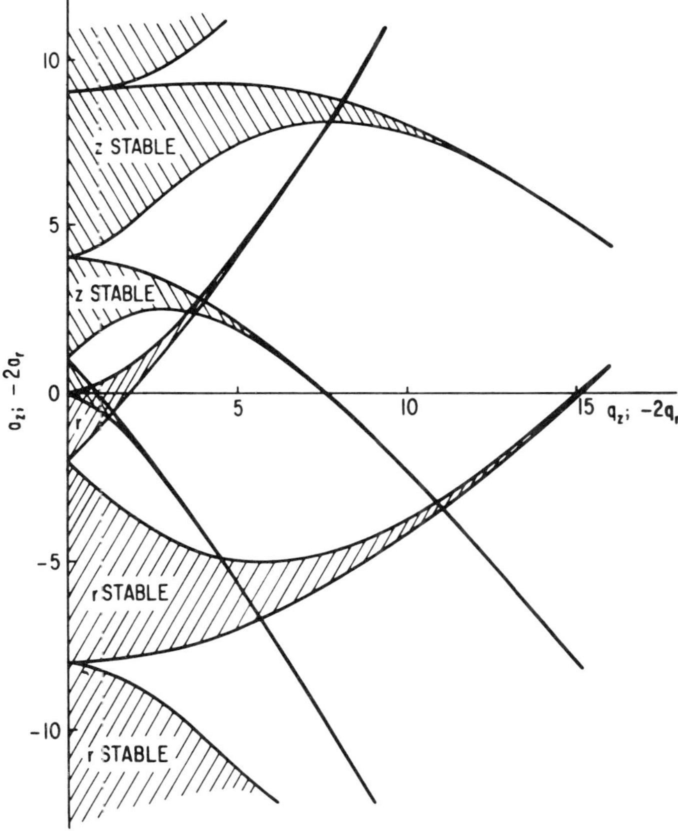

FIG 69. Mathieu diagram for the quadrupole trap obtained by superimposing two one-dimensional diagrams that differ by the factor -2.

stable operation where the amplitudes of ion motion are finite. However, the amplitudes are too large for most practical devices in all except the area closest to the origin. The stable region of particular interest is shown in detail in Fig. 70 with some of the iso-β lines. The significance of the iso-β lines

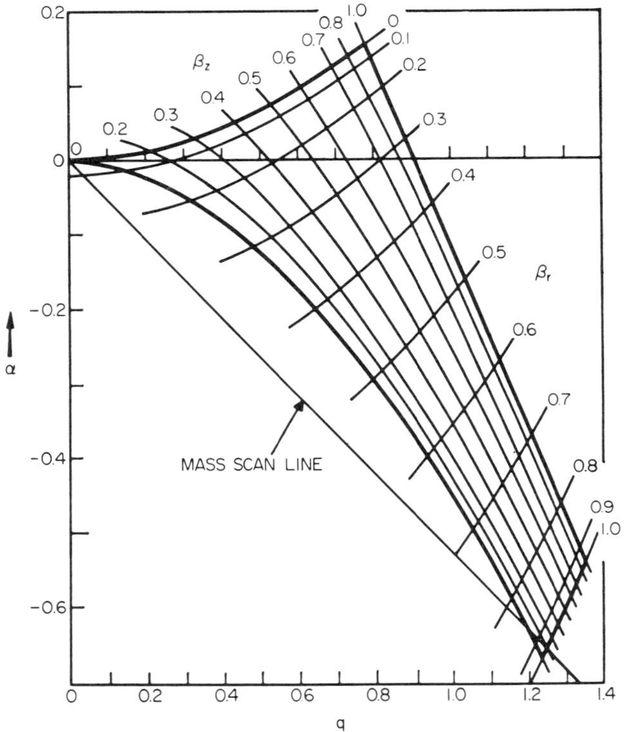

FIG. 70. Detail of the Mathieu diagram, showing the region usually employed. The iso-β lines are related to the characteristic ion motions in the r and z directions. A scan line for mass-selective ion storage is also shown.

in understanding ion motion was discussed and illustrated in Section II. If the fundamental frequency of ion motion in a particular direction is ω_0, then

$$\beta = 2\omega_0/\omega \qquad (66)$$

The limits of stability are the two sets of lines $\beta_{r,z} = 0$ and $\beta_{r,z} = 1$. Ions may be contained in all coordinate directions when a and q are within the stable region, provided the maximum amplitude of oscillation is less than the dimensions of the device. As discussed in Section II the maximum amplitude is a function of the initial conditions—the initial position and energy of the ion and the phase of the rf field at ion formation. Generally speaking, ions must be produced within the device, although schemes might be devised for introducing some ions from the exterior. If the ions are formed by electron bombardment, the initial velocity is essentially thermal except for some fragment ions. The effect of initial field phase depends on the (a, q) conditions and is particularly important in determining the boundaries of the stable region for a device of finite size (Section V-B-4). For a finite device, the boundaries of Fig. 70 apply only to one initial phase of the rf field.

2. Mass Analysis

Several different techniques have been used to detect ions stored in the trap. One method is to detect and measure the ions of a particular m/e by sensing the characteristic motion associated with their particular position in the (a, q) diagram $(9, 10, 71)$. This can be done with an auxiliary oscillatory circuit tuned to the fundamental frequency of the ion motion and determining the power absorption or damping in that circuit as the ions oscillate in resonance. (Ions have also been detected by the voltages they induce without any auxiliary rf, but this is not a sensitive method.) The auxiliary field may be applied between the two end caps so that it acts principally in the z direction and is tuned to ω_0^z (9). The auxiliary voltage must be small to prevent the ions from being rapidly lost to the walls. In this method the characteristic β_z is chosen in the center of the stable region where ions are easily stored. Other ions may simultaneously have stable trajectories. This is a disadvantage because the presence of one ion species may influence the buildup in concentration of another species.

A second method $(17, 70, 73)$ of using the trap is to build up the ion concentration for an appropriate time and then to pulse the ions from the trap through a perforated cap electrode to an ion collector, which is usually an electron multiplier. To make this method mass-selective, it is necessary that only a single m/e species be stored. This is accomplished by operating with a scan line passing close to the apex of the stability diagram (see Fig. 70) in a manner analogous to that of the mass filter. This reduces the problem of one ionic species influencing another, since ions of only one e/m ratio are stable and stored at any given time. The theoretical treatment of the resolution of the mass filter presented in Section III may be applied to this second method of operating the ion cage. Section V-B-4 treats the resolution based on a computer simulation of ion trajectories.

3. Resolution Using Resonant Detection of Ions

The detection using an auxiliary rf circuit is closely related to the use of a resonant field for isotope separation with the mass filter (Section III-E). The theory developed there can be adapted to explore the limits on the resolution in the ion cage.

The equation of motion in the z direction in the presence of the auxiliary field is

$$(d^2z/d\xi^2) + (a_z - 2q_z \cos 2\xi)z = b \cos (\alpha\xi + \delta) \tag{67}$$

where

$$b = -4V_{\text{Res}}/m\omega_0^2 z_0^2, \qquad \alpha = 2\omega_{\text{Res}}/\omega_0 \tag{68}$$

V_{Res} is the auxiliary voltage with frequency ω_{Res}, and δ is a phase constant. The solutions of interest are those where $\alpha = \beta_z$. In this case (81) the unperturbed solution is multiplied by a factor, z_{env}, proportional to the time the resonant field is applied. The oscillation has a linearly increasing amplitude, Eq. (44). The perturbing envelope is given by

$$z_{\text{env}} = (b/2W)\xi \tag{69}$$

where W is the Wronskian determinant for the disturbed Mathieu functions.

When the ions are near, but not at, resonance and

$$\alpha = \beta \pm (\Delta\beta/2), \qquad \text{where} \quad \Delta\beta \ll 1 \tag{70}$$

the maximum amplitude of motion in the z direction is increased and there is a beating motion of frequency $\Delta\beta/4\pi$. The envelope of the trajectory is multiplied by

$$z_{\text{env}} = (2b/W \, \Delta\beta) \sin (\Delta\beta/4)\xi \tag{71}$$

of which Eq. (69) is a special case. The maximum beat amplitude is $2b/W \, \Delta\beta$. Ions that are close to resonance and which gain sufficient amplitude during the beat to strike the electrodes will absorb energy from the field and attenuate the resonance circuit. The attenuation, therefore, has a line width as a function of β which limits the resolving power. The half-width may be estimated by using some simple assumptions. It is assumed that initially the ions have maximum amplitudes uniformly distributed in the z direction between 0 and z_0. Half the ions will be lost when the amplitude of the perturbation factor z_{env} reaches $z_0/2$ (that is, when $\Delta\beta = 4b/z_0 W$). For ions *exactly* at resonance and starting at $z = 0$, there is a characteristic time for ion loss, τ_l (ions with $z_0 \neq 0$ will be lost more quickly), given by $\tau_l = 4Wz_0/b\omega$. Hence,

$$\Delta\beta = \frac{16}{\omega\tau_l}$$

and

$$\left(\frac{\beta}{\Delta\beta}\right) = \frac{\omega_{\text{Res}}\tau_l}{8} = \frac{\pi}{4}\frac{\tau_l}{\tau_{\text{res}}} \tag{72}$$

where τ_{res} is the period of the resonating field. The relationship between $(M/\Delta M)$ and $(\beta/\Delta\beta)$ can be deduced from the stability diagram. According to Fischer (10), $(M/\Delta M) = 1.07 \, (\beta/\Delta\beta)$ for $\beta = 0.6$ when near the q axis, and $(M/\Delta M) = 0.84\tau_l/\tau_{\text{res}}$. The resolving power is proportional to the number of resonance periods an ion originating in the $z = 0$ plane can undergo. If the amplitude of V_{res} is decreased, the resolving power increases. An upper limit is reached when the ion loss from resonance is not due to collection at the wall but is due to a collision (ion–ion or ion–neutral). There is no point in making τ_l greater than the mean (unperturbed) lifetime of the ions. When ion-neutral collisions are dominant, the limit to the attainable resolution is pressure-dependent.

4. Resolution Using Mass-Selective Storage: Computer Simulation

The operation of the ion trap using mass-selective ion storage with a mass scan line (Fig. 70) through the apex of the stability diagram has been simulated (*17*) by numerical integration of the equations of motion. In this method of operation the trap is mass scanned by fixing ω and V/U, and by slowly varying the magnitude of V so that species of differing m/e move one by one into and through the stable region. Ions in the stable region build up in concentration and are periodically pulsed from the trap into a detector. Each mass peak consists of a number of ion pulses, the envelope of the pulses giving the peak shape. The storage time between pulses is chosen according to the ion formation rate, lower rates requiring longer storage times and lower scan speeds so that the pulsed output signal remains high. This method sacrifices knowledge of the time variation during the storage time, in return for the ease of measurement obtained by integration of the ion formation rate.

The computer simulation was concerned mainly with the influence of the initial position and velocity of the ions and the phase of the rf field at ion formation on the stability diagram for a device of finite size. The resolution and its associated relative sensitivity could then be determined for any scan line. Linear devices, such as the monopole and mass filter, are frequently limited in resolution by the number of cycles an ion takes to pass through them. In the three-dimensional quadrupole the ions can be stored for long periods, except perhaps at high pressures where ion-neutral scattering may limit storage. The resolution limits are therefore those imposed by the perfection of the field or the measurable sensitivity.

For the simulation, it was convenient to choose a relatively short storage time. Generally $(200/\pi)$ rf cycles was chosen. Figure 71(a) shows computed stability limits for other storage times when stability is arbitrarily defined as a maximum amplitude of less than ten times the initial displacement in both the r and z directions. The choice of $200/\pi$ is seen to be a good compromise. To determine the effect of the distribution in the position of ion formation on the peak shapes, the stability apex was computed for a number of different stability criteria, as shown in Fig. 71(b). These results are for the initial phase of the rf field that gives the smallest amplitude of oscillation. For this phase only, peak shapes shown in Fig. 72 were deduced for various assumptions

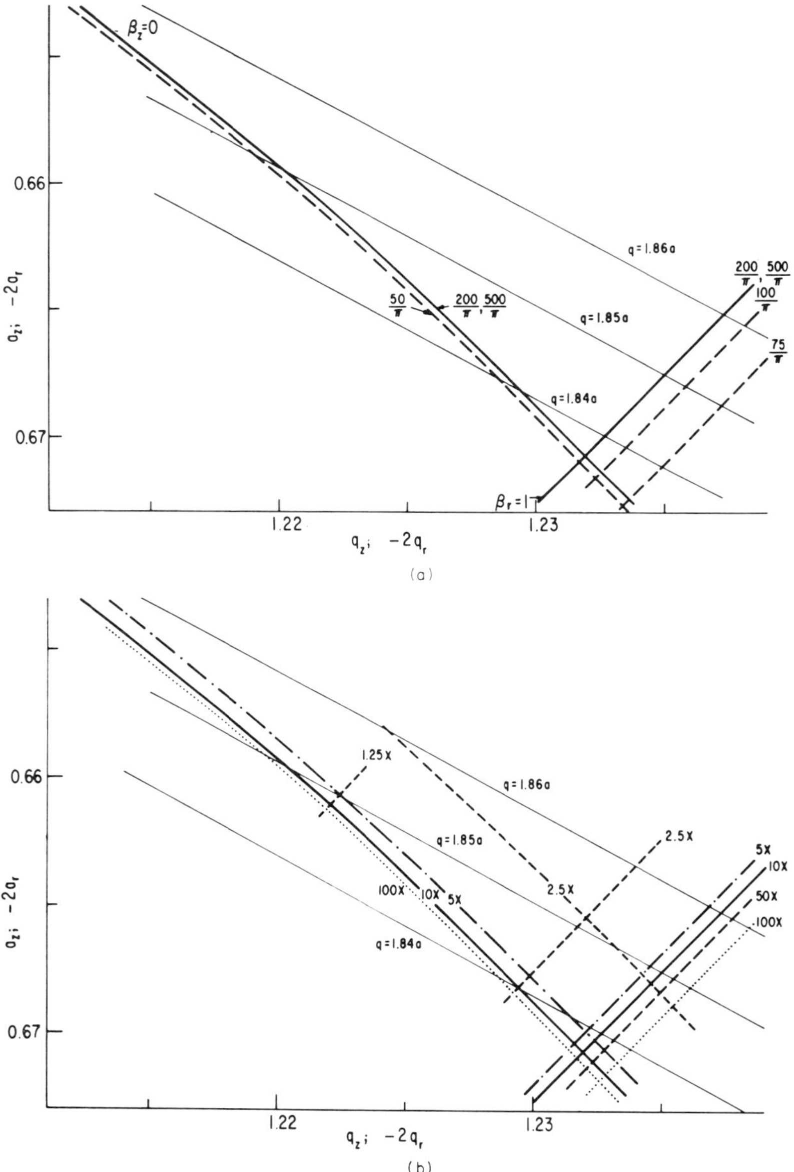

FIG. 71. Tip of the stability diagram used for mass-selective storage as computed from ion trajectories (17): (a) for different storage times, given in rf cycles when stability is defined as less than ten times the initial displacement (with no initial velocity); (b) for different stability criteria and a fixed storage time of $200/\pi$ cycles.

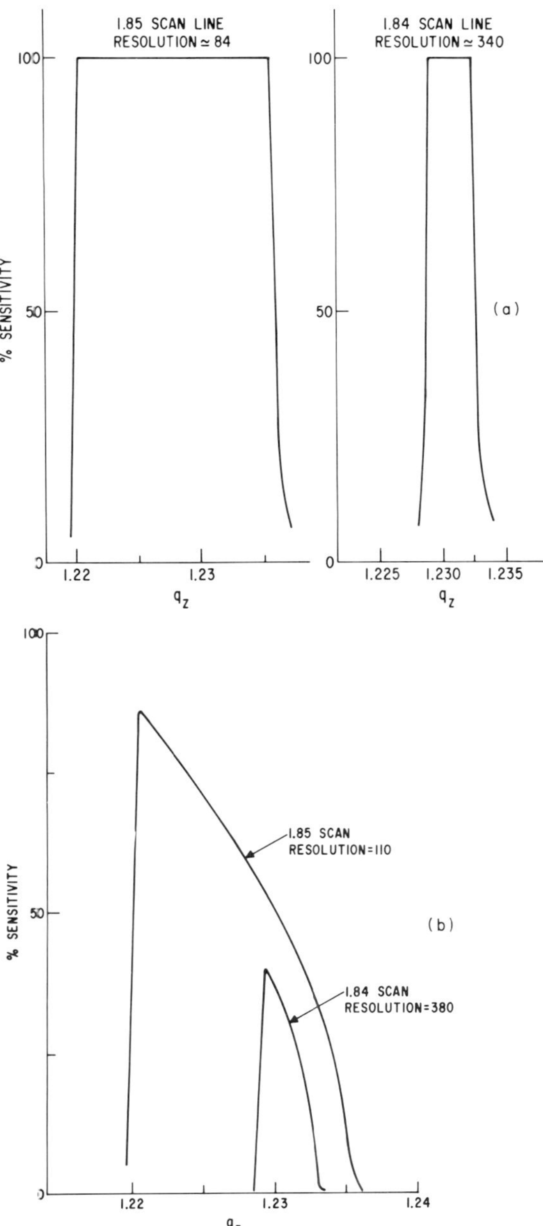

FIG. 72. Computed peak shapes derived from the results of Fig. 71 and various assumptions on the possible initial positions of ion formation (17): (a) uniform formation in central one-tenth in both r and z directions; (b) uniform formation for all values of r, but only for the central one-tenth in the z direction.

regarding the allowed initial positions of ion formation. In (a) the ion formation is uniform but occurs only in the central one-tenth of the device in both r and z directions. That is, all ions are stored until $u_M/u_0 > 10$. In (b) the allowed positions of ion formation include all values of r but only the central one-tenth in the z direction. This more closely corresponds to experimental situations where the electron beam is perpendicular to the z axis. The main factor influencing the peak shape, resolution, and sensitivity is the dependence of the stability boundaries on the phase of the rf field at ion formation. As the q/a ratio is decreased for higher resolution the "acceptance" (10, 71) of the device decreases. The computed boundaries as a function of initial phase are shown in Fig. 73, using the 10 × stability criterion. Ions to the left

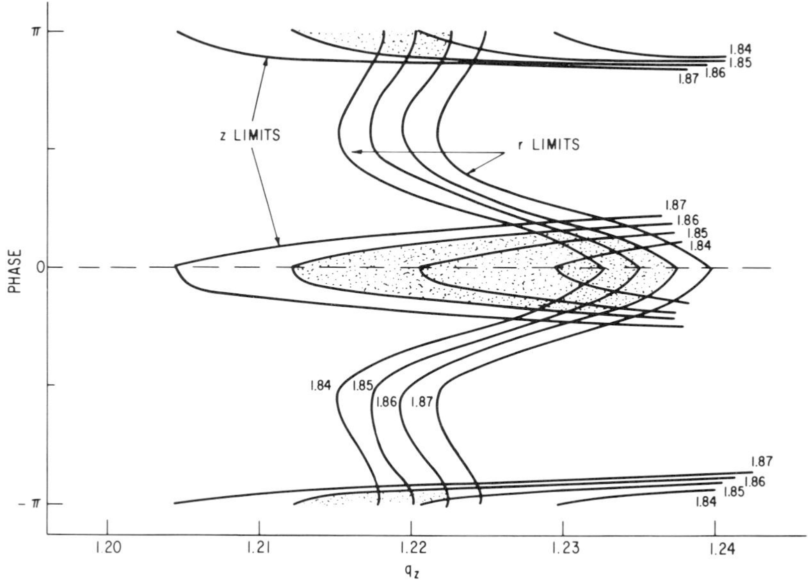

FIG. 73A. For $200/\pi$ cycles and the 10 × stability criterion, the stability limits are given for various a/q scan lines as a function of the phase of the rf field at ion formation and the q value. Those phases stable in both r and z directions are contained between pairs of contours, as indicated by the shading for the $q = 1.86a$ line (17).

of the r contours are stable in the r direction; ions to the right of the z contours are stable in the z direction. For a particular scan line, as illustrated by the shading for the $q = 1.86a$ line, the phases for stability in all directions are those between the contours. Summing the phases gives the peak shapes of Fig. 73B. The dashed curves are for the case when ion formation is allowed

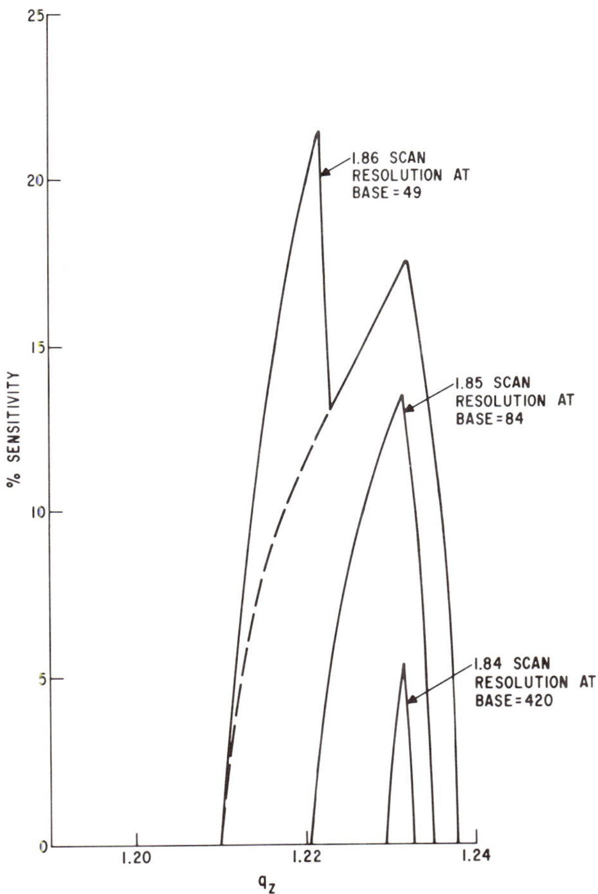

FIG. 73B. Peak shapes for various scan lines deduced from Fig. 73A.

for only half the rf cycle. The peak asymmetries are opposite to those likely to result from the initial positions of ion formation [Fig. 72(b)]. The relationship between resolution and sensitivity due to the phase factors *only* is given in Fig. 74.

It is interesting to consider effects due to initial ion velocity, such as might be found with some fragment ions. Figure 75 shows the stability apex computed for various initial velocities in both z and r directions for a single phase of the rf field, a time of $200/\pi$ cycles and a device with a maximum allowable amplitude of 10^{-2} meter. The broken lines are those for $u_M/u_0 = 10$. The initial velocities are in terms of meters per radian of the applied rf field and the

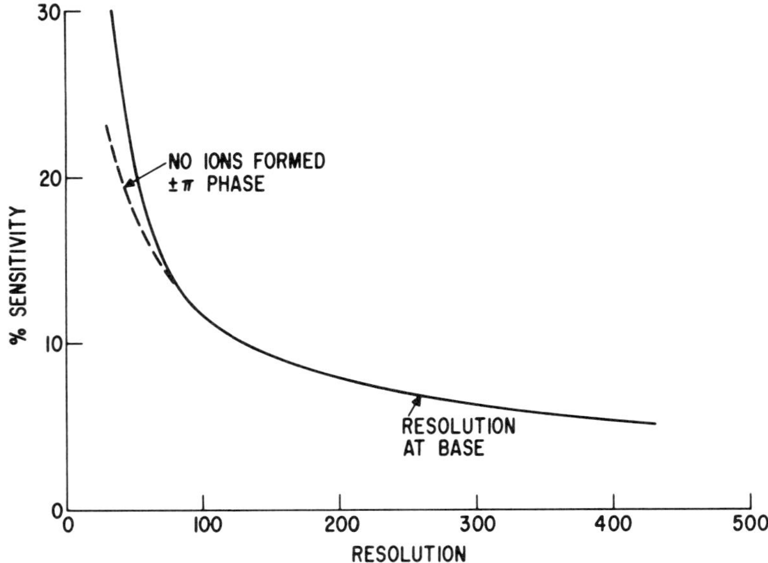

FIG. 74. Computed resolution versus sensitivity curve based on the results of Fig. 73 (*17*).

initial displacements are assumed to be zero. With higher frequency fields, the voltages for a given m/e and q value are larger, and a larger initial velocity can be tolerated. For an applied frequency of 1 MHz and an ion of 30 amu, the velocities illustrated correspond to about 0.03 and 0.13 eV. When operating the device with any appreciable resolution, ions with initial kinetic energy will be discriminated against. This causes the cracking patterns to differ from those in conventional mass spectrometers.

C. Experimental

1. Performance as a Mass Spectrometer

a. Resonant detection of ions. Use of resonant detection of ions has been reported by Fischer (*10*) and Rettinghaus (*71*). Their detection techniques were closely related, but their operation of the ion trap was somewhat different. Some of the theoretical considerations concerning resolution in this mode of operation were dealt with in Section V-B.

Fischer's device had an r_0 value of 2 cm and the electrodes were machined hyperboloids with the separation between the endcaps accurate to 5×10^{-3} cm. The curvature was correct within 10^{-2} cm up to the radius r_0. The elec-

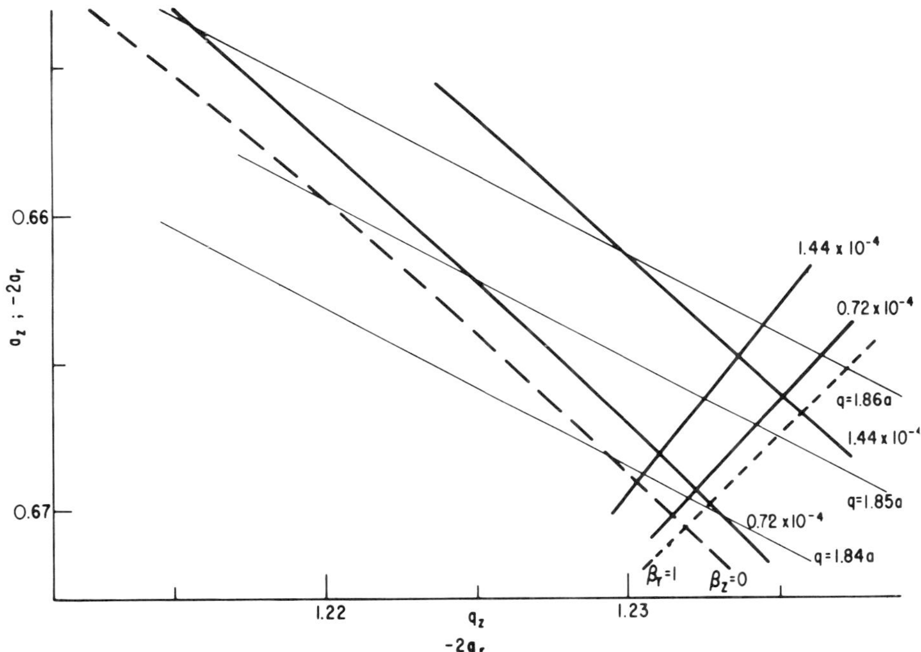

FIG. 75. Stability tip for ions formed with initial kinetic energy at the center of the device. The usual 10 × criterion is shown as broken lines. The device size was taken as 10^{-2} meters. The initial kinetic energies are given in meters per radian of the applied field (*17*).

tron filament was placed behind one of the cap electrodes. A small magnetic field in the z direction was sometimes used to increase the electron path length. (The ions were only slightly affected.) The main rf frequency was 500 kHz with a power capability of 30 watts and maximum amplitude of 1000 volts. A schematic diagram of the circuitry is given in Fig. 76. The resonant detection frequency was chosen as 150 kHz (or $\beta_z = 0.6$). The output of the auxiliary generator was applied across a 5 $M\Omega$ resistor and across half the pure resistance of the resonator tuned to 150 kHz. The voltage across the resonator was then proportional to its resistance and inversely proportional to the attenuation. This voltage was amplified, demodulated, and used as the y input to an oscilloscope display. The resonance amplitude was normally in the 50- to 200-mV range ($\sim 0.1\%$ of main rf amplitude V). The effect of the demodulation stage is illustrated by Fig. 77.

The mass spectrum was scanned by applying a sawtooth signal as part of the dc voltage on the ring electrode. The sawtooth was also used for the x axis of the oscilloscope. As different ions passed through their $\beta_z = 0.6$ line, their number was measured by the 150 kHz attenuation. Figure 78(a) shows

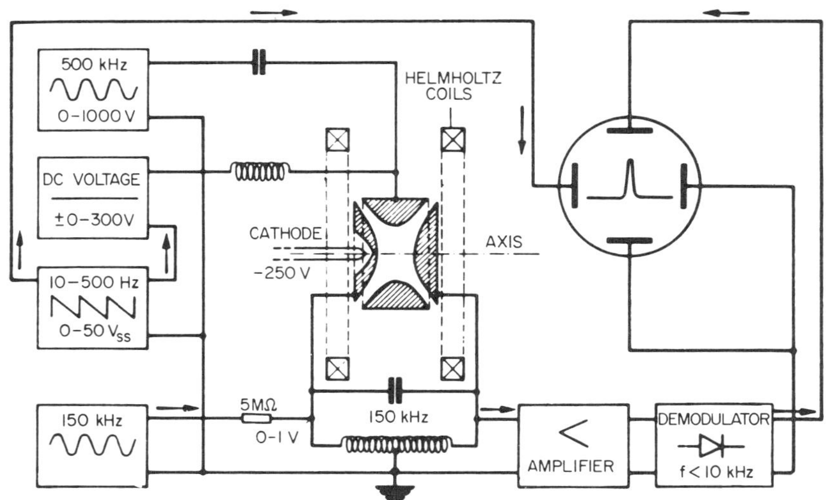

FIG. 76. Circuit used by Fischer for detection of ions by using the damping of an auxiliary rf circuit tuned to the fundamental frequency of ion motion (*10*).

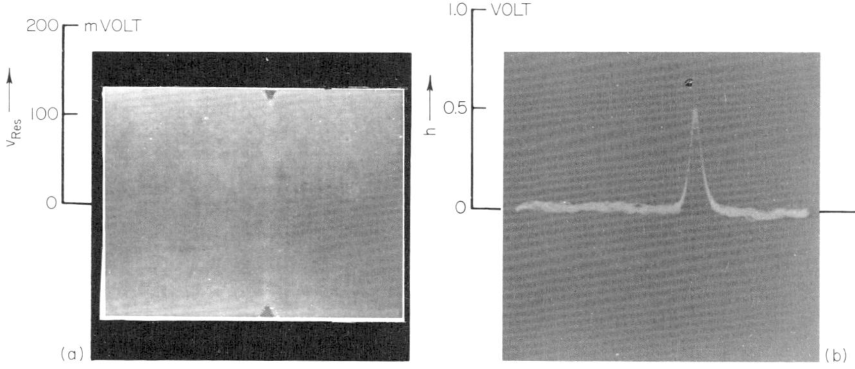

FIG. 77. Illustration of the ion signal (a) before and (b) after demodulation in the circuit of Fig. 76 (*10*).

the stability regions for $M = 18$ and 28 plotted in U-V space, with the application of a sawtooth sweep, and Fig. 78 B illustrates the resulting oscilloscope display. Note that the conditions at detection are not the same for ions of different m/e. The r trajectories are quite different and the behavior of the ions during the remainder of the sweep depends on the particular m/e. This is not a desirable situation. For example, $M = 18$ is unstable during measurement of $M = 28$, but $M = 28$ is stable during most of the sweep. Ion-ion interactions

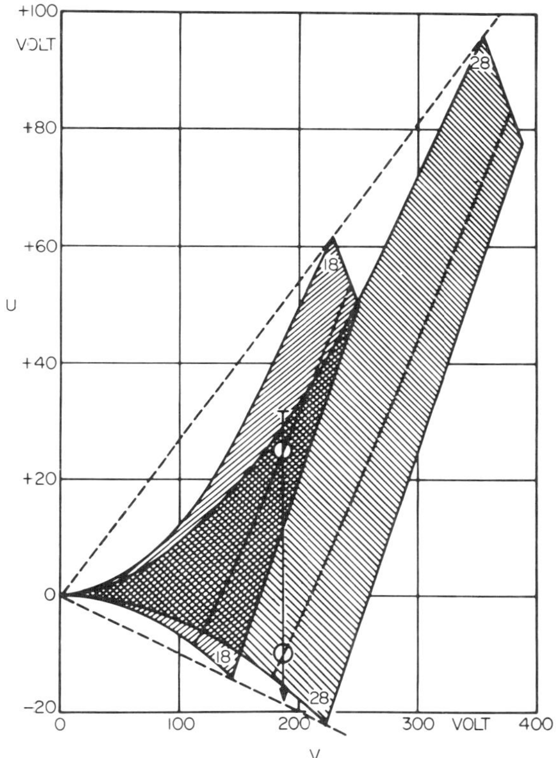

FIG. 78 A.   Stability diagram in $U-V$ space, showing stable regions for $M = 18$ and $M = 28$ and their relation to a sawtooth voltage used to sweep the mass range.

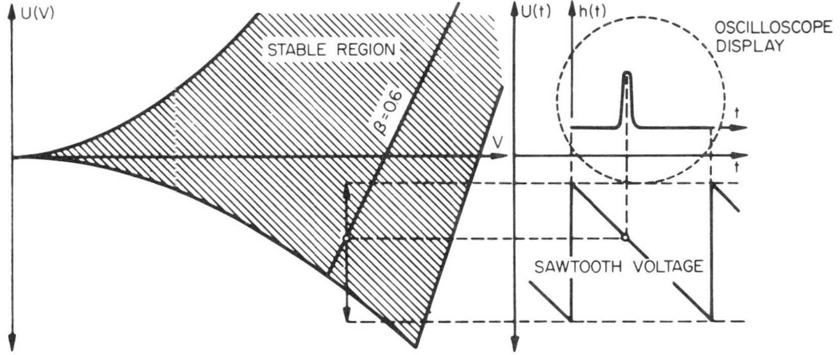

FIG. 78 B.   Illustration of how the spectrum of Fig. 78 A is displayed (10).

might affect the observed results, and the different ionic species have different buildup times. The method has a limited mass range for a particular value of V. The trap was operated in a continuous manner, setting V for the desired mass range and continuously ionizing and storing ions as the spectrum was scanned.

The resolving power achieved was $M/\Delta M = 85$ as shown by the separation of the krypton isotopes in Fig. 79. The spectra were recorded at the

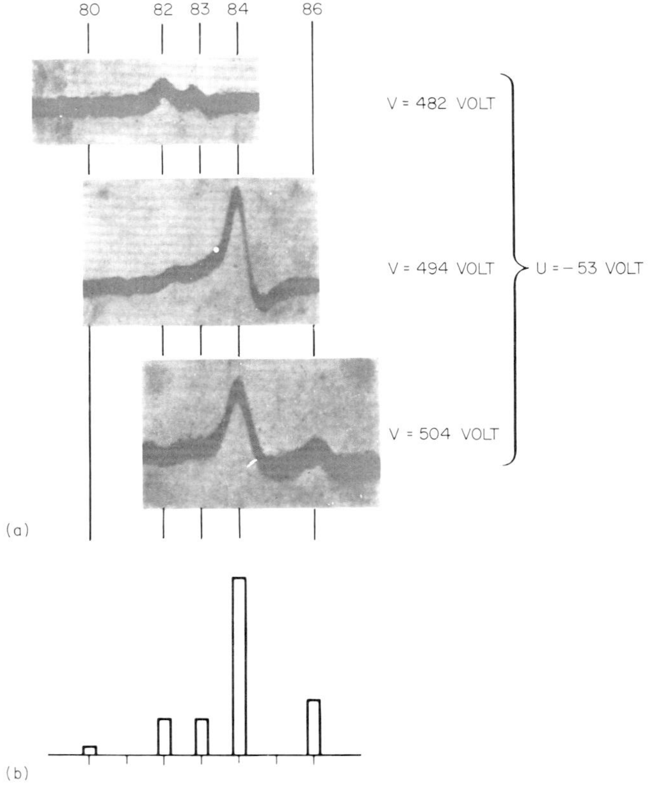

FIG. 79. (a) Krypton spectra obtained by Fischer (10) near the limit of stability for several settings of the rf amplitude V. For $V = 482$ volts, $M = 84$ was beyond the stable region. (b) True isotopic distribution.

edge of the stability region and only there was such resolution attainable. With the $V = 482$-volt operation, $M = 84$ was beyond the stable region and was not observed. The pressure was 3×10^{-6} torr. At a pressure of 3×10^{-5} torr of heptane the resolving power was about 20. Fischer suggested that at lower pressures the maximum resolving power would be greater, owing to the increase in the mean collision time with neutrals, and that the sensitivity could be increased by decreasing the scanning rate and thereby increasing the available ion storage time. The smallest detectable partial pressure was about 2×10^{-8} torr (equivalent to 2×10^4 ions/cm^3 in the trap), but this was limited only by the inability to reduce the total pressure below 10^{-6} torr.

Rettinghaus (71) used a more approximate geometry and encountered problems due to field distortions. The parameter r_0 was 1.2 cm and the electrodes were spherical rather than hyperbolic. The electron filament was placed behind a hole in the ring electrode. The circuitry for ion detection is shown in Fig. 80. The main rf frequency was 1.6 MHz and the detection was

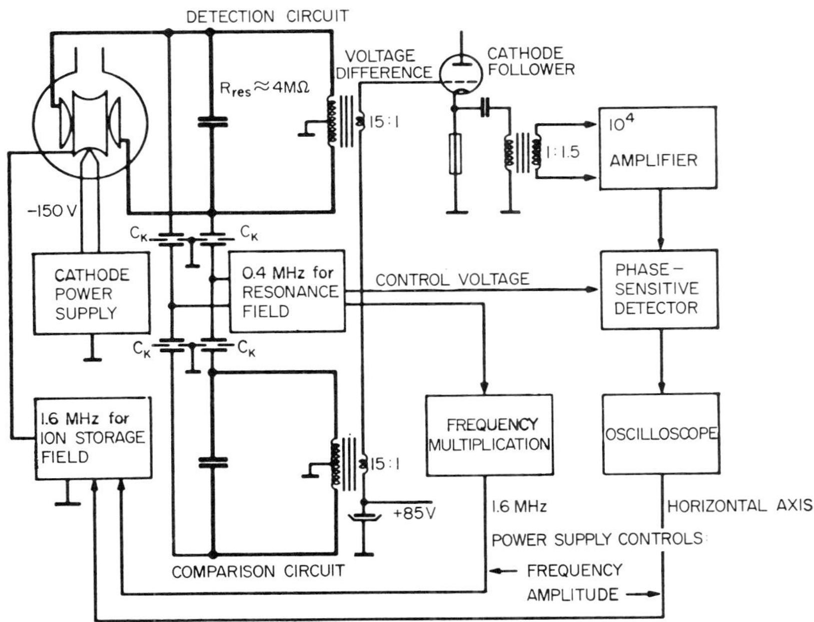

FIG. 80. Sensitive circuit used by Rettinghaus (71) for resonant detection of the stored ions.

at $\beta_z = 0.5$, perhaps an unfortunate choice because certain field faults give nonlinear effects at precisely that value of β (see Section VI). The trap without ions was balanced by a matching circuit. When ions were in resonance, the voltage difference was amplified and measured with a phase-sensitive detector. This method was very sensitive, and about four ions in the cage were claimed as the minimum detectable signal. The ion trap was operated with no applied dc voltage ($a = 0$). An initial value of the rf amplitude V was chosen to correspond to the lowest mass in the range being detected. The electron beam was switched on for a chosen ionization time, usually 5 sec. The value of V was then swept at a rate equivalent to about 1 amu/sec with the electron beam switched off. As ions of higher mass passed through the $\beta = 0.5$ point, their concentration was measured. This method suffers from the same objections as Fischer's method in that different species are treated in different ways and different ions may interact with each other as a result of ion-ion scattering or space-charge effects.

Rettinghaus achieved minimum detectable pressures in the 10^{-13} torr range and a maximum resolution of 300, which was apparently limited by field faults. Due to the long ion storage period, (partly necessary as the result of a low ionization rate) there were interesting, but undesirable, interactions between ions and neutrals. For example, Rettinghaus observed the formation of HCO^+ from CO^+. Ion-neutral collisions are discussed later.

b. Mass selective storage. Dawson and Whetten (*70, 72, 73*) used accurately machined devices and operated with mass selective storage. The dimensions varied from an r_0 of 5.6×10^{-3} meter to 2.8×10^{-2} meter. The electron filament was positioned behind the ring electrode with a grid so that the electron beam could be cut off a short time before the stored ions were pulsed from the trap so as to avoid interference from soft X-rays, metastables, or unstable ions. The electron multiplier output was amplified and fed through a gated circuit to a peak-reading amplifier so that the peak envelopes could be displayed on an X–Y recorder. The gate was adjusted to coincide with the ion signal from the pulse. The maximum observable resolution was about 300 with the smallest device and 1000 with the largest. The limitation appeared to be the perfection of the field geometry. The minimum detectable partial pressure was about 10^{-13} torr, but this was a function of the total pressure due to some associated noise problems.

The devices have been operated at rf frequencies between 500 kHz and 1.8 MHz. The storage is better at higher frequencies, but the power requirements become large because the power depends on the fifth power of the frequency. At 500 kHz, the power is low and all-solid-state power supplies can be constructed.

The storage times were typically about 10 msec for a pressure of 10^{-8} torr. The peak shapes were poor, with peaks split into doublets or triplets. This was due to the occurrence of nonlinear resonances because of the presence of field faults (some of the experimental data are given in Section VI). The problems could be avoided by applying a small bias between the two end caps, the bias being a fixed percentage (usually about 5%) of the dc voltage applied to the ring electrode. This gave good peak shapes at the expense of some loss in sensitivity. The bias shifts the z stability boundary to higher q values and moves the apex of the stability diagram away from the region where nonlinear resonances occur.

Recently Dawson, Hedman, and Whetten (*73a*) operated a device employing electrodes of 20×20 stainless steel mesh. The mesh was crudely shaped to approximate the hyperboloid surfaces. With a size of $z_0 = 2$ cm, the resolution was about 75.

2. Maximum Ion Density and Space Charge

Fischer (*10*) estimated the saturated ion density in his trap from the dc shift, ΔU, in the stability diagram due to the space charge of the ions. He used the relationship

$$\Delta U = \rho r_0{}^2/4\varepsilon \tag{73}$$

where ε is the dielectric constant and ρ is the ion density. The maximum ion density was found to be 4×10^6 ions/cm^3 for krypton. The maximum ion density for a given (a, q) was proportional to the rf amplitude V, and therefore was greater for high m/e or high frequency or for a large device.

Dehmelt and Major (*77*), using a trap with $r_0 = 3.5$ cm, 1 MHz frequency, $a = 0$, and a low q, found for $H_2{}^+$ a maximum ion density of about 10^5 ions/cm^3. This was also the concentration estimated by Dawson and Whetten (*82*) for $M = 28$, with $r_0 = 1.4$ cm and a frequency of 1.8 MHz. More recent measurements by Dawson and Whetten have indicated that ion densities can be as great as 10^8 ions/cm^3 (*82a*).

Space charge provides the eventual limitation to the ion concentration. Fischer treated space charge by assuming an equal distribution over the entire field, with the effect being that of an additional voltage, ΔU, acting to defocus the ions in all directions. ΔU is given by Eq. (73). The stable regions are altered as in Fig. 81. Dehmelt (*80*) has taken a different approach, assum-

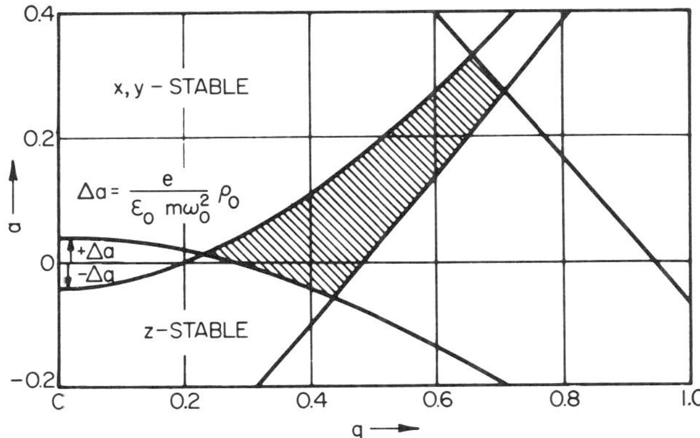

FIG. 81. The effect of space charge on the stability diagram, according to Fischer (*10*).

ing that the ions arrange themselves in the bottom of a pseudo-potential well so that they exactly cancel the field. He derives an expression

$$\rho_{max} = \frac{3eV^2}{16\pi m\omega^2 z_0^4} \tag{74}$$

The charge density is again a uniform one.

3. Ion-Loss Processes

A number of processes of ion loss can occur to limit the buildup in concentration of ions in the trap before the space-charge limit is reached. These will be discussed in turn. There is the possibility of ion-ion scattering, ion-neutral scattering, and (under certain special conditions) it is found that the trap can self-empty by an interaction of the ion cloud with the main rf field.

a. Ion-ion scattering. Fischer (*10*) found that introduction of a second gas, whose ions were stable, during the observation of another species caused both a shift in the resonance (additional space charge) and a reduction and broadening of the resonance. The effect was not due to ion-neutral effects, since the second gas had to be in the region of stability. This suggested that ion-ion collisions might be important. Dehmelt (*80*) calculated the ion-ion collision time as 10^{-3} sec for protons at room temperature when the ion density was at its theoretical maximum in Major's trap (*77*). He used the formula for a uniform plasma:

$$t_c = 0.78 \times 10^7 A^{1/2}(kT_i/e)^{3/2}\rho^{-1}Z^{-4}(\ln \Lambda)^{-1} \tag{75}$$

where A is atomic weight, Z is ionic charge, T_i is the ion temperature, and Λ is the ratio of the cutoff distance for coulombic interaction of an ion with its neighbors to the distance of closest approach. Dehmelt pointed out that such collisions cannot lead to a net energy adsorption from the field for a homogeneous rf field. In an inhomogeneous field this is no longer strictly true. In any case, since the trap is of finite dimensions, ion-ion collisions can lead to ion loss.

Taking into account the experimental ion density in Major's trap (1/30 of theoretical maximum) and the average ion energy, the hot-ion self-collision time was calculated from (Eq. 75) to be about 50 sec. The experimental mean lifetimes were less than this except at the highest vacuum, presumably indicating that ion-ion collisions were not the main scattering process.

Operating with $a = 0$, Dawson and Whetten (*70*) have studied the kinetics of ion loss from the trap and found that at very low pressures, the ion loss process is bimolecular. That is, the ion-loss rate is proportional to the square of the number of ions in the trap. Figure 82 shows some of their results. Ions were formed for 10 sec by electron bombardment. The filament was then

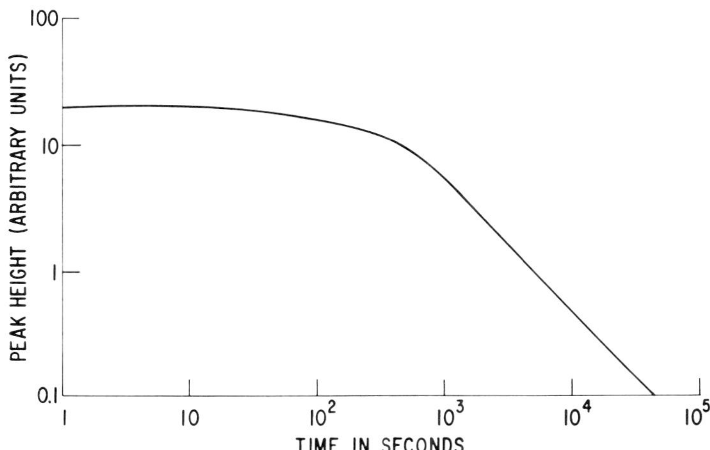

FIG. 82. Ion storage at 3×10^{-10} torr. The electron beam was switched on for 10 sec to fill the trap. The number of ions remaining in the trap was measured as a function of time and is shown on a log-log plot. The a value was zero and q was about 0.7 for mass 28 (70).

switched off, and after appropriate storage times the number of ions remaining was measured.

The slope of -1 on the log–log plot indicates a bimolecular process. The pressure during the storage was 3×10^{-10} torr, and some ions could be measured several days after formation. At higher pressures, ion storage kinetics tend to become more complex, but a transition to ion-neutral loss where the loss rate is proportional to the number of ions is often seen at higher gas pressures (82).

b. *Ion-neutral scattering.* Clear evidence of ion-neutral effects is contained in Rettinghaus' results (71), particularly in the presence of $M = 29$ in the mass spectra. Figure 83 presents the $29/(28 + 29)$ ratio as a function of the storage time of the ions before measurement. $M = 29$ was presumably HCO^+ formed by the reaction of CO^+ with background gases. The total pressure was about 2×10^{-9} torr. Rettinghaus observed mean ion lifetimes (the time to lose half the ions) as long as 20 min.

Fischer (10) determined the mean lifetime from the rate of buildup of ions in the trap, assuming that the ion-loss kinetics were first order in ion concentration. The data at different electron currents were not entirely consistent with this assumption. For nitrogen at 6×10^{-6} torr, he found a maximum mean lifetime of 1.5×10^{-2} sec. From the gas-kinetic collision cross section he estimated a collision time of 4×10^{-4} sec, suggesting that each ion underwent 40 collisions on the average before being lost.

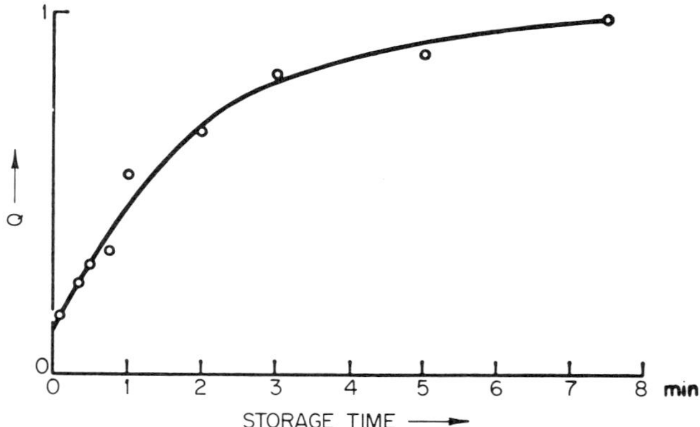

FIG. 83. The ratio $29/(28 + 29) = Q$ as a function of the storage time of the ions before measurement, as found by Rettinghaus (71). Mass 29 is HCO^+ formed by an ion-molecule reaction of CO^+. The pressure was 2×10^{-9} torr.

If the ion loss is by ion-ion rather than ion-neutral scattering, the concept of a mean lifetime (time for loss of half the ions) is no longer very useful, since it is dependent on the ion concentration.

In considering ion-neutral scattering it is important to distinguish between several possible processes with differing probabilities.

(1) Resonant charge exchange with the parent gas: This is a very probable process. The fast ion is replaced by a thermal ion at the moment of collision. If the process occurs near the wall of the trap, the newly formed ion may have a maximum amplitude of motion such that it is lost to the wall. Collisions near the center of the device could be stabilizing.

(2) Other charge exchange and ion-molecule reactions: These will lead to different ion species, which may or may not be stored dependent on their (a, q) value and therefore on their e/m. Cross sections for these processes and their energy dependence vary widely.

(3) Elastic collisions that change momenta of the ions: These might be important for ion loss where the neutral atoms are much heavier than the ions. Collisions between heavy ions and light atoms should not be so important as was experimentally observed by Wuerker *et al.* (11) for metal dust particles in air (see Section V-D), and by Huggett and Menasian for Hg^+ in helium [according to Dehmelt (80)]. For Hg^+ in krypton and neon, Dawson and Whetten (82) found that the heavy ions are stabilized by the presence of the light gas.

c. Self-emptying. Both Fischer (10) and Whetten and Dawson (73) have found phenomena where, when the ion concentration is high, the ion cloud becomes unstable and (with continuous ion formation) the ion concentration oscillates. The instabilities are separated by time periods of 10^{-4} to 10^{-1} sec.

Fischer attributed these effects to the shift in operating conditions with space charge, but was unable to explain the oscillatory effect. Dawson and Whetten observed partial self-ejection of ions from the trap in the z direction into their ion collector when the ion concentration reached a critical value, and only for certain β_z lines in the stability diagram. The time between the pulses of self-ejected ions was proportional to the ion formation rate. For a given ion formation rate, the self-ejection time was proportional to the square of the applied rf frequency. Figure 84 shows a photograph of the self-ejection of

FIG. 84. Oscilloscope trace showing spontaneous emptying of the ion trap in the z direction whenever the ion concentration builds up to a critical value. The self-emptying occurs only at certain β_z positions. The large pulses are due to a voltage pulse applied to the end cap to empty the trap. The time between self-emptying pulses is 6×10^{-4} sec *(73)*.

$M = 28$ occurring between the intervals when the trap was emptied by pulsing the perforated end cap. The self-ejection may be due to plasma oscillations when the ion concentration is such that the plasma frequency equals the applied rf frequency. The plasma frequency is given by

$$\omega_p = (4\pi\rho e^2/m)^{1/2} \tag{76}$$

where ρ is the ion density. Such a plasma oscillation can only lead to ion loss when the collective oscillation leads to net increases in amplitude for individual ions. Along the lines $\beta_z = \frac{1}{2}$ and $\beta_z = \frac{2}{3}$, individual ions oscillate in the z direction with a fundamental frequency that is a subharmonic of the applied rf frequency, and it is along such lines that the spontaneous instabilities were observed. Plasma resonance, however, requires that a relatively large ion density be present. A frequency of 500 kHz requires $n = 1.6 \times 10^8$ ions/cm^3.

d. Effect of ion loss kinetics on mass spectra. Consider operation of the trap with continuous formation of ions at a rate K_1. For ion-neutral scattering, the ion loss rate is $K_2 N$, where N is the number of ions present. The ion concentration builds up at a rate of

$$dN/dt = K_1 - K_2 N \quad \text{or} \quad N = (K_1/K_2)(1 - 1/e^{K_2 t}) \tag{77}$$

After sufficient time, if the ion concentration is not first limited by space charge, we have $N_\infty = K_1/K_2$. If two ions (e.g., isotopes) are formed at rates that differ, but are lost by processes with the same rate constant K_2, then at any time the concentrations of the two ions in the trap will be proportional to their formation rates.

If only ion-ion scattering occurs, with a loss rate of $K_3 N^2$, the equations are

$$dN/dt = K_1 - K_3 N^2 \quad \text{and} \quad N = (K_1/K_3)^{1/2}(1 - 2/[e^{2(K_1 K_3)^{1/2} t} + 1]) \tag{78}$$

At short times, the ion concentrations are proportional to their formation rates. As the storage time is increased, the relative concentration of the lesser species increases. At very large times, $N_\infty = (K_1/K_3)^{1/2}$. To compare ion formation rates directly, relatively short times must be used. On the other hand, at longer times, a minor species will continue to increase in concentration and will be more easily detected.

If a single species is considered at different pressures, taking $K_1 = K_1'P$, and $K_2 = K_2'P$, then, for ion-neutral scattering, N_∞ is independent of pressure. For ion-ion scattering, N_∞ will depend on $P^{1/2}$.

The proper use of the trap requires knowledge of the details of the scattering processes that are occurring.

D. Non-Mass-Spectrometric Uses

The containment of micron-sized charged particles of iron and aluminum was examined by Wuerker *et al.* (*11, 83*). The tube was mounted with the z axis vertical, and gravitational forces were counteracted by applying a potential between the cap electrodes. Single particle trajectories in the rz plane could be examined and photographed using reflected light. Figure 85 shows the kind of Lissajous figures obtained by observing a single particle in the rz plane when the β_z/β_r ratios were 2 : 1, 1 : 1, and 1 : 2. Many particle containment was also examined, and it was found that by having a background gas pressure of about 10^{-3} torr, the motion of the charged particles was damped. The particles then formed stable arrays, the time to form the stable array being an inverse function of the pressure. The stable arrays are compressed by increasing q, individual particle motions getting larger, until the array "dissolves" at some critical q. This q is greater the smaller the number of

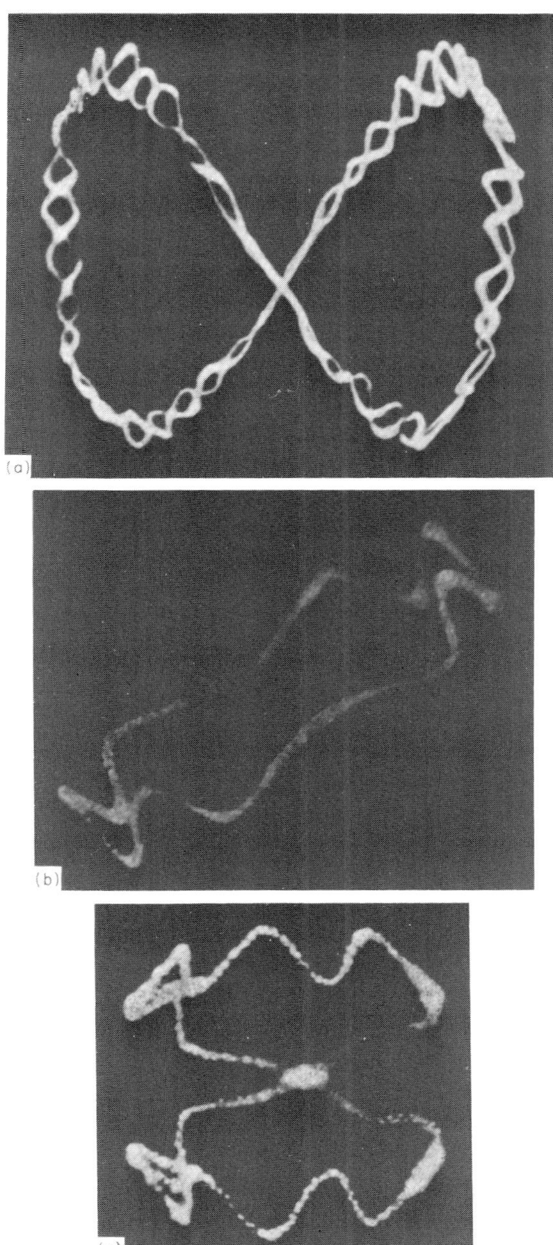

FIG. 85. Lissajous figures (*11*) obtained by viewing the trajectory of a macroscopic trapped particle in the r-z plane: (a) $\beta_z/\beta_r = 2$, $a_z = 0$, $q_z = 0.232$; (b) $\beta_z/\beta_r = 1$, $a_z = -0.0643$, $q_z = 0.502$; (c) $\beta_z/\beta_r = 0.5$, $a_z = 0.102$, $q_z = 0.502$.

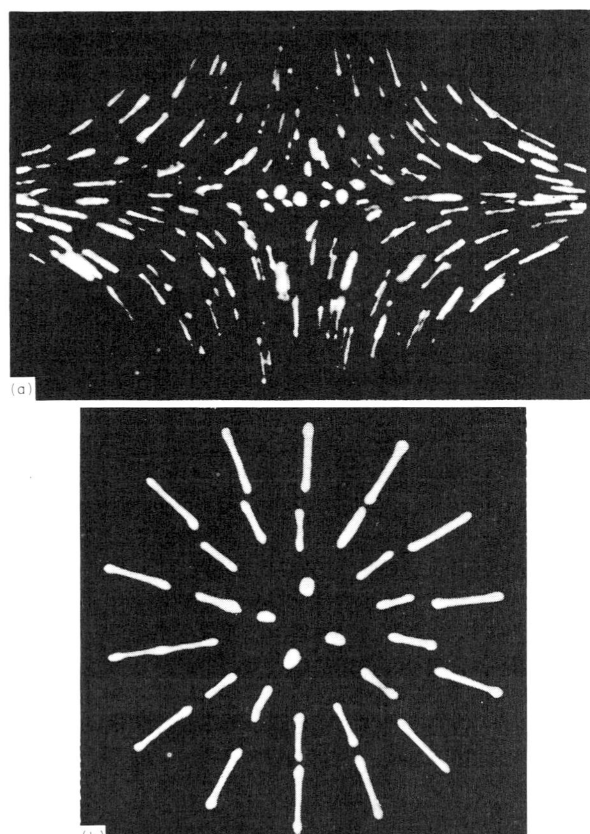

FIG. 86. (a) View of r–z plane for a many body suspension of macroscopic charged particles (*11*). The array has been "crystallized" by damping collisions with background gas at 10^{-3} torr. (b) View of r–θ plane for a "crystallized" suspension of 32 charged particles.

particles. Figure 86(a) shows an example of a many particle stable array in the rz plane, the lines representing the limit of oscillation of the individual particles. Figure 86(b) shows a suspension of 32 particles viewed in the $r\theta$ plane. The possible relevance of these observations to ion containment is intriguing, especially since the stabilization of mercury ions in a trap by the presence of a high neon pressure (10^{-2} torr) has been observed (*80, 82*). The containment of large charged particles was proposed for use in an accelerometer (*84*). It has also been used to study the acceleration of particles due to laser-induced vaporization of the material (*85*).

Dehmelt first suggested using trapped ions for magnetic resonance experiments (74) and later carried out the first successful experiments (75). He^+ ions were contained in a quadrupole rf trap and bombarded with polarized cesium atoms. The helium ions were rapidly polarized by spin exchange collisions. The polarization was then detected by observing the helium ion lifetime, since the charge exchange cross section of

$$Cs + He^+ \rightarrow Cs^+ + He^* \tag{79}$$

depends on the relative orientations of the spins of the valence electrons. If they are antiparallel, the energy defect is much smaller (the He^* is singlet). Related experiments were attempted by Burnham and Kleppner (79). The method was also applied successfully to a precision determination of hfs separation of the $(He^3)^+$ ion in the 8 GHz region (86).

Dehmelt *et al.* (76, 80) have used linearly polarized photons to produce protons by photodissociation of H_2^+. The cross section is dependent on the orientation of the H_2^+. The magnetic resonance of H_2^+ was monitored by measuring the H^+ ion concentration. Jefferts (78) has used a similar technique to study rf transitions characteristic of the interaction between the electron spin and the molecular rotation in parahydrogen.

Another interesting use of quadrupole traps has been to contain electrons. This can be done by increasing the frequency and size parameter z_0 to compensate for the much lower mass (83). A magnetic field parallel to the z axis may be used to facilitate containment (10, 87). Electrons may also be trapped with the combination of an electrostatic quadrupole field and a magnetic field (88).

E. Advantages

The three-dimensional quadrupole ion trap, by virtue of its storage capability, possesses some unique properties. The ions have an exceedingly long path length even in a small device, so the number of rf cycles an ion spends in the field is not a limitation on the resolution. The limitations are the perfection of the field and the sensitivity of ion detection. Because of the path length, the rf frequency may be reduced with a corresponding decrease in the power requirements, since rf power varies with f^5. Other advantages of storage are that the electron beam may be cut off before ion measurement, to prevent unwanted secondary effects, and integration of the ion formation rate permits operation at very low pressures.

Physical processes occurring during ion storage may distort the measured abundances of ionic species unless the user is aware of their existence and takes suitable precautions. These processes include charge exchange and saturation effects (Section V-C-2 and 3). They are more important at long storage times and high pressures. More work is necessary to explore fully the properties of the ion trap before it can be most effectively used. However, it does show considerable promise as an experimental tool in ion and molecular physics.

VI. Nonlinear Resonances in Quadrupole Fields

A. Introduction

The preceding sections have treated ion motion in perfect quadrupole fields. In practice, field imperfections are always present to some degree. Electrodes that are circular rather than hyperbolic in cross section are in widespread use. The rf field may have harmonics in addition to the fundamental frequency, and small electrode spacing errors are likely. Local deviations from perfect fields may be relatively unimportant, but systematic errors whose influence can build up over the entire flight path of the ions are very significant. These errors are particularly important in high resolution devices and in the ion trap, where the ions have a long path length in the field.

Systematic imperfections in the fields cause nonlinear resonances that are observed as peak-splitting and peak-shape distortions. Some ions that would normally have (a, q) positions within the stable region are found to have unstable trajectories. Nonlinear resonances occur at narrow lines within the stable region.

Nonlinear effects have been considered by von Busch and Paul (89) for the mass filter, using an analytical approach. They investigated the effects experimentally in a device 3 meters in length. This work will be described first. Some of the mass filter results can be applied directly to the monopole. The effect of field errors on focusing properties has been studied by computation of ion trajectories (90). Finally, results for the quadrupole ion trap are discussed. More theoretical and experimental data (72, 73) are available for this case than for the others and there is excellent confirmation of the theory.

B. The Mass Filter

1. Theory

The potential in the mass filter can be expressed as

$$\Phi = \sum_N A_N(r/r_0)^N \cos N(\phi - \phi_N)[U - \sum_n b_n V \cos (n\omega/2)(t - t_n)] \quad (80)$$

where r and ϕ are polar coordinates, U and V are the dc and rf potentials, and ω is the angular rf frequency. For an ideal quadrupole field, only the term with $N = n = 2$ is retained, and the equation can be readily transformed to Cartesian coordinates. Taking $A_2 = 1$, $b_2 = 1$,

$$\Phi = (1/r_0^2)(x^2 - y^2)[U - V \cos \omega(t - t_2)] \tag{81}$$

If only slight deviations from quadrupole field are considered, then $A_N \ll A_2$, $b_n \ll b_2$ and the terms that have both $N \neq 2$ and $n \neq 2$ can be neglected. The remaining two sets of terms are of different types. Terms with $N = 2, n > 2$ are the result of harmonics in the rf voltage; that is, of time-varying errors. Terms with $N > 2$, $n = 2$ result from mechanical errors in the quadrupole field geometry. For example, when $N = 3, n = 2$, the third-order term by itself represents a pure hexapole field. In Cartesian coordinates,

$$\Phi_3 = (A_3/r_0^3)(3yx^2 - y^3)[U - V \cos \omega(t - t_2)] \tag{82}$$

Laplace's equation is still obeyed, but the restoring force on an ion is no longer linearly dependent on its displacement from the center. There is also a coupling between the x and y oscillations.

From the general theory of two-dimensional harmonic oscillators, one can deduce (89) that the higher order terms ($N \neq 2$, $n = 2$) produce sum resonances at those conditions where

$$(\beta_x/2)K + (N - K)(\beta_y/2) = 1 \tag{83}$$

K can have the values N, $N - 2$, $N - 4 \ldots$. Since resonance lines of the form $\beta_x + (N - 2)(\beta_y/2) = 1$ pass through the apex of the stable region, they can result in peak splitting in the mass filter by causing ion trajectories to become unstable.

Third-order (hexapole) distortions cause resonance lines at

$$\beta_x = \frac{2}{3}, \qquad \left(\frac{\beta_x}{2}\right) + \beta_y = 1 \tag{84}$$

and

$$\beta_y = \frac{2}{3}, \qquad \left(\frac{\beta_y}{2}\right) + \beta_x = 1$$

Since the third-order distortion is asymmetric, there are two possible hexapole distortions.

Fourth-order resonances are at

$$\beta_x = \tfrac{1}{2}, \qquad \beta_x + \beta_y = 1, \qquad \beta_y = \tfrac{1}{2} \tag{85}$$

Some of the resonance lines are shown in Fig. 87.

The terms with $n > 2$ would be important if β_x or β_y could be equal to n, but such values lie outside the stable region anyway. However, odd harmonics of $\omega/2$ can make the trajectories unstable when $\beta_x = \tfrac{1}{2}$ or $\beta_y = \tfrac{1}{2}$.

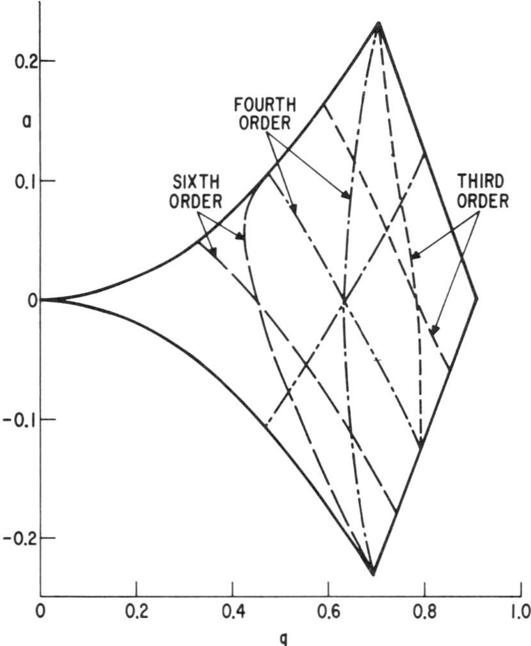

FIG. 87. Stability diagram for the mass filter and monopole, showing the nonlinear resonance lines of third, fourth, and sixth order. Of the two possible third-order resonances, only one is shown. Only two of the four sixth-order lines are shown.

This analytical theory predicts the position at which resonances occur, but not their magnitude for a given amount of distortion. The magnitude can be determined, however, using computer simulation (Section VI-D) (*90*). Computer calculations have recently been made for third- and fourth-order distortions in the mass filter (*90*).

2. Experimental Observations of Nonlinear Resonances

von Busch and Paul (*89*) used an isotope separator (*13*) for an experimental investigation of nonlinear resonances. The field length was 3 meters, the rf frequency was 2.6 MHz, and r_0 was 1.5 cm. Measurements made on sodium ions are shown in Fig. 88. The transmitted ion current is given as a function of q when $a = 0$. Resonance dips can be seen at $\beta_x = \beta_y = \frac{1}{3}, \frac{1}{2}$, and $\frac{2}{3}$. When $a \neq 0$, then $\beta_x \neq \beta_y$ and the $\beta = \frac{1}{3}$ dip splits into two weak intensity minima. These appeared to be the lines $2\beta_x + \beta_y = 1$ and $\beta_x + 2\beta_y = 1$ rather than $\beta_x = \frac{1}{3}$ and $\beta_y = \frac{1}{3}$. It appeared that some but not all the sixth-order resonances were excited. The dip at $\beta = \frac{1}{2}$ suggested a fourth-order resonance;

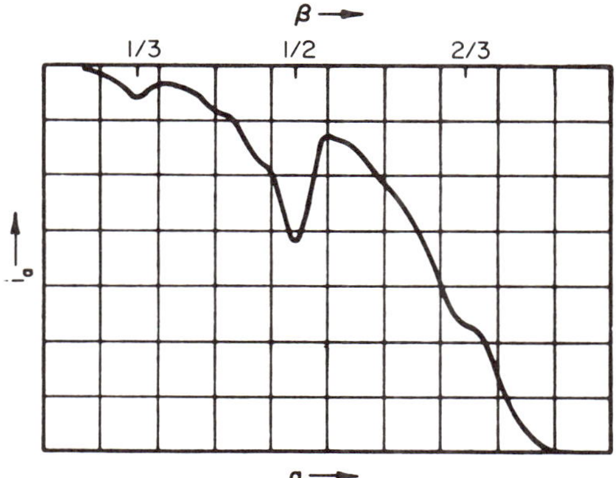

FIG. 88. Experimental observations (89) with a 3-meter mass filter of the ion transmission as a function of q when $a = 0$. The ions spend 135 rf periods in the field. q varies from 0.35 to 0.95.

however, for $a \neq 0$, there was splitting into $\beta_x = \frac{1}{2}$ and $\beta_y = \frac{1}{2}$, but not $\beta_x + \beta_y = 1$. The approximation of hyperbolic electrodes by using circular cylinders should give a sixth-order potential as the lowest interference term (89).

The rf generator used in the experiment employed frequency doubling that could introduce the subharmonic at half the frequency, so resonance minima were expected at $\beta_{x,y} = \frac{1}{2}$. However, no changes in resonance intensity were found when frequency doubling was eliminated, and the cause of the $\beta_{x,y} = \frac{1}{2}$ resonances has not been determined.

Spacing errors in the quadrupole rods give rise to multipole terms of arbitrary order. The resonance at $\beta_{x,y} = \frac{2}{3}$ may be a third-order resonance, but it was so weak that splitting could not be measured for $a \neq 0$.

Levine and Tobias (91) observed peak splitting that was associated with a buildup of deposits along the rods after operating for long times. The deposits may alter the potential so as to cause nonlinear resonances of the ions. Hein (92) has discussed experimental observations relating to the presence and extent of undesired precursor peaks.

C. The Monopole

Nonlinear resonances analogous to those in the mass filter can occur in the monopole. Low order sum resonances, however, are important only if the monopole is operated at the apex of the stability region. Higher order sum resonance lines, and lines such as $\beta_x = \frac{1}{3}$, intersect the $\beta_y = 0$ boundary in positions where the monopole might be operated. For monopoles operated in the conventional manner (Section IV), there would be dips in the ion transmission at the resonance lines. These dips would move to the low mass side of the peak as the ion mass is lowered. High mass ions spend a longer time in the field and therefore the effects would be more noticeable. A simple change in the a/q ratio would move operation away from the resonance line.

However, distortions will affect the focusing properties as well. The computer simulation of monopole performance described in Section IV has been extended to include third order distortions (90). Since only one quadrant of the mass filter field is used, there are four possible ways in which third-order distortions can appear. Two are symmetric about the $x = 0$ axis and correspond to positive and negative third order terms, the other two being asymmetric.

In the computer simulation, initial ion displacements of 10^{-3} meter were assumed in the y direction and 5×10^{-4} meter in the x direction. The ions stayed in the rf field for 20 cycles. Focusing properties in the y direction only were considered. Small asymmetric third order distortions have little effect because the ions oscillate rapidly about the x axis. For the symmetric third-order distortions, a positive weighting factor A_3 causes ions that enter at phases of the rf field, such that they have a large maximum y displacement, to be more strongly focused than those entering at other phases (β_y is larger). Since the monopole operated in the conventional manner is poorly focused (see Fig. 53), this type of distortion can be advantageous. A negative weighting factor A_3 gives additional defocusing and is very undesirable. Figure 89(a) shows peak shapes deduced from y trajectories for A_3/r_0 equal to 0.5, and Fig. 89(b) shows those for $A_3/r_0 = -0.5$. These are relatively small distortions (less than 1% at the maximum displacement). The ions stay only 20 cycles in the field, but the distortions have important effects because of the utilization of focusing properties. As can be seen, a positive distortion is actually beneficial.

D. The Quadrupole Ion Trap

1. Theory

The quadrupole ion trap has been treated in detail (72, 73), both by computer simulation and experimentally, for geometrical distortions that retain the rotational symmetry. Third- and fourth-order potential terms are of the form

$$\Phi_3 = (\tfrac{1}{4}z_0^2)(A_3/z_0)(3r^2z - 2z^3)(U - V\cos\omega t) \tag{86}$$

$$\Phi_4 = (\tfrac{1}{4}z_0^2)(A_4/z_0^2)(r^4 + 2z^4 - 8r^2z^2)(U - V\cos\omega t) \tag{87}$$

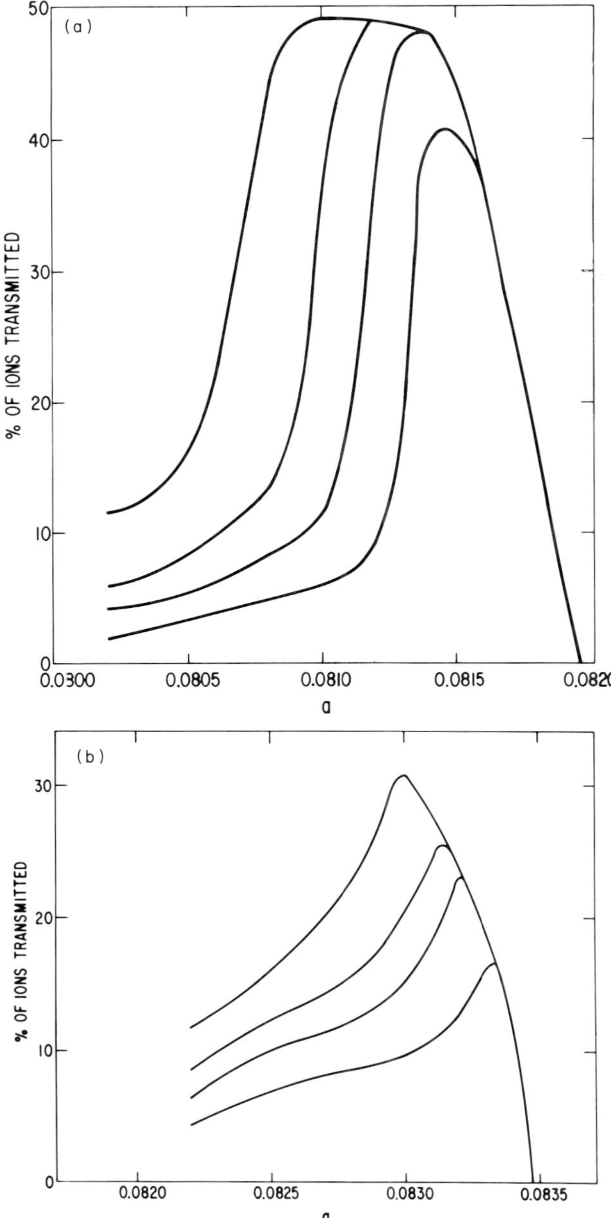

FIG. 89. Computed peak shapes (*90*) in a monopole with third-order distortions that are symmetrical about the $x = 0$ axis. Ions stay 20 cycles within the field. The scan line is $a = 0.2q$. The initial y displacement is 10^{-3} meters, the initial x displacement 5×10^{-4} meters. Only y focusing has been considered. (a) Distortion weighting factor $A_3/r_0 = 0.5$; (b) $A_3/r_0 = -0.5$. The peak shapes correspond to exit slit widths of 2, 3, 4, and 6×10^{-3} meters.

The resonance lines are the same as those given in Section VI-B for the x and y directions in the mass filter. A large degree of third order distortion would be expected from an asymmetric, incorrect spacing of the end cap electrodes, and fourth order distortion from a symmetric incorrect spacing.

2. Computer Simulation

The computer simulation was done by computation of trajectories, assuming initial displacements of 10^{-3} meters in both r and z directions, and assuming a maximum displacement before ion loss of 10^{-2} meters. Only one phase of the rf field at ion formation was considered, and only a third order distortion. The storage time was $200/\pi$ rf cycles. Figure 90 shows peak shapes for several mass scan lines with A_3 taken as $10z_0/3$. This is a very large distortion (about equivalent to a 5% error in the z spacing of the caps). The resonance dip increases at high resolution, and multiple splitting from a single resonance is possible. With less distortion, the device can be operated at lower q/a values. The computed resonance lines obtained by plotting the dips of Fig. 90 in (a, q) space correspond to the line $\beta_z/2 + \beta_r = 1$ as predicted by the theory, although the exact position of the line is very slightly dependent on the degree of distortion. Large distortions make it difficult to achieve high resolution, since they change the shape of the stability tip (the tip is "blunted").

The results have also been plotted to give the number of cycles before ion loss at the resonance position as a function of the third order weighting factor A/z_0 for scan lines of differing "nominal" (without distortion) resolution. This is shown in Fig. 91. The data almost fall on straight lines on the log-log plot. The position of the lines in the A direction is roughly inversely proportional to the resolution. One can therefore estimate the permissible geometrical error for a given performance. High resolution requires long storage times and is therefore a particularly demanding requirement.

3. Experimental Observations

Peaks with multiple splitting have been observed in devices of several different sizes (73), the effects being more severe in the smallest size ($z_0 = 4 \times 10^{-3}$ meters). The extent of the splitting is changed by changing the spacing of the end caps (74), as would be predicted. Examples of the $M = 28$ peak at various resolutions are given in Fig. 92. The resonance dips are large and narrow. Slow scanning is required to prevent their presence from being masked by the response of the recording system. The storage time was 0.1 sec at 1 MHz frequency so that some ions were in the field for 10^5 cycles.

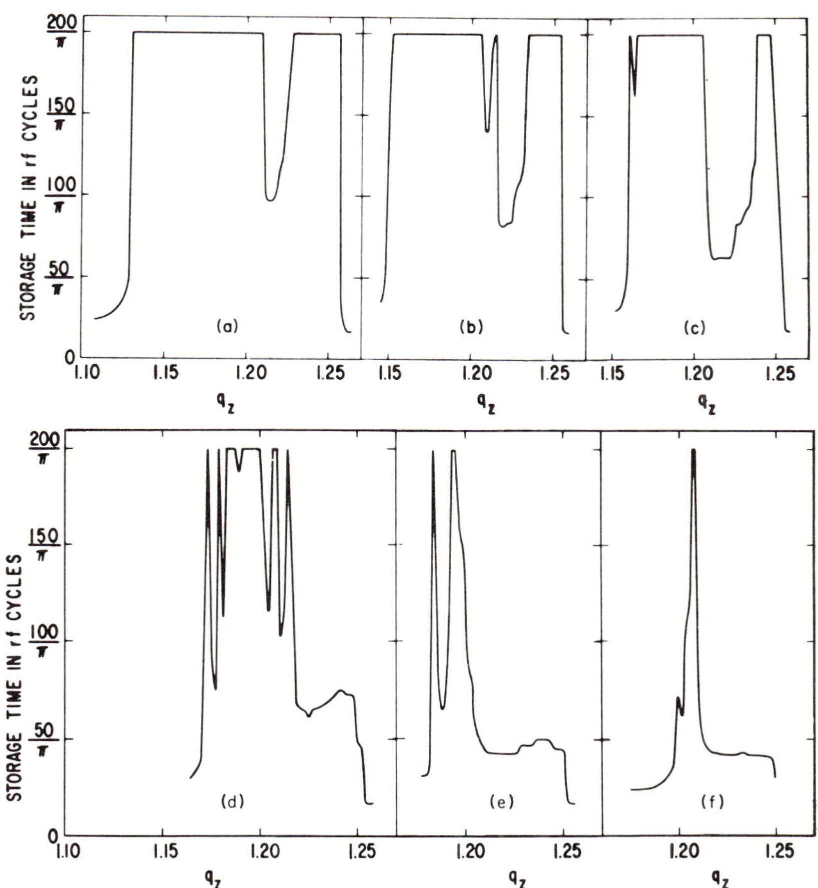

FIG. 90. Computed peak shapes (72) for the quadrupole ion trap for various scan lines when the third-order distortion weighting factor A_3 was $10z_0/3$. The scan lines were $q/a = :$ (a) 2.25, (b) 2.0, (c) 1.9875, (d) 1.975, (e) 1.9625, and (f) 1.95.

The position of the experimental resonances has been plotted on the stability diagram in Fig. 93. The major resonances fall where predicted by the theory. Many minor dips are also present, and these have not been explained.

It was found empirically (70) that a small bias between the two end caps, usually a few percent of the dc voltage between the ring and the caps, gave improved peak shapes. This has been explained by both experimental observations and computer simulation to be the result of a shifting of the z stability boundary to higher q values, with little shifting in the resonance line locations. A stability tip free from resonances is obtained. The higher the resolution, the smaller the bias that is required. The use of the bias is

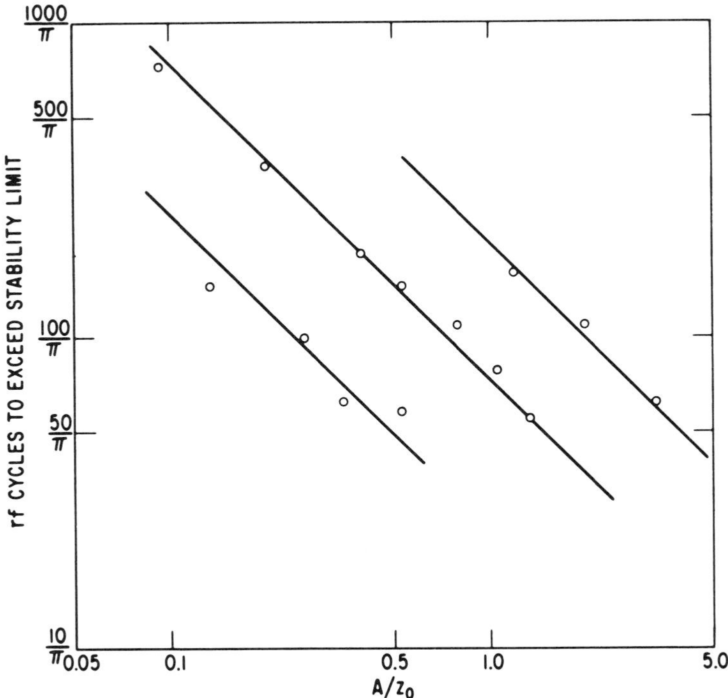

FIG. 91. A log-log plot (72) of the number of rf cycles taken to exceed the 10^{-2}-meter stability limit at the position of the third-order resonance as a function of the third-order weighting factor. The lines are for scans of nominal resolution of 10, 26, and 52, reading from right to left.

accompanied by a decrease in sensitivity. This is the result of a broadening of the z stability boundary, with its location becoming strongly dependent on the position of ion formation. The loss of sensitivity due to the position factor *only* was computed for various percentages of bias as a function of the resolution and is given in Fig. 94. This provides a qualitative picture of the likely deterioration in sensitivity as the result of the use of a bias voltage to avoid poor peak shapes. It appears likely that this biasing technique could be adapted for use with quadrupole mass filters.

VII. Summary and Future Trends

In the preceding sections we have described in some detail the achievements to date in rf quadrupole field mass spectroscopy. The evolution of these instruments has been extremely rapid, reflecting inherent advantages such as

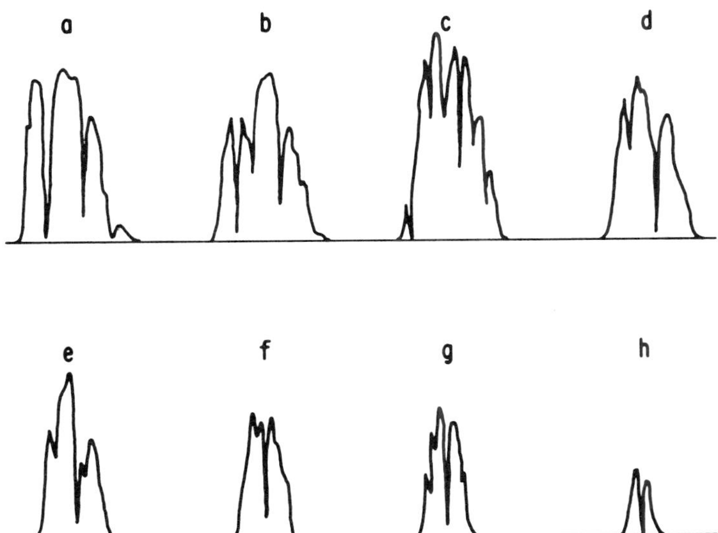

FIG. 92. Experimental peak shapes (72) in the ion trap with storage for 10^5 rf cycles. The peak is $M = 28$ at a pressure of 10^{-9} torr. The q/a value decreases from (a) to (h). Resolution for (h) was about 18.

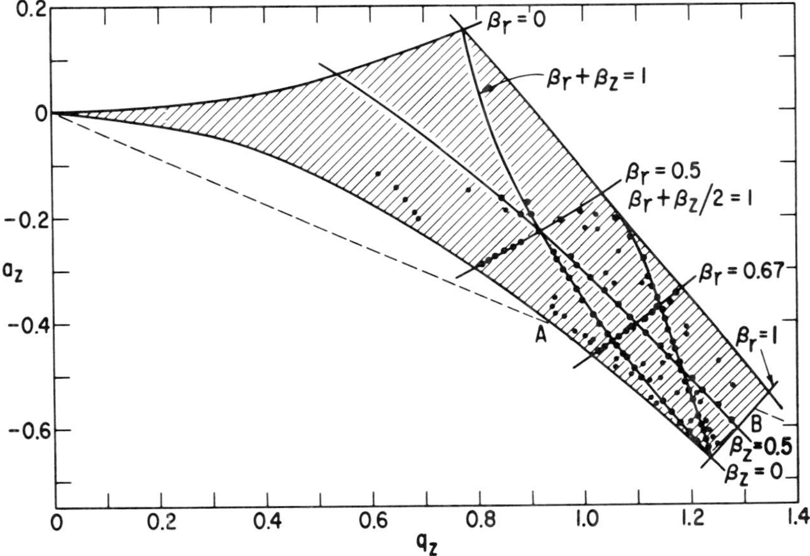

FIG. 93. Stability diagram for the ion trap showing the predicted resonance lines for third- and fourth-order distortions. The points correspond to experimentally observed dips in the ion peaks (73).

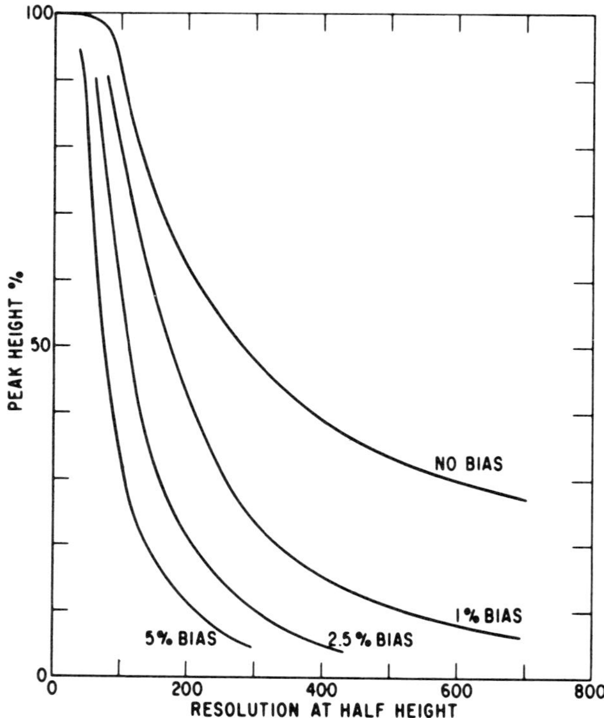

FIG. 94. The resolution versus sensitivity computed to show the decrease in sensitivity when a bias voltage is used between the end caps in order to obtain a good peak shape. Only the spread in position of ion formation was considered (72).

their small size and weight and the absence of a magnetic field. Of the many types of dynamic mass spectrometers [Blauth (1), for example, lists 47] only the quadrupole mass filter, the monopole, and the time-of-flight spectrometer are in general use at the present time.

The role of quadrupole mass spectroscopy has been greatest in the field of partial pressure analysis and in special applications. However, the performance of commercial instruments is still improving, and their use is extending into routine chemical analysis—especially in combination with gas chromatography.

What are the limitations in the present form of these devices? One limitation is that the rf power increases with the square of the ionic mass, and higher mass ranges can be obtained only by use of high power or low rf frequencies or very small radius instruments. Low frequencies sacrifice resolution, except with inconveniently long instruments, especially because a limit is being

TABLE V

Commercially Available Spectrometers

Manufacturer	Model	Resolution	Sensitivity (amp/torr)	Minimum partial pressure (torr)	Dynamic (range)	Maximum scan rate (sec/amu)	Mass range	Number of ranges
Quadrupole mass filters								
Balzers	QMG101	100	1000	10^{-13}	—	2×10^{-4}	1–400	(2)
CEC	21–440	2M	10	5×10^{-14}	10^6	5×10^{-4}	1–300	(1)
Centronic	Q.806	80	—	10^{-12}	—	5×10^{-4}	1–100	(1)
EAI	Quad 300	2M	10	—	—	10^{-3}	1–500	(3)
Extranuclear	—	Various						
Granville	750	2M	100	5×10^{-15}	—	5×10^{-4}	1–750	(3)
Phillips								
Varian/EAI	Quad 250A	2M	100	10^{-15}	10^6	5×10^{-4}	1–800	(3)
Varian/EAI	Quad 150A	2M	40	2×10^{-14}	—	10^{-3}	1–300	(2)
Varian	QRGA	100	—	2×10^{-13}	—	2×10^{-4}	1–250	(2)
Varian Mat	AMP-3	200	100	10^{-13}	—	10^{-3}	1–200	(1)
Monopoles								
General Electric	600	2M	100	10^{-14}	10^6	5×10^{-4}	1–600	(1)
General Electric	300	2M	50	10^{-13}	—	10^{-3}	1–300	(1)
VEECO	SPI-10	2M at $M = 50$	—	10^{-10}	—	10^{-3}	1–200	(2)

reached to further lowering of the ion energy. A small radius presents difficult problems in maintaining the accuracy of machining and of alignment, and
sacrifices some sensitivity. Development is already at the stage where circular
cylindrical rods are being replaced with hyperbolic rods to improve the performance. The quadrupole ion trap offers some potential advantages in that
ions may be stored for many rf cycles in a small volume, and for high masses
a low rf frequency may be used. Our knowledge of ion storage properties is
still too incomplete to predict how useful this property will be. It is probable
that highly accurate quadrupole fields will be necessary to achieve high
performance. Ion storage may also be useful when the rate of ion formation is
very low, such as at extremely low partial pressures.

Further evolution of the quadrupole and monopole may come in different
methods of utilizing the fields. The effective length of the instrument could
be increased by multiple traverses of one or more devices. Alternatively the
rods might be bent into a closed circle configuration. One might use to better
advantage the focusing properties of quadrupole fields, as suggested by Lever
(16) in his exact focusing monopole. An exact focusing four-rod structure
should also be feasible. The study of particle motion in hexapole and higher
order fields may lead to fruitful developments in other directions.

All these factors suggest a continuing, but perhaps slower, evolution of
quadrupole field devices.

Appendix: Commercially Available Instruments

To illustrate the state of the art, Table V provides a list of some of the
commercially available rf quadrupole mass spectrometers with their adver-
tized performance at the time of writing. There is an emphasis on U.S. manufacturers, and the list is not necessarily complete.

Different definitions of resolution are used by different manufacturers and
therefore resolutions are not necessarily directly comparable and may vary
with the mass range. Sensitivities depend on the resolution and usually on the
electron multiplier gain. The maximum sensitivity is not usually available at
the fastest scan rate.

References

1. E. W. Blauth, in " Dynamic Mass Spectrometers." Elsevier, Amsterdam, 1966.
2. F. M. J. Pichanick, in " Methods of Experimental Physics" (L. Marton, ed.-in-chief),
   Vol. 4A, "Atomic Sources and Detectors" (V. W. Hughes and H. L. Schultz, eds.).
   Academic Press, New York, 1967.
3. E. D. Courant, M.S. Livingston, and H. S. Snyder, *Phys. Rev.* **88**, 1190 (1952).
4. W. Paul and H. Steinwedel, *Z. Naturforsch.* **8a**, 448 (1953).
5. J. P. Blewett, *Phys. Rev.* **88**, 1197 (1952).

6. W. Paul and H. Steinwedel, German Patent No. 944, 900, June 28, (1956).
7. R. F. Post, *in* Univ. of California Rad. Lab. Rept. UCRL-2209 (1953).
8. W. Paul, O. Osberghaus, and E. Fischer, *Forschungsber. Wirtsch. Verkehrsministeriums Nordrheim Westfalen* **415** (1955).
9. K. Berkling, "Diplomarbeit." Physik. Inst. Univ. Bonn, Germany, 1956.
10. E. Fischer, *Z, Fhysik* **156**, 26 (1959).
11. R. F. Wuerker, H. Shelton, and R. V. Langmuir, *J. Appl. Phys.* **30**, 342 (1959).
12. U. von Zahn, *Rev. Sci. Instr.* **34**, 1 (1963).
13. W. Paul, H. P. Reinhard, and U. von Zahn, *Z. Physik* **152**, 143 (1958).
14. N. W. McLachlan, *in* "Theory and Application of Mathieu Functions." Oxford Univ. Press, London and New York, 1951.
15. Computation Lab. Natl. Bur. Std., "Tables Relating to Mathieu Functions." Columbia Univ. Press, New York, 1951.
16. R. F. Lever, *IBM J. Res. and Dev.* **10**, 26 (1966).
17. P. H. Dawson and N. R. Whetten, *J. Vac. Sci. and Technol.* **5**, 1 (1968).
18. W. Paul and M. Raether, *Z. Physik* **140**, 262 (1955).
19. I. E. Dayton, F. C. Shoemaker, and R. F. Mozley, *Rev. Sci. Instr.* **25**, 485 (1954).
20. W. M. Brubaker, *Colloq. Spectrosc. Intern., 9th, Lyon, France, June* (1961). Not generally available.
21. U. von Zahn, Thesis, Univ. of Bonn, Bonn, Germany (1956).
22. P. Marchand and P. Marmet, *Can. J. Phys.* **42**, 1914 (1964).
23. K. Ogata and H. Matsuda, *Phys. Rev.* **89**, 27 (1953).
24. J. Mattauch and R. Bieri, *Z. Naturforsch.* **9a**, 303 (1954).
25. G. V. Schiersledt, H. Ewald, H. Liebl, and G. Sauermann, *Z. Naturforsch.* **11a**, 216 (1956).
26. W. M. Brubaker, *Intern. Instr. Conf. Proc. 5th, Stockholm, Sweden, Sept., 1960.*
27. W. M. Brubaker, *in* "Advances in Mass Spectrometry," Vol. 4. Elsevier, Amsterdam, 1968.
28. W. M. Brubaker and J. Tuul, *Rev. Sci. Instr.* **35**, 1007 (1964).
29. R. S. Narcisi and A. D. Bailey, *in* "Space Research" (D. G. King-Hele, P. Muller, and G. Righini, eds.), Vol. V. North-Holland, Amsterdam, 1965.
30. G. F. Sauter, R. A. Gerber, and H. J. Oskam, *Rev. Sci. Instr.* **37**, 572 (1966).
31. H. Böhm and K. G. Günther, *Vakuum-Tech.* **7**, 192 (1965).
32. H. Böhm and K. G. Günther, *Z. Angew. Phys.* **17**, 553 (1964).
33. K. G. Günther and W. Hänlein, *in Trans. Vacuum Symposium and Second International Congress, 8th, 1961,* 573. Pergamon Press, New York, 1962.
34. P. Blum and F. L. Torney, *Rev. Sci. Instr.* **38**, 1404 (1967).
35. P. A. Redhead, *Can. J. Phys.* **37**, 1260 (1959).
36. P. Marchand, C. Paquet, and P. Marmet, *Rev. Sci. Instr.* **37**, 1702 (1966).
37. D. F. Munro, *Rev. Sci. Instr.* **38**, 1532 (1967).
38. M. S. Story, *in Natl. Vacuum Symp. Am. Vacuum Soc., 14th, Kansas City, Missouri, Oct, 1967.* Not generally available. For abstract, see *J. Vac. Sci. and Techn.* **4**, 326 (1967).
38a. A. R. Fairbairn, *Rev. Sci. Instr.* **40**, 380 (1969).
39. U. von Zahn, *Z. Physik* **168**, 129 (1962).
40. U. von Zahn, S. Gebauer, and W. Paul, *in Ann. Conf. Mass Spectrometry, 10th, New Orleans, La., June, 1962.* Not generally available.
41. M. Mosharrafa and H. J. Oskam, *Physica* **32**, 1759 (1966).
42. W. M. Brubaker, *in Ann. Conf. Mass Spectrometry and Allied Topics, 16th, Pittsburgh, Pennsylvania, May, 1968.* Not generally available.

43. W. M. Brubaker, *in Natl. Vacuum Symp. Am Vacuum Soc., 14th, Kansas City, Missouri, Oct., 1967.* Not generally available. For abstract see *J. Vac. Soc. and Techn.* **4,** 326 (1967).

44. G. O. Brink, *Rev. Sci. Instr.* **37,** 857 (1966).

44a. J. A. Burt, *Rev. Sci. Instr.* **40,** 358 (1969).

45. F. von Busch and W. Paul, *Z. Physik* **164,** 581 (1961).

46. W. Paul and H. P. Reinhard, *Proc. Symp. Isotope Separation, Amsterdam, 1957.*

47. M. C. Paul, *Vacuum* **15,** 239 (1965).

48. R. A. Gerber, G. F. Sauter, and H. J. Oskam, *Physica* **32,** 2173 (1966).

49. G. F. Sauter, R. A. Gerber, and H. J. Oskam, *Physica* **32,** 1921 (1966).

50. J. M. Madson and H. J. Oskam, *Phys. Letters* **25(A),** 407 (1967).

51. M. Mosharrafa and H. J. Oskam, *Proc. Ann. Conf. Mass Spectronomy and Allied Topics, 12th, Montreal, Canada, June, 1964.* Not generally available.

52. H. G. Bennewitz and R. Wedemeyer, *Z. Physik* **172,** 1 (1963).

53. C. K. Crawford, *J. Vac. Sci. and Technol.* **5,** 131 (1968).

54. D. E. Golden, G. Sinnott, and R. N. Varney, *Phys. Rev. Letters,* **20,** 239 (1968).

55. D. Gutman, A. J. Hay, and R. L. Belford, *J. Phys. Chem.* **70,** 1786 (1966).

56. D. Gutman, R. L. Belford, A. J. Hay, and R. Pancirov, *J. Phys. Chem.* **70,** 1793 (1966).

57. A. S. Gilmour, Jr., and F. A. Giori, *Ann. Conf. Mass Spectrometry and Allied Topics, 13th, St. Louis, Missouri, May, 1965.* Not generally available.

58. U. von Zahn and H. Tatarczyk, *Phys. Letters* **12,** 190 (1964).

59. H. Tatarczyk and U. von Zahn, *Z. Naturforsch.* **20a,** 1708 (1965).

60. R. Gradewald, *Ann. Physik* **20,** 1 (1967).

60a. M. Mosharrafa, D. Witsoe, R. Patterson, and W. G. Kubicek, in *Ann. Conf. Mass Spectrometry and Allied Topics, 17th, Dallas, Texas, May, 1969.* Not generally available.

61. K. Mauersberger, D. Müller, D. Offermann, and U. von Zahn, *in* " Space Research " (R. L. Smith-Rose, ed.), Vol. VII, p. 1150. North Holland, Amsterdam, 1967.

62. J. B. Hudson and R. L. Watters. *IEEE Trans.* **IM-15,** No. 3, 94 (1966).

63. J. B. Hudson, *Natl. Symp. Am. Vacuum Soc., Kansas City, Missouri, Oct., 1967.* Not generally available. For abstract, see *J. Vac. Sci. and Technol.* **4,** 325 (1967).

64. P. H. Dawson and N. R. Whetten, *Rev. Sci. Instr.* **39,** 1417 (1968).

65. P. H. Dawson and N. R. Whetten, *J. Vac. Sci. and Technol.* **6,** 97 (1969).

66. S. J. Lins and M. C. Paul, *Rev. Sci. Instr.* **35,** 1084 (1964).

67. R. E. Grande, R. L. Watters, and J. B. Hudson, *J. Vac. Sci. and Technol.* **3,** 329 (1966).

68. J. Gross, D. Offermann, and U. von Zahn, *in* " Space Research " (A. P. Mitra, L. G. Jacchia, W. S. Newman. eds.), Vol. VIII. North Holland, Amsterdam, 1968.

69. G. Hartmann, K. Mauersberger, and D. Müller, *in* " Space Research" (A. P. Mitra, L. G. Jacchia, W. S. Newman, eds.), Vol. VIII. North Holland, Amsterdam, 1968.

70. P. H. Dawson and N. R. Whetten, *J. Vac. Sci. and Technol.* **5,** 11 (1968).

71. G. Rettinghaus, *Z. Angew. Phys.* **22,** 321 (1967).

71a. E. W. Purcell and H. C. Torey, *Phys. Rev.* **69,** 37 (1946).

72. P. H. Dawson and N. R. Whetten, *J. Mass Spect. and Ion Phys.* **2,** 45 (1969)

73. N. R. Whetten and P. H. Dawson, *J. Vac. Sci. and Technol.* **6,** 100 (1969).

73a. P. H. Dawson, J. Hedman, and N. R. Whetten. *Rev. Sci. Instr.* In press.

74. H. G. Dehmelt, *Phys. Rev.* **103,** 1125 (1956).

75. H. G. Dehmelt and F. G. Major, *Phys. Rev. Letters* **8,** 213 (1962).

76. C. B. Richardson, K. B. Jefferts, and H. G. Dehmelt, *Phys. Rev.* **165,** 80 (1968).

77. F. G. Major and H. G. Dehmelt, *Phys. Rev.* **170,** 91 (1968).

78. K. B. Jefferts, *Phys. Rev. Letters* **20,** 39 (1968).

79. D. C. Burnham and D. Kleppner, *Bull. Am. Phys. Soc., Series II,* **11,** 70 (1966).

80. H. G. Dehmelt, *Advan. At. Mol. Phys.* **3**, 53 (1968).

81. G. Kotowski, *Z. Angew. Math. Mech.* **23**, 213 (1943).

82. P. H. Dawson and N. R. Whetten, *Die Naturwissenschaften* **3**, 109 (1969).

82a. P. H. Dawson and N. R. Whetten, unpublished.

83. R. F. Wuerker, H. M. Goldenberg, and R. V. Langmuir, *J. Appl. Phys.* **30**, 441 (1959).

84. D. B. Langmuir, R. V. Langmuir, H. Shelton, and R. F. Wuerker, U.S. Patent No. 3,065,640, November 27, 1962.

85. R. W. Waniek and P. J. Jarmuz, *Appl. Phys. Letters* **12**, 52, (1968).

86. E. N. Fortson, F. G. Major, and H. G. Dehmelt, *Phys. Rev. Letters* **16**, 221 (1966).

87. T. W. Karrass and E. Lindman, *J. Appl. Phys.* **36**, 18 (1965).

88. H. Kleinpoppen and J. D. Schumann, *Z. Angew. Phys.* **22**, 152 (1967).

89. F. von Busch and W. Paul, *Z. Physik* **164**, 588 (1961).

90. P. H. Dawson and N. R. Whetten, *J. Mass Spect. and Ion Phys.* In press.

91. L. P. Levine and J. Tobias, *Rev. Sci. Instr.* **36**, 1894 (1965).

92. R. A. Hein, *First Ann. Symp., New England Section of American Vacuum Society, Boston, Mass., April, 1969.* Not generally available.

Theory of the Generation of Harmonics and Combination Frequencies in a Plasma*

MAHENDRA SINGH SODHA

*Physics Department, Indian Institute of Technology
New Delhi, India*

AND

PREDHIMAN KRISHAN KAW†

*Plasma Physics Laboratory, Princeton University
Princeton, New Jersey*

* Work supported partially by U.S. Environmental Science Services Administration and partially by U.S. Atomic Energy Commission.

† A large portion of this work was done when this author was at Indian Institute of Technology, New Delhi, India.

187

In this paper, the authors have reviewed the work done on the theory of generation of harmonics and combination frequencies in the current density in a plasma, emphasizing the kinetic approach to these problems rather than the elementary phenomenological approach considered by earlier reviewers (*1*). Starting from the Boltzmann transfer equation for electrons, explicit expressions are derived for the harmonic and combination frequency components of the current density due to the various different mechanisms. The effects of the external magnetic fields and the plasma oscillations are also taken into account in some cases. Wherever the kinetic treatments are not available, the elementary approach is presented. These expressions for the current density components have been used for studying in detail two phenomena connected with the nonlinear propagation of electromagnetic waves in plasma, viz.,

1. The nonlinear growth of the harmonic and combination frequency waves in a plasma.

2. The magnitudes of the harmonic and combination frequency components in the reflected wave from a semiinfinite plasma-free space interface.

To get an idea of the variation of the harmonic and combination frequency components in the current density and in the reflected wave, with the relevant plasma parameters, numerical results are presented at many stages. The absolute orders of magnitudes of these components are also evaluated in most of the cases.

I. Introduction

In recent years a number of papers on generation of harmonics and combination frequencies of electromagnetic waves in a plasma have appeared in the literature. A common characteristic of almost all these investigations is their concentration on a single mechanism, to the extent that other mechanisms proposed by different workers to explain these phenomena are not even mentioned. This has led to improper and somewhat erroneous correlation of various investigations (especially between theory and experiment) and the incorrect conclusion that little theoretical work has been done in this field. Another feature of most of these investigations is the limitation of scope to the generation of harmonic and combination frequency components in the current density. The equally important feature of the problem, viz., the application of these results to predict the growth of the harmonic and combination frequency waves in a plasma (in which high intensity electromagnetic waves are propagating) has not been considered in most of these investigations.

In a series of papers published recently, the authors and their collaborators have evaluated the harmonic and combination frequency components in the current density to include the various relevant mechanisms, using the technique of Boltzmann's transfer equation. These results have been used to predict the growth of the harmonic and combination frequency components in high intensity electromagnetic waves propagating through a plasma. The magnitude of these components in the reflected parts of strong electromagnetic waves incident on a plasma-free space interface has also been evaluated. This paper presents a review of the theoretical work on the various mechanisms proposed for the generation of the harmonic and combination frequencies in a plasma. In this review, the authors have emphasized the kinetic approach (based on Boltzmann's transfer equation) to the problem rather than the elementary phenomenological approach, adopted in another recent review of this field by Wetzel and Tang (1).

The first significant theoretical analysis of the phenomenon of harmonic generation in a plasma appears to have been made by Margenau and Hartman (2), in which they investigated the time dependence of the various parts of the distribution function of electron velocities in the presence of an alternating electric field. Based on this analysis an explicit expression for the third harmonic component of current density was obtained by Rosen (3); this expression was used by Sodha and Palumbo (4) to study the growth of the third harmonic component in an electromagnetic wave propagating in a plasma. Following a technique similar to that of Margenau and Hartman (2), Vilenskii (5, 6), Fain (7), Gurevich (8), Ginzburg and Gurevich (9), Wetzel (10), Silin (11), Chiyoda (12), Krenz (13), Gupta (14, 15), Mittal and Kaw (16), Sodha and Kaw (17–24), and Paranjape (25) have carried out the investigations more rigorously, and extension has been made in some papers to include the effect of inhomogenities, magnetic fields, external dc electric fields, etc. The problem of the generation of combination frequencies has also been considered in some of the investigations. A quasihydrodynamic approach to these problems has been made by Murphy (26), Tang (27) and Desloge and Coleman (28), and a purely phenomenological hydrodynamic model has been followed by Wetzel (29, 30), Blachier et al. (31), Whitmer and Barrett (32, 33), Barrett et al. (34), Krenz (35), Krenz and Kino (36, 37), Smerd (38), Taylor (39), Visvanathan (40), Rydbeck (41), etc.

The generation of harmonics and combination frequencies due to the periodic changes in the electron density, arising because of the periodic variation of the electrical vector, has been investigated by Ginzburg (42), Baird and Coleman (43), Moriyama and Sumi (44), and Chiyoda and Tamaru (45). Forsterling and Wüster (46, 47) have used a phenomenological approach to investigate the generation of harmonics around plasma resonances in an inhomogeneous plasma, whereas a more rigorous kinetic theory technique has been used for a similar problem by Moriyama and Sumi (48, 49).

The first experimental investigation of the generation of harmonics in a plasma was carried out by Uenohara *et al.* (*50*). Further investigations by other workers were motivated by two different objectives, viz.,

1. Improvement in the performance of plasma as a high frequency electromagnetic wave generator from a device point of view, e.g., the work of Froome (*51–54*), Bierrum and Walsch (*55, 56*), Swan (*57*), etc.

2. Understanding of the physics of harmonic generation and possible verification of the proposed theories, e.g., the investigations of Baird and Coleman (*43*), Hill and Tetenbaum (*58*), Dreicer (*59*), Tetenbaum *et al.* (*60*), Stern (*61*), Green (*62*), etc.

In this review, the authors have not attempted any correlation between the experimental and theoretical results because they believe with Wetzel and Tang (*1*) that the nonlinear processes underlying the production of harmonics and combination frequencies usually occur together; since none of the theories of the present day even claims to take all the nonlinearities into account, such an attempt is not worth while at the present stage.

II. Boltzmann Transfer Equation

A. Formulation

To investigate the phenomenon of the generation of harmonics and combination frequencies in the current density in a plasma, one has to analyze the time dependence of the various components of the distribution function of electron velocities. The contribution of the ions to the current density is negligible because of their relatively large mass and hence low mobility. However, the ions and neutral molecules affect the electron velocity distribution because of the randomizing electron collisions caused by them. Moreover, the velocity distribution for either of the heavier species may be assumed to be Maxwellian, corresponding to the temperature of the gas. This is justified because neither of them is appreciably affected by the external electric fields and because the change in the kinetic energy of these particles in collisions with electrons is negligible as a result of the large mass ratio involved.

If $f(v_x, v_y, v_z, x, y, z, t)$ denotes the distribution function of electron velocities, then $f\, dv_x\, dv_y\, dv_z\, dx\, dy\, dz$ gives the average number of electrons in a small volume element $dv_x\, dv_y\, dv_z\, dx\, dy\, dz$ of the six-dimensional phase space. For investigating problems in which the concept of a distribution function of electron velocities is useful, the starting point is the Boltzmann's transfer equation, viz.,

$$\frac{\partial f}{\partial t} + \mathbf{v} \cdot \nabla f + \boldsymbol{\alpha}' \cdot \nabla_v f = \left(\frac{\partial f}{\partial t}\right)_c \tag{2.1}$$

where $\mathbf{v}$ is the electron velocity, $\boldsymbol{\alpha}'$ is the acceleration of electrons, t is the time, ∇ and ∇_v are respectively the gradients in the position and velocity spaces, and $(\partial f/\partial t)_c$ is the rate of change of f due to collisions. The first term on the left side arises because of the time dependent or transient nature of the external electric field. The second term denotes the effect of the density and temperature gradients and the third term covers the effects of the external electric and magnetic fields that accelerate the electrons.

The following discussion will be limited to the case in which the electron density gradient, the temperature gradient, and the electrical vector lie in a plane, and in which a static magnetic field is applied perpendicular to this plane. Thus, one may have

$$E_Z = 0, \qquad \frac{\partial N}{\partial z} = \frac{\partial T}{\partial z} = 0$$

This case corresponds to the propagation of a plane-polarized electromagnetic wave along the direction of the external magnetic field; the consideration of the general case, when the magnetic field is not along the direction of propagation, adds little to an understanding of the physics of the phenomena under study.

The acceleration $\boldsymbol{\alpha}'$ of electrons due to an electrical vector $\mathbf{E}$ and a magnetic field $\mathbf{B}$ is given by

$$-\boldsymbol{\alpha}' = \mathbf{a} + \mathbf{v} \times \boldsymbol{\omega} \tag{2.2}$$

where $\mathbf{a} = e\mathbf{E}/m$, $\boldsymbol{\omega} = e\mathbf{B}/mc$, e and m are the electronic charge and mass, respectively, and c is the velocity of light in vacuum. For the present case, $a_z = 0$ and $\omega_x = \omega_y = 0$.

In the nonequilibrium stationary states of interest to the phenomenon of electromagnetic wave propagation in plasmas, the random velocity of the electron is much larger than its directed (drift) velocity; in this case it is convenient to expand the distribution function of electron velocities in a series of spherical harmonics in velocity space. The spherical harmonic expansion can in turn be readily converted to a fully symmetric base-tensor form (62a), viz.,

$$f = \sum Y_{lms} f^l_{ms} = \sum_l \frac{f^l_i \, v^l}{v^l} \tag{2.3}$$

where the symbol l stands for the lth-order dot product. Substituting Eqs. (2.2) and (2.3) in (2.1), limiting ourselves to the particular case mentioned above, viz.,

$$\partial T/\partial z = \partial N/\partial z = 0, \qquad E_z = 0, \qquad B_x = B_y = 0$$

(neglecting f^3 and other higher order terms in the expansion of f), proceeding as in Shkarofsky *et al.* (62a) (remembering that the f^l are irreducible base tensors; i.e., give zero by the operation of contraction), and following

Ginzburg and Gurevich (9) for the value of the collision term, one obtains the following system of linked equations:

$$\frac{\partial f^0}{\partial t} + \frac{v}{3}\left(\frac{\partial f_x{}^1}{\partial x} + \frac{\partial f_y{}^1}{\partial y}\right) = \frac{1}{3v^2}\frac{\partial}{\partial v}\left[v^2(a_x f_x{}^1 + a_y f_y{}^1)\right] + \left(\frac{\partial f^0}{\partial t}\right)_c \qquad (2.4\text{a})$$

$$\frac{\partial f_x{}^1}{\partial t} + v f_x{}^1 + \omega_B f_y{}^1 = -v\frac{\partial f^0}{\partial x} + a_x\frac{\partial f^0}{\partial v} - \frac{2v}{5}\left(\frac{\partial f_{xx}^2}{\partial x} + \frac{\partial f_{yx}^2}{\partial y}\right)$$

$$+ \frac{2}{5v^3}\frac{\partial}{\partial v}\left[v^3(a_x f_{xx}^2 + a_y f_{yx}^2)\right] \qquad (2.4\text{b})$$

$$\frac{\partial f_y{}^1}{\partial t} + v f_y{}^1 - \omega_B f_x{}^1 = -v\frac{\partial f^0}{\partial y} + a_y\frac{\partial f^0}{\partial v} - \frac{2v}{5}\left(\frac{\partial f_{xy}^2}{\partial x} + \frac{\partial f_{yy}^2}{\partial v}\right)$$

$$+ \frac{2}{5v^3}\frac{\partial}{\partial v}\left[v^3(a_y f_{yy}^2 + a_x f_{xy}^2)\right] \qquad (2.4\text{c})$$

$$\frac{\partial f_{xx}^2}{\partial t} + v f_{xx}^2 - 2\omega_B f_{yx}^2 = -\frac{v}{3}\left(2\frac{\partial f_x{}^1}{\partial x} - \frac{\partial f_y{}^1}{\partial y}\right) + \frac{v}{3}\frac{\partial}{\partial v}(2a_x f_x{}^1 - a_y f_y{}^1)$$

$$(2.4\text{d1})$$

$$\frac{\partial f_{yy}^2}{\partial t} + v f_{yy}^2 + 2\omega_B f_{xy}^2 = -\frac{v}{3}\left(2\frac{\partial f_y{}^1}{\partial y} - \frac{\partial f_x{}^1}{\partial x}\right) + \frac{v}{3}\frac{\partial}{\partial v}(2a_y f_y{}^1 - a_x f_x{}^1)$$

$$(2.4\text{d2})$$

$$\frac{\partial f_{xy}^2}{\partial t} + v f_{xy}^2 + \omega_B(f_{xx}^2 - f_{yy}^2) = -\frac{v}{2}\left(\frac{\partial f_x{}^1}{\partial y} + \frac{\partial f_y{}^1}{\partial x}\right) + \frac{v}{2}\frac{\partial}{\partial v}(a_x f_y{}^1 + a_y f_x{}^1)$$

$$(2.4\text{e})$$

where $f_{xy}^2 = f_{yx}^2$, $\omega_B = \omega_z = eB_0/mc$, $v \equiv v(v)$ is the elastic collision frequency of electrons with heavy particles and $(\partial f^0/\partial t)_c$ denotes the effect of collisions on the isotropic part of the distribution function of electron velocities. In most of the work carried out to date on magnetoplasmas, some of the terms in the above equations have been neglected; the numerical results presented for magnetoplasmas should therefore be treated with caution.

It is perhaps pertinent to point out here that the retention of terms up to f^2 only is a necessary and sufficient condition for obtaining correct lowest order magnitudes of the generated harmonic and combination frequency components in the current density, provided one limits one's analysis up to the third harmonic and the second-order combination frequencies ($\omega_1 \pm 2\omega_2$). Thus, for higher harmonics and higher order combination frequency terms, $f^3, f^4, \ldots,$etc., may also become important. The proof of this statement for third harmonic generation in a homogeneous isotropic plasma is given in Appendix F.

In the following subsection, we shall try to summarize briefly the time dependence of the various parts of the distribution function under different conditions by deriving some recurrence relations between them in a manner first given by Margenau and Hartman (2). These recurrence relations could be used as such to investigate the harmonic and combination frequency components in the current density. However, we shall prefer to start *ab initio* in each case, using Eqs. (2.4a–e). The following subsection helps only in writing down the time dependence of various parts of the distribution function.

B. Time Dependence of the Components of Distribution Function

The time-dependent nature of the isotropic part of the distribution function of electron velocities in a uniform plasma, in the presence of a sufficiently strong low frequency microwave field, and the resulting generation of harmonics and combination frequencies in the current density, was first analyzed in their pioneering paper by Margenau and Hartman (2). They investigated in detail the time dependence of the various parts of the distribution function by choosing a general Fourier expansion for f and utilizing the Boltzmann's transfer equation for further analysis. In this subsection, we have extended their treatment to include the effect of an additional dc electric field and another alternating electric field of different frequency.

Consider a uniform plasma placed under the simultaneous influence of a dc electric field and two microwave fields of frequencies ω_1 and ω_2, all directed along the x axis; i.e.,

$$E_x = E_d + E_1 \exp(i\omega_1 t) + E_2 \exp(i\omega_2 t)$$
$$E_y = E_z = 0 \tag{2.5}$$

It is assumed further for simplicity that the E_1 field is sufficiently weak so that it does not give any higher order components of the current density. The quantities E_1 and E_2 may be assumed to be complex to account for both the amplitude and the phase of the electrical vector. The Boltzmann's transfer equation for electrons in terms of the variable $\Delta_x = v_x/v$ may be written as

$$\frac{\partial f}{\partial t} - \{a_d + a_1 \exp(i\omega_1 t) + a_2 \exp(i\omega_2 t)\}$$

$$\left[\Delta_x \frac{\partial f}{\partial v} + \frac{1 - \Delta_x^2}{v} \frac{\partial f}{\partial \Delta_x}\right] = \left(\frac{\partial f}{\partial t}\right)_c \tag{2.6}$$

where $a_d = eE_d/m$, $a_1 = eE_1/m$, and $a_2 = eE_2/m$. A very general expansion of the distribution function (taking account of the fact that the E_1 field is weak) is

$$f = \sum_{k=0}^{\infty} P_k(\Delta_x) f^k(v, t)$$

$$= \sum_{k=0}^{\infty} P_k(\Delta_x)\Big\{ f_0^k + f_{11}^k \exp(i\omega_1 t) + \sum_{m=1}^{\infty} f_{2m}^k \exp(im\omega_2 t)$$

$$+ \sum_{n=1}^{\infty} [f_{1n}^{k+} \exp\{i(\omega_1 + n\omega_2)t\} + f_{1n}^{k-} \exp\{i(\omega_1 - n\omega_2)t\}]\Big\} \tag{2.7}$$

The electron density is then given by

$$N = 4\pi \int_0^{\infty} \Big[f_0^0 + \sum_m f_{2m}^0 \exp(im\omega_2 t)$$

$$+ \sum_n \{f_{1n}^{0+} \exp\{i(\omega_1 + n\omega_2)t\} + f_{1n}^{0-} \exp\{i(\omega_1 - n\omega_2)t\}\}\Big] v^2 \, dv \tag{2.8}$$

and the current density by

$$J = - (4\pi e/3) \int_0^{\infty} v^3 \Big[f_0^1 + f_{11}^1 \exp(i\omega_1 t) + \sum_m f_{2m}^1 \exp(im\omega_2 t)$$

$$+ \sum_n \{f_{1n}^{1+} \exp\{i(\omega_1 + n\omega_2)t\} + f_{1n}^{1-} \exp\{i(\omega_1 - n\omega_2)t\}\}\Big] dv \tag{2.9}$$

Substituting (2.7) in (2.6) and using the orthogonal properties of the Legendre polynomials as well as the Fourier terms, one obtains the following set of recurrence relations:

For $m, n > 1$:

$$\left(\frac{\partial f^k_{2,m}}{\partial t}\right)_c = im\omega_2 f^k_{2,m} - \frac{k+1}{2k+3} v^{-(k+2)} \frac{d}{dv}\left[v^{k+2}\left(a_d f^{k+1}_{2,m} + \frac{a_2}{2} f^{k+1}_{2,m-1}\right.\right.$$

$$\left.+ \frac{\tilde{a}_2}{2} f^{k+1}_{2,m+1} + \frac{\tilde{a}_1}{2} f^{k+1,+}_{1,m} + \frac{a_1}{2} \tilde{f}^{k+1,-}_{1,m}\right)\right]$$

$$- \frac{k}{2k-1} v^{k-1} \frac{d}{dv}\left[v^{1-k}\left(a_d f^{k-1}_{2,m} + \frac{a_2}{2} f^{k-1}_{2,m-1}\right.\right.$$

$$\left.\left.+ \frac{\tilde{a}_1}{2} f^{k-1,+}_{1,m} + \frac{a_1}{2} \tilde{f}^{k-1,-}_{1,m} + \frac{\tilde{a}_2}{2} f^{k-1}_{2,m+1}\right)\right] \tag{2.10a}$$

$$\left(\frac{\partial f^{k,+}_{1,n}}{\partial t}\right)_c = i(\omega_1 + n\omega_2) f^{k,+}_{1,n} - \frac{k+1}{2k+3} v^{-(k+2)}$$

$$\times \frac{d}{dv}\left[v^{k+2}\left(a_d f^{k+1,+}_{1,n} + \frac{a_1}{2} f^{k+1}_{2,n} + \frac{a_2}{2} f^{k+1,+}_{1,n} + \frac{\tilde{a}_2}{2} f^{k+1,+}_{1,n+1}\right)\right]$$

$$- \frac{k}{2k-1} v^{k-1} \frac{d}{dv}\left[v^{1-k}\left(a_d f^{k-1,+}_{1,n} + \frac{a_1}{2} f^{k-1}_{2,n}\right.\right.$$

$$\left.\left.+ \frac{a_2}{2} f^{k-1,+}_{1,n-1} + \frac{\tilde{a}_2}{2} f^{k-1,+}_{1,n+1}\right)\right] \tag{2.10b}$$

$$\left(\frac{\partial f^{k,-}_{1,n}}{\partial t}\right)_c = i(\omega_1 - n\omega_2) f^{k,-}_{1,n} - \frac{k+1}{2k+3} v^{-(k+2)}$$

$$\times \frac{d}{dv}\left[v^{k+2}\left(a_d f^{k+1,-}_{1,n} + \frac{a_1}{2} \tilde{f}^{k+1}_{2,n} + \frac{\tilde{a}_2}{2} f^{k+1,-}_{1,n-1} + \frac{a_2}{2} f^{k+1,-}_{1,n+1}\right)\right]$$

$$- \frac{k}{2k-1} v^{k-1} \frac{d}{dv}\left[v^{1-k}\left(a_d f^{k-1,-}_{1,n} + \frac{a_1}{2} \tilde{f}^{k-1}_{2,n}\right.\right.$$

$$\left.\left.+ \frac{\tilde{a}_2}{2} f^{k-1,-}_{1,n-1} + \frac{a_2}{2} f^{k-1,-}_{1,n+1}\right)\right] \tag{2.10c}$$

For $m, n = 1$:

$$\left(\frac{\partial f^k_{2,1}}{\partial t}\right)_c = i\omega_2 f^k_{2,1} - \frac{k+1}{2k+3} v^{-(k+2)} \frac{d}{dv}\left[v^{k+2}\left(a_d f^{k+1}_{2,1} + a_2 f^{k+1}_0\right.\right.$$

$$\left.\left.+ \frac{\tilde{a}_2}{2} f^{k+1}_{2,2} + \frac{\tilde{a}_1}{2} f^{k+1,+}_{1,2} + \frac{a_1}{2} \tilde{f}^{k+1,-}_{1,2}\right)\right] - \frac{k}{2k-1} v^{k-1} \frac{d}{dv}$$

$$\times \left[v^{1-k}\left(a_d f^{k-1}_{2,1} + a_2 f^{k-1}_0 + \frac{\tilde{a}_2}{2} f^{k-1}_{2,2} + \frac{\tilde{a}_1}{2} f^{k-1,+}_{1,2} + \frac{a_1}{2} \tilde{f}^{k-1,-}_{1,2}\right)\right] \tag{2.11a}$$

$$\left(\frac{\partial f_{1,1}^{k}}{\partial t}\right)_{c} = i\omega_{1} f_{1,1}^{k} - \frac{k+1}{2k+3} v^{-(k+2)} \frac{d}{dv}\left[v^{k+2}\left(a_{d} f_{1,1}^{k+1} + a_{1} f_{0}^{k+1}\right.\right.$$

$$\left.\left. + \frac{\tilde{a}_{2}}{2} f_{1,1}^{k+1,+} + \frac{a_{2}}{2} f_{1,1}^{k+1,-}\right)\right] - \frac{k}{2k-1} v^{k-1}$$

$$\times \frac{d}{dv}\left[v^{1-k}\left(a_{d} f_{1,1}^{k-1} + a_{1} f_{0}^{k-1} + \frac{\tilde{a}_{2}}{2} f_{1,1}^{k-1,+} + \frac{a_{2}}{2} f_{1,1}^{k-1,-}\right)\right]$$

$$(2.11b)$$

$$\left(\frac{\partial f_{1,1}^{k,+}}{\partial t}\right)_{c} = i(\omega_{1} + \omega_{2}) f_{1,1}^{k,+} - \frac{k+1}{2k+3} v^{-(k+2)}$$

$$\times \frac{d}{dv}\left[v^{k+2}\left(a_{d} f_{1,1}^{k+1,+} + \frac{a_{1}}{2} f_{2,1}^{k+1} + \frac{\tilde{a}_{2}}{2} f_{1,2}^{k+1,+} + \frac{a_{2}}{2} f_{1,1}^{k+1}\right)\right]$$

$$- \frac{k}{2k-1} v^{k-1} \frac{d}{dv}\left[v^{1-k}\left(a_{d} f_{1,\cdot1}^{k-1,+} + \frac{a_{1}}{2} f_{2,1}^{k-1}\right.\right.$$

$$\left.\left. + \frac{\tilde{a}_{2}}{2} f_{1,2}^{k-1,+} + \frac{a_{2}}{2} f_{1,1}^{k-1}\right)\right]$$

$$(2.11c)$$

$$\left(\frac{\partial f_{1,1}^{k,-}}{\partial t}\right)_{c} = i(\omega_{1} - \omega_{2}) f_{1,1}^{k,-} - \frac{k+1}{2k+3} v^{-(k+2)}$$

$$\times \frac{d}{dv}\left[v^{k+2}\left(a_{d} f_{1,1}^{k+1,-} + \frac{a_{1}}{2} \tilde{f}_{2,1}^{k+1} + \frac{a_{2}}{2} f_{1,2}^{k+1,-} + \frac{\tilde{a}_{2}}{2} f_{1,1}^{k+1}\right)\right]$$

$$- \frac{k}{2k-1} v^{k-1} \frac{d}{dv}\left[v^{1-k}\left(a_{d} f_{1,1}^{k-1,-} + \frac{a_{1}}{2} \tilde{f}_{2,1}^{k-1}\right.\right.$$

$$\left.\left. + \frac{a_{2}}{2} f_{1,2}^{k-1,-} + \frac{\tilde{a}_{2}}{2} f_{1,1}^{k-1}\right)\right]$$

$$(2.11d)$$

For, $m = 0$:

$$\left(\frac{\partial f_{0}^{k}}{\partial t}\right)_{c} = - \frac{k+1}{2k+3} v^{-(k+2)} \frac{d}{dv}\left[v^{k+2}\left(a_{d} f_{0}^{k+1} + \frac{\tilde{a}_{1}}{2} f_{1,1}^{k+1} + \frac{\tilde{a}_{2}}{2} f_{2,1}^{k+1}\right)\right]$$

$$- \frac{k}{2k-1} v^{k-1} \frac{d}{dv}\left[v^{1-k}\left(a_{d} f_{0}^{k-1} + \frac{\tilde{a}_{1}}{2} f_{1,1}^{k-1} + \frac{\tilde{a}_{2}}{2} f_{2,1}^{k-1}\right)\right]$$

$$(2.12)$$

If the electron density, following Margenau and Hartman (2), is assumed to be constant in time, then

$$4\pi \int_0^\infty v^2 f_0{}^0 \, dv = N$$

$$\int_0^\infty f_{2,m}^0 v^2 \, dv = \int_0^\infty f_{1,n}^{0\pm} v^2 \, dv = 0 \qquad m, n > 0$$

As an aid to further discussion, one should keep in mind that the terms with two superscripts (e.g., $f_{1,1}^{k+1,+}, f_{1,1}^{k+1,-}$, etc.) are the combination frequency components of the distribution function, while those with only one superscript correspond to the fundamental or harmonic terms. Consider first the case where there is no dc electric field. In this case, the recurrence relations are such that they connect in the same equation only those harmonic or fundamental terms $f_{2,m}^k$ for which the sum of upper and lower indices $(k + m)$ is either odd or even, and only those combination frequency terms f_{1n}^{k+} for which $(k + n)$ is exactly the reverse or, respectively, even and odd. Now, for a finite electron density N, $f_0{}^0$ has to be finite. Since the sum of the upper and lower indices for $f_0{}^0$ is zero, it will connect only with those harmonic terms f_{2m}^k for which $(k + m)$ is even and only those combination frequency terms f_{1n}^{k+} for which $(k + n)$ is odd. Thus, for an electron density independent of time, one may immediately write the following time dependence (for the present case in the absence of a dc electric field):

$$f^0 = f_0{}^0 + f_{22}^0 \exp(2i\omega_2 t) + f_{11}^{0+} \exp\{i(\omega_1 + \omega_2)t\}$$
$$+ f_{11}^{0-} \exp\{i(\omega_1 - \omega_2)t\} + \cdots$$

$$f^1 = f_{11}^1 \exp(i\omega_1 t) + f_{21}^1 \exp(i\omega_2 t) + f_{23}^1 \exp(3i\omega_2 t)$$
$$+ f_{12}^{1+} \exp\{i(\omega_1 + 2\omega_2)t\} + f_{12}^{1-} \exp\{i(\omega_1 - 2\omega_2)t\} + \cdots$$

and

$$f^2 = f_0{}^2 + f_{22}^2 \exp(2i\omega_2 t) + f_{11}^{2+} \exp\{i(\omega_1 + \omega_2)t\}$$
$$+ f_{11}^{2-} \exp\{i(\omega_1 - \omega_2)t\} + \cdots$$

When both microwave fields are strong, it can be shown in a similar manner that the distribution function has the following components (the symbols have been changed slightly for future convenience):

$$f^0 = f_0{}^0 + f_{12}^0 \exp(2i\omega_1 t) + f_{22}^0 \exp(2i\omega_2 t)$$
$$+ f_{12}^{0+} \exp\{i(\omega_1 + \omega_2)t\} + f_{12}^{0-} \exp\{i(\omega_1 - \omega_2)t\} + \cdots \qquad (2.13a)$$

$$f^1 = f_{11}^1 \exp(i\omega_1 t) + f_{13}^1 \exp(3i\omega_1 t) + f_{21}^1 \exp(i\omega_2 t) + f_{23}^1 \exp(3i\omega_2 t)$$
$$+ f_{12}^{1+'} \exp\{i(\omega_1 + 2\omega_2)t\} + f_{12}^{1-'} \exp\{i(\omega_1 - 2\omega_2)t\}$$
$$+ f_{21}^{1+'} \exp\{i(\omega_2 + 2\omega_1)t\} + f_{21}^{1-'} \exp\{i(\omega_2 - 2\omega_1)t\} + \cdots \qquad (2.13b)$$

and

$$f^2 = f_0{}^2 + f_{12}^2 \exp(2i\omega_1 t) + f_{22}^2 \exp(2i\omega_2 t)$$
$$+ f_{12}^{2+} \exp\{i(\omega_1 + \omega_2)t\} + f_{12}^{2-} \exp\{i(\omega_1 - \omega_2)t\} + \cdots \quad (2.13c)$$

Physically, these expansions show that two microwave fields alone can generate only odd harmonics and second-order combination frequencies $(\omega_1 \pm 2\omega_2, \omega_2 \pm 2\omega_1)$ in the current density; moreover, no direct component of the current density is predicted (as is to be expected). Rosen (3) used an expansion similar to the one given above for investigating the third harmonic generation in a plasma; Vilenskii (5, 6) and Gurevich (8) used it for analyzing the generation of combination frequencies in a plasma. However, the importance of the terms with frequencies $(\omega_1 \pm \omega_2)$ in f^0 and f^2, in determining the magnitude of the combination frequency components in the current density, was first pointed out by Sodha and Kaw (19).

In the presence of a dc electric field, it is seen that the recurrence relations are such that they connect in the same equation the harmonic functions f_{2m}^k having odd $(k + m)$ to those with even $(k + m)$ and to the combination frequency functions $f_{1,n}^{k;+}$ with both odd and even $(k + n)$. This means that even for the case of a time-independent electron density, one has to choose

$$f^0 = f_0{}^0 + f_{11}^0 \exp(i\omega_1 t) + f_{12}^0 \exp(2i\omega_1 t) + f_{13}^0 \exp(3i\omega_1 t)$$
$$+ f_{21}^0 \exp(i\omega_2 t) + f_{22}^0 \exp(2i\omega_2 t) + f_{23}^0 \exp(3i\omega_2 t)$$
$$+ f_{12}^{0+} \exp\{i(\omega_1 + \omega_2)t\} + f_{12}^{0-} \exp\{i(\omega_1 - \omega_2)t\}$$
$$+ f_{12}^{0+\prime} \exp\{i(\omega_1 + 2\omega_2)t\} + f_{12}^{0-\prime} \exp\{i(\omega_1 - 2\omega_2)t\}$$
$$+ f_{21}^{0+\prime} \exp\{i(\omega_2 + 2\omega_1)t\} + f_{21}^{0-\prime} \exp\{i(\omega_2 - 2\omega_1)t\} + \cdots \quad (2.14a)$$

$$f^1 = f_0{}^1 + f_{11}^1 \exp(i\omega_1 t) + f_{12}^1 \exp(2i\omega_1 t) + f_{13}^1 \exp(3i\omega_1 t)$$
$$+ f_{21}^1 \exp(i\omega_2 t) + f_{22}^1 \exp(2i\omega_2 t) + f_{23}^1 \exp(3i\omega_2 t)$$
$$+ f_{12}^{1+} \exp\{i(\omega_1 + \omega_2)t\} + f_{12}^{1-} \exp\{i(\omega_1 - \omega_2)t\}$$
$$+ f_{12}^{1+\prime} \exp\{i(\omega_1 + 2\omega_2)t\} + f_{12}^{1-\prime} \exp\{i(\omega_1 - 2\omega_2)t\}$$
$$+ f_{21}^{1+\prime} \exp\{i(\omega_2 + 2\omega_1)t\} + f_{21}^{1-\prime} \exp\{i(\omega_2 - 2\omega_1)t\} + \cdots \quad (2.14b)$$

$$f^2 = f_0{}^2 + f_{11}^2 \exp(i\omega_1 t) + f_{12}^2 \exp(2i\omega_1 t) + f_{13}^2 \exp(3i\omega_1 t)$$
$$+ f_{21}^2 \exp(i\omega_2 t) + f_{22}^2 \exp(2i\omega_2 t) + f_{23}^2 \exp(3i\omega_2 t)$$
$$+ f_{12}^{2+} \exp\{i(\omega_1 + \omega_2)t\} + f_{12}^{2-} \exp\{i(\omega_1 - \omega_2)t\}$$
$$+ f_{12}^{2+\prime} \exp\{i(\omega_1 + 2\omega_2)t\} + f_{12}^{2-\prime} \exp\{i(\omega_1 - 2\omega_2)t\}$$
$$+ f_{21}^{2+\prime} \exp\{i(\omega_2 + 2\omega_1)t\} + f_{21}^{2-\prime} \exp\{i(\omega_2 - 2\omega_1)t\} \quad (2.14c)$$

Thus, in the presence of a dc field, even harmonics as well as the first-order combination frequencies $(\omega_1 \pm \omega_2)$ will be generated in addition to the usual odd harmonics and the combination frequencies $(\omega_1 \pm 2\omega_2)$, $(\omega_2 \pm 2\omega_1)$; moreover, a dc component of the current is also present, as is to be expected. Such an expansion was first used by Sodha and Kaw (23) in their investigation of the generation of harmonics and combination frequencies in the presence of an external dc electric field.

In the above analysis, the diffusion terms in the Boltzmann's equation have not been taken into account, thereby limiting the treatment to uniform plasmas only. Vilenskii (5, 6) and Wetzel (10) have investigated in detail the effect of the interaction of an electron density gradient with a microwave field on the time dependence of the first two components of the distribution function. When two microwave fields are present in an inhomogeneous plasma, it can be shown that an expansion similar to the one derived above for a homogeneous plasma in the presence of a dc field, Eq. (2.14) should be used. Such an expansion has been used by Sodha and Kaw (21) and Gupta (14, 15) for studying the second harmonic and the sum-and-difference frequency generation in the current density in an inhomogeneous plasma. The simpler problem of second harmonic generation in an inhomogeneous plasma, in the absence and presence of magnetic fields, had earlier been analyzed by Chiyoda (12) and by Sodha and Kaw (20, 22).

III. Harmonic and Combination Frequency Components in Current Density

The current density in a plasma is given by

$$\mathbf{J} = -e \int_{-\infty}^{+\infty} \int_{-\infty}^{+\infty} \int_{-\infty}^{+\infty} \mathbf{v} f \, dv_x \, dv_y \, dv_z \tag{3.1}$$

To obtain explicit expressions for the harmonic and combination frequency components of the current density, therefore, one should evaluate the various frequency components of $f^0, f^1, f^2, \ldots$, etc., by using Eqs. (2.4a–e) and the expected time dependence discussed in Section II-B.

In further discussion, the homogeneous and inhomogeneous plasmas will be treated separately. Further, among homogeneous plasmas also, a distinction will be made between neutral plasmas and plasmas with induced excessive ionization. The effect of a magnetic field will also be discussed in some cases.

A. Homogeneous Neutral Plasmas

By a neutral plasma, we mean one in which there are no excess free charges of any kind and the Laplace's equation is satisfied at all points.

1. Two Microwave Fields Only

Consider first the case of a plasma placed under the influence of an electric field

$$\mathbf{E} = \mathbf{E}_1 \exp(i\omega_1 t) + \mathbf{E}_2 \exp(i\omega_2 t) \tag{3.2}$$

directed along the x direction. Equations (2.4a–e) for the present case will have the forms

$$\frac{\partial f^0}{\partial t} = \frac{a}{3v^2} \frac{\partial}{\partial v}(v^2 f^1) + \left(\frac{\partial f^0}{\partial t}\right)_c \tag{3.3a}$$

$$\frac{\partial f^1}{\partial t} + vf^1 = a\left[\frac{\partial f^0}{\partial v} + \frac{2}{5v^3}\frac{\partial}{\partial v}(v^3 f^2)\right] \tag{3.3b}$$

$$\frac{\partial f^2}{\partial t} + vf^2 = \frac{2v}{3}a\frac{\partial}{\partial v}\left(\frac{f^1}{v}\right) \tag{3.3c}$$

where $a = a_1 \exp(i\omega_1 t) + a_2 \exp(i\omega_2 t)$ and the expected time dependence of f^0, f^1, and f^2 are given by Eq. (2.13).

Vilenskii (5, 6) considered the simultaneous influence of a weak field of frequency ω_1 and a strong field of frequency ω_2 on a plasma which led to the generation of combination frequencies $(\omega_1 \pm 2\omega_2)$. He did not however realize the importance of the contributions arising from the higher order term f^2; this was pointed out by Gurevich (8). Sodha and Kaw (19) showed that the terms with frequencies $(\omega_1 \pm \omega_2)$ in f^0 and f^2 also give appreciable contributions to the combination frequency components of current density; these terms had been neglected by Vilenskii (5, 6) and Gurevich (8). Rosen (3) derived an explicit expression for the third-harmonic component in the current density without taking the important contribution of f^2 terms into account. Sodha and Kaw (17) investigated the third harmonic generation with the f^2 terms; however, they used a nonconvergent expansion for the distribution function, which led to some erroneous results. Krenz (13) has recently given a correct theoretical investigation of the third harmonic generation, taking account of f^2 terms. In what follows, the authors have given a comprehensive treatment that considers all important terms and have derived explicit expressions for the third-harmonic and combination frequency components of the current density.

Substituting Eq. (2.13) in (3.3a–c), equating the time-independent terms and the coefficients of various frequency terms on both sides of each resulting equation, and solving, one obtains explicit expressions for the different frequency components of f^0, f^1, f^2, etc. (For details please see Appendix A). Using the approximations

$$f^1_{13}, f^1_{23}, f^{1\pm'}_{12}, f^{1\pm'}_{21} \ll f^1_{11} \quad \text{or} \quad f^1_{21}$$

and

$$f_{12}^0, f_{22}^0, f_{12}^{0\pm}, f_{12}^2, f_{22}^2, f_{12}^{2\pm} \ll f_0^0$$

which correspond to the assumption that the magnitudes of the harmonic and combination frequency components of the current density are much smaller than those of the fundamentals, one obtains finally for the components of f^1, the following equations:

$$f_{11}^1 = \frac{a_1}{v + i\omega_1} \frac{\partial f_0^0}{\partial v}, \qquad f_{21}^1 = \frac{a_2}{v + i\omega_2} \frac{\partial f_0^0}{\partial v} \tag{3.4a,b}$$

$$
\begin{aligned}
f_{13}^1 = \frac{a_1^3}{v + 3i\omega_1} &\left[\frac{1}{24i\omega_1} \frac{\partial}{\partial v} \left\{ \frac{1}{v^2} \frac{\partial}{\partial v} \left(\frac{v^2}{v + i\omega_1} \frac{\partial f_0^0}{\partial v} \right) \right\} \right. \\
&+ \frac{1}{15} \frac{\partial}{\partial v} \left\{ \frac{v}{v + 2i\omega_1} \frac{\partial}{\partial v} \left(\frac{1}{v + i\omega_1} \cdot \frac{1}{v} \frac{\partial f_0^0}{\partial v} \right) \right\} \\
&+ \left. \frac{1}{5(v + 2i\omega_1)} \frac{\partial}{\partial v} \left(\frac{1}{v + i\omega_1} \cdot \frac{1}{v} \frac{\partial f_0^0}{\partial v} \right) \right]
\end{aligned}
\tag{3.4c}
$$

$$
\begin{aligned}
f_{23}^1 = \frac{a_2^3}{v + 3i\omega_2} &\left[\frac{1}{24i\omega_2} \frac{\partial}{\partial v} \left\{ \frac{1}{v^2} \frac{\partial}{\partial v} \left(\frac{v^2}{v + i\omega_2} \frac{\partial f_0^0}{\partial v} \right) \right\} \right. \\
&+ \frac{1}{15} \frac{\partial}{\partial v} \left\{ \frac{v}{v + 2i\omega_2} \frac{\partial}{\partial v} \left(\frac{v}{v + i\omega_2} \cdot \frac{1}{v} \frac{\partial f_0^0}{\partial v} \right) \right\} \\
&+ \left. \frac{1}{5(v + 2i\omega_2)} \frac{\partial}{\partial v} \left(\frac{1}{v + i\omega_2} \cdot \frac{1}{v} \frac{\partial f_0^0}{\partial v} \right) \right]
\end{aligned}
\tag{3.4d}
$$

$$
\begin{aligned}
f_{12}^{1+\prime} = \frac{a_1 a_2^2}{v + i(\omega_1 + 2\omega_2)} &\left[\frac{1}{24i\omega_2} \frac{\partial}{\partial v} \left\{ \frac{1}{v^2} \frac{\partial}{\partial v} \left(\frac{v^2}{v + i\omega_2} \frac{\partial f_0^0}{\partial v} \right) \right\} \right. \\
&+ \frac{1}{15} \frac{\partial}{\partial v} \left\{ \frac{v}{v + 2i\omega_2} \frac{\partial}{\partial v} \left(\frac{1}{v + i\omega_2} \cdot \frac{1}{v} \frac{\partial f_0^0}{\partial v} \right) \right\} \\
&+ \frac{1}{5(v + 2i\omega_2)} \frac{\partial}{\partial v} \left(\frac{1}{v + i\omega_2} \cdot \frac{1}{v} \frac{\partial f_0^0}{\partial v} \right) \\
&+ \frac{1}{12i(\omega_1 + \omega_2)} \frac{\partial}{\partial v} \left\{ \frac{1}{v^2} \frac{\partial}{\partial v} \left[v^2 \left(\frac{1}{v + i\omega_1} + \frac{1}{v + i\omega_2} \right) \frac{\partial f_0^0}{\partial v} \right] \right\} \\
&+ \frac{1}{15} \frac{\partial}{\partial v} \left\{ \frac{v}{v + i(\omega_1 + \omega_2)} \frac{\partial}{\partial v} \left[\frac{1}{v} \left(\frac{1}{v + i\omega_1} + \frac{1}{v + i\omega_2} \right) \frac{\partial f_0^0}{\partial v} \right] \right\} \\
&+ \left. \frac{1}{5[v + i(\omega_1 + \omega_2)]} \frac{\partial}{\partial v} \left\{ \frac{1}{v} \left(\frac{1}{v + i\omega_1} + \frac{1}{v + i\omega_2} \right) \frac{\partial f_0}{\partial v} \right\} \right]
\end{aligned}
\tag{3.4e}
$$

$$
\begin{aligned}
f_{12}^{1-\prime} = \frac{a_1 \tilde{a}_2{}^2}{v + i(\omega_1 - 2\omega_2)} \Bigg[&- \frac{1}{24 i \omega_2} \frac{\partial}{\partial v} \left\{ \frac{1}{v^2} \frac{\partial}{\partial v} \left(\frac{v^2}{v - i\omega_2} \frac{\partial f_0{}^0}{\partial v} \right) \right\} \\
&+ \frac{1}{15} \frac{\partial}{\partial v} \left\{ \frac{v}{v - 2i\omega_2} \frac{\partial}{\partial v} \left(\frac{1}{v - i\omega_2} \cdot \frac{1}{v} \frac{\partial f_0{}^0}{\partial v} \right) \right\} \\
&+ \frac{1}{5(v - 2i\omega_2)} \frac{\partial}{\partial v} \left(\frac{1}{v - i\omega_2} \cdot \frac{1}{v} \frac{\partial f_0{}^0}{\partial v} \right) \\
&+ \frac{1}{12 i(\omega_1 - \omega_2)} \frac{\partial}{\partial v} \left\{ \frac{1}{v^2} \frac{\partial}{\partial v} \left[v^2 \left(\frac{1}{v + i\omega_1} + \frac{1}{v - i\omega_2} \right) \frac{\partial f_0{}^0}{\partial v} \right] \right\} \\
&+ \frac{1}{15} \frac{\partial}{\partial v} \left\{ \frac{v}{v + i(\omega_1 - \omega_2)} \frac{\partial}{\partial v} \left[\frac{1}{v} \left(\frac{1}{v + i\omega_1} + \frac{1}{v - i\omega_2} \right) \frac{\partial f_0{}^0}{\partial v} \right] \right\} \\
&+ \frac{1}{5[v + i(\omega_1 - \omega_2)]} \frac{\partial}{\partial v} \left\{ \frac{1}{v} \left(\frac{1}{v + i\omega_1} + \frac{1}{v - i\omega_2} \right) \frac{\partial f_0{}^0}{\partial v} \right\} \Bigg]
\end{aligned}
$$

$$\tag{3.4f}$$

$$
\begin{aligned}
f_{21}^{1+\prime} = \frac{a_2 a_1{}^2}{v + i(\omega_2 + 2\omega_1)} \Bigg[&\frac{1}{24 i \omega_1} \frac{\partial}{\partial v} \left\{ \frac{1}{v^2} \frac{\partial}{\partial v} \left(\frac{v^2}{v + i\omega_1} \frac{\partial f_0{}^0}{\partial v} \right) \right\} \\
&+ \frac{1}{15} \frac{\partial}{\partial v} \left\{ \frac{v}{v + 2i\omega_1} \frac{\partial}{\partial v} \left(\frac{1}{v + i\omega_1} \cdot \frac{1}{v} \frac{\partial f_0{}^0}{\partial v} \right) \right\} \\
&+ \frac{1}{5(v + 2i\omega_1)} \frac{\partial}{\partial v} \left(\frac{1}{v + i\omega_1} \cdot \frac{1}{v} \frac{\partial f_0{}^0}{\partial v} \right) \\
&+ \frac{1}{12 i(\omega_1 + \omega_2)} \frac{\partial}{\partial v} \left\{ \frac{1}{v^2} \frac{\partial}{\partial v} \left[v^2 \left(\frac{1}{v + i\omega_1} + \frac{1}{v + i\omega_2} \right) \frac{\partial f_0{}^0}{\partial v} \right] \right\} \\
&+ \frac{1}{15} \frac{\partial}{\partial v} \left\{ \frac{v}{v + i(\omega_1 + \omega_2)} \frac{\partial}{\partial v} \left[\frac{1}{v} \left(\frac{1}{v + i\omega_1} + \frac{1}{v + i\omega_2} \right) \frac{\partial f_0{}^0}{\partial v} \right] \right\} \\
&+ \frac{1}{5[v + i(\omega_1 + \omega_2)]} \frac{\partial}{\partial v} \left\{ \frac{1}{v} \left(\frac{1}{v + i\omega_1} + \frac{1}{v + i\omega_2} \right) \frac{\partial f_0{}^0}{\partial v} \right\} \Bigg]
\end{aligned}
$$

$$\tag{3.4g}$$

and

$$
\begin{aligned}
f_{21}^{1-\prime} = \frac{a_2 \tilde{a}_1{}^2}{v + i(\omega_2 - 2\omega_1)} \Bigg[&- \frac{1}{24 i \omega_1} \frac{\partial}{\partial v} \left\{ \frac{1}{v^2} \frac{\partial}{\partial v} \left(\frac{v^2}{v - i\omega_1} \frac{\partial f_0{}^0}{\partial v} \right) \right\} \\
&+ \frac{1}{15} \frac{\partial}{\partial v} \left\{ \frac{v}{v - 2i\omega_1} \frac{\partial}{\partial v} \left[\frac{1}{v - i\omega_1} \cdot \frac{1}{v} \frac{\partial f_0{}^0}{\partial v} \right] \right\} \\
&+ \frac{1}{5(v - 2i\omega_1)} \frac{\partial}{\partial v} \left\{ \frac{1}{v - i\omega_1} \cdot \frac{1}{v} \frac{\partial f_0{}^0}{\partial v} \right\}
\end{aligned}
$$

$$- \frac{1}{12i(\omega_1 - \omega_2)} \frac{\partial}{\partial v} \left\{ \frac{1}{v^2} \frac{\partial}{\partial v} \left[v^2 \left(\frac{1}{v - i\omega_1} + \frac{1}{v + i\omega_2} \right) \frac{\partial f_0{}^0}{\partial v} \right] \right\}$$

$$+ \frac{1}{15} \frac{\partial}{\partial v} \left\{ \frac{v}{v - i(\omega_1 - \omega_2)} \frac{\partial}{\partial v} \left[\frac{1}{v} \left(\frac{1}{v - i\omega_1} + \frac{1}{v + i\omega_2} \right) \frac{\partial f_0{}^0}{\partial v} \right] \right\}$$

$$+ \frac{1}{5[v - i(\omega_1 - \omega_2)]} \frac{\partial}{\partial v} \left\{ \frac{1}{v} \left(\frac{1}{v - i\omega_1} + \frac{1}{v + i\omega_2} \right) \frac{\partial f_0{}^0}{\partial v} \right\}$$

$$(3.4h)$$

where $f_0{}^0$ is to be evaluated from the equation

$$- \frac{1}{3v^2} \frac{\partial}{\partial v} \left[v^2 \left(\frac{\tilde{a}_1}{2} f_{11}^1 + \frac{\tilde{a}_2}{2} f_{21}^1 \right) \right] = \frac{m}{Mv^2} \frac{\partial}{\partial v} (vv^3 f_0{}^0) + \frac{kT}{Mv^2} \frac{\partial}{\partial v} \left(vv^2 \frac{\partial f_0{}^0}{\partial v} \right)$$

$$(3.5)$$

in which the value of $(\partial f^0 / \partial t)_c$ for a Lorentzian plasma derived by Desloge and Matthysse (63) has been used; k is the Boltzmann's constant, T is the temperature of the gas, and M is the mass of a molecule.

An examination of Eqs. (3.4a–h) reveals that the expression for $f_{23}^1/a_2{}^3$ can be obtained from that of $f_{13}^1/a_1{}^3$ by replacing ω_1 by ω_2; the expression for $f_{12}^{1-'}/a_1\tilde{a}_2{}^2$ can be obtained from that of $f_{12}^{1+'}/a_1a_2{}^2$ by replacing ω_2 with $-(\omega_2)$. Also, expressions for $f_{12}^{1+'}/a_2 a_1{}^2$ and $f_{21}^{1-'}/a_2\tilde{a}_1{}^2$ can be obtained from those of $f_{12}^{1+'}/a_1a_2{}^2$ and $f_{12}^{1-'}/a_1\tilde{a}_2{}^2$ by interchanging the positions occupied by ω_1 and ω_2.

Using Eqs. (2.3), (2.13), (3.1), and (3.4a–h)—keeping in mind that all the f_y^k are zero (because the external electric fields are all along the x direction only) and that $\Delta_x = v_x/v$—and following Ginzburg and Gurevich (9), one obtains for the harmonic and combination frequency components of the current density:

$$\mathbf{J}_{13} = (m/M)\alpha A_{13} \mathscr{E}_1{}^2 \mathbf{E}_1 \exp(3i\omega_1 t) \tag{3.6a}$$

$$\mathbf{J}_{23} = (m/M)\alpha A_{23} \mathscr{E}_2{}^2 \mathbf{E}_2 \exp(3i\omega_2 t) \tag{3.6b}$$

$$\mathbf{J}_{12}^{+'} = (m/M)\alpha A_{12}^{+'} \mathscr{E}_2{}^2 \mathbf{E}_1 \exp\{i(\omega_1 + 2\omega_2)t\} \tag{3.6c}$$

$$\mathbf{J}_{12}^{-'} = (m/M)\alpha A_{12}^{-'} \tilde{\mathscr{E}}_2{}^2 \mathbf{E}_1 \exp\{i(\omega_1 - 2\omega_2)t\} \tag{3.6d}$$

$$\mathbf{J}_{21}^{+'} = (m/M)\alpha A_{21}^{+'} \mathscr{E}_1{}^2 \mathbf{E}_2 \exp\{i(\omega_2 + 2\omega_1)t\} \tag{3.6e}$$

$$\mathbf{J}_{21}^{-'} = (m/M)\alpha A_{21}^{-'} \tilde{\mathscr{E}}_1{}^2 \mathbf{E}_2 \exp\{i(\omega_2 - 2\omega_1)t\} \tag{3.6f}$$

where $\alpha = e^2 E_{00}^2 M/6m^2\omega_0{}^2 kT$, $\mathscr{E}_1 = E_1/E_{00}$, $\mathscr{E}_2 = E_2/E_{00}$,

$$A_{13} = - \frac{kT}{m} \cdot \frac{\omega_p{}^2\omega_0{}^2}{2\pi} \left[\int_0^\infty v^2 f_0{}^0 \, dv \right]^{-1} \int_0^\infty v^3 \left(\frac{f_{13}^1}{a_1{}^3} \right) dv \tag{3.7a}$$

$$A_{12}^{+'} = - \frac{kT}{m} \cdot \frac{\omega_p^2 \omega_0^2}{2\pi} \left[\int_0^\infty v^2 f_0^{\ 0} \, dv \right]^{-1} \int_0^\infty v^3 \left(\frac{f_{12}^{1+'}}{a_1 a_2^{\ 2}} \right) dv \qquad (3.7b)$$

in which $\omega_p = (4\pi N e^2/m)^{1/2}$ is the plasma frequency, N is the electron density, E_{00} is an arbitrary normalizing field, ω_0 is an arbitrary frequency, and the symbol $\sim$ over any letter denotes its complex conjugate. The expressions for the remaining A can be obtained from those of A_{13} and $A_{12}^{+'}$ in a manner outlined above for the components of $f^1/a_p^{\ 2} a_q$, $(p, q = 1, 2)$. It should be noted from Eqs. (3.4a–h) that $f^1/a_p^{\ 2} a_q$ will all be independent of a_p or a_q and hence of the electric fields. Further, it is seen that they have been evaluated in terms of $f_0^{\ 0}$; the first step in evaluating the integrals (3.7a, b) is therefore to determine the form of $f_0^{\ 0}$ from Eq. (3.5). Following Sodha and Palumbo (64), the solution to Eq. (3.5) may be written as

$$f_0^{\ 0} \propto \exp(-u^2) \left[1 + \alpha \omega_0^2 \int_0^{u^2} \left(\frac{\mathscr{E}_1 \tilde{\mathscr{E}}_1}{v^2 + \omega_1^{\ 2}} + \frac{\mathscr{E}_2 \tilde{\mathscr{E}}_2}{v^2 + \omega_2^{\ 2}} \right) du^2 \right] \qquad (3.8)$$

where $u = (m/2kT)^{1/2} v$ is the dimensionless electron velocity, and it has been assumed that

$$\left[\alpha \omega_0^2 \int_0^{u^2} \left(\frac{\mathscr{E}_1 \tilde{\mathscr{E}}_1}{v^2 + \omega_1^{\ 2}} + \frac{\mathscr{E}_2 \tilde{\mathscr{E}}_2}{v^2 + \omega_2^{\ 2}} \right) du^2 \right]^2 \ll 1$$

(this assumption is not very stringent at moderately high electric fields). From Eq. (3.8) one notes that $f_0^{\ 0}$ is non-Maxwellian in form; the departure from an equilibrium Maxwellian form (viz., $\exp(-u^2)$) is of the order of α times the Maxwellian component itself. In further discussion, however, we shall assume $f_0^{\ 0}$ to be Maxwellian corresponding to the temperature of the gas; this assumption is justified because the contributions to the harmonic and combination frequency components of current density, arising because of the departures from an equilibrium Maxwellian form of $f_0^{\ 0}$ will be of the order of $(m/M)\alpha^2$, and these may be neglected in comparison to the contributions of the order of $(m/M)\alpha$ arising because of the Maxwellian part of $f_0^{\ 0}$. Further, taking $f_0^{\ 0}$ to be Maxwellian removes the limitation of our treatment with regard to the Lorentzian nature of the plasmas, and it may be applied to non-Lorentzian plasmas as well.

The integrals expressed in Eqs. (3.7a, b) may be put in closed forms in either of the following cases:

Case I. $\omega_1 \gg v$, $\omega_2 \gg v$, $|\omega_1 - 2\omega_2| \gg v$ and $|\omega_2 - 2\omega_1| \gg v$.

Case II. $v \gg 3\omega_1$, $v \gg 3\omega_2$, and $v = v_0 u''$, where v_0 is the collision frequency corresponding to $u = 1$ or $v = (2kT/m)^{1/2}$. The expressions of interest are given in Appendix A.

It is noted that for $n = 0$,

$$A_{13} = A_{23} = A_{12}^{\pm\prime} = A_{21}^{\pm\prime} = 0 \tag{3.9}$$

which means that all harmonics and combination frequency components will vanish for an electron collision frequency independent of electron velocity. Rosen (3) obtained the same result for the third harmonic only by his approximate treatment, in which he had not taken the higher order asymmetrical terms f^2 into account. This conclusion regarding the vanishing of these nonlinear high frequency components for a velocity-independent collision frequency seems also to be consistent with the general observation that the nonlinear effects of any form manifest themselves for homogeneous plasmas if and only if there are departures from a velocity-independent collision frequency. When the electron collision frequency is proportional to electron velocity (in other words, when the collision cross section is independent of energy or when $n = 1$), the expressions for the A are considerably simplified and one has

$$A_{13} = \frac{\omega_p^2 \omega_0^2}{\omega_1^3 \pi} \left[\frac{1}{\pi^{1/2}} \left(\frac{v_0^3}{54\omega_1^3} - \frac{1}{30} \frac{v_0}{\omega_1} \right) - \frac{23i}{720} \frac{v_0^2}{\omega_1^2} \right] \tag{3.10a}$$

$$\begin{aligned}
A_{12}^{+\prime} = \frac{\omega_p^2 \omega_0^2}{\pi(\omega_1 + 2\omega_2)^2} &\left[\frac{1}{\pi^{1/2}} \left\{ v_0^3 \left[\frac{2}{3} \frac{\omega_1^2 + \omega_2^2}{\omega_1^2 \omega_2^2 (\omega_1 + \omega_2)^2} + \frac{1}{6\omega_2^4} \right] \right. \right. \\
&\left. - v_0 \left(\frac{3}{5\omega_1\omega_2} + \frac{3}{10\omega_2^2} \right) \right\} \\
&\left. - \frac{3}{4} iv_0^2 \left\{ \frac{1}{3\omega_1\omega_2(\omega_1 + \omega_2)} + \frac{3}{5} \cdot \frac{\omega_1^2 + \omega_2^2}{\omega_1^2 \omega_2^2 (\omega_1 + \omega_2)^2} + \frac{23}{60\omega_2^3} \right\} \right]
\end{aligned}$$

$$\tag{3.10b}$$

Expressions for the other A can be obtained from these.

From Eqs. (3.6a–f) one notes that the harmonic and combination frequency components of the current density are of the order of $m/Ma\mathscr{E}_p^2 E_q$ $(p, q = 1, 2)$ or $(e^2/6m\omega_0^2 kT)E_p^2 E_q$. In order to get an idea of the order of magnitude of these high frequency components in the current density and their variations with the various important parameters, some calculations have been made for $n = 1$, i.e., for the case when the electron collision frequency is proportional to the electron velocity; it will be assumed that the field with frequency ω_2 is weak so that the nonlinearities arise only because of the field with frequency ω_1, and only the frequencies $3\omega_1$, $\omega_2 + 2\omega_1$ and $\omega_2 - 2\omega_1$ are generated. It is noted at the outset that all A are directly proportional to ω_p^2 and v_0; therefore one expects the magnitude of the harmonic and the combination frequency components in the current density to increase with the electron density and the electron collision frequency.

TABLE IA

VARIATIONS OF j_{13}, $j_{21}^{+\prime}$, AND $j_{21}^{-\prime}$ WITH v_0/ω_0 FOR $\omega_0 = \omega_1$, $\omega_2/\omega_1 = 3$

v_0/ω_0	j_{13}	$j_{21}^{+\prime}$	$j_{21}^{-\prime}$
0.001	7.528×10^{-5}	4.500×10^{-5}	2.257×10^{-4}
0.003	2.257×10^{-4}	1.355×10^{-4}	6.770×10^{-4}
0.010	7.528×10^{-4}	4.512×10^{-4}	2.257×10^{-3}
0.030	2.260×10^{-3}	1.355×10^{-3}	6.755×10^{-3}
0.100	7.668×10^{-3}	4.550×10^{-3}	2.200×10^{-2}

Table IA illustrates the variation of

$$j_{13} = \frac{24m\pi\omega_0^3 kT}{\omega_p^2 e^2} \left| \frac{J_{13}}{E_1^3 \exp(3i\omega_1 t)} \right| \tag{3.11a}$$

$$j_{21}^{+\prime} = \frac{24m\pi\omega_0^3 kT}{\omega_p^2 e^2} \left| \frac{J_{21}^{+\prime}}{E_2 E_1^2 \exp\{i(\omega_2 + 2\omega_1)t\}} \right| \tag{3.11b}$$

$$j_{21}^{-\prime} = \frac{24m\pi\omega_0^3 kT}{\omega_p^2 e^2} \left| \frac{J_{21}^{-\prime}}{E_2 \tilde{E}_1^2 \exp\{i(\omega_2 - 2\omega_1)t\}} \right| \tag{3.11c}$$

with v_0/ω_0 for $\omega_0 = \omega_1$ and $\omega_2/\omega_1 = 3$. One notes that the magnitude of j_{21}^{-} is always greater than that of $j_{21}^{+\prime}$. The variations with the collision frequency are appreciable. Table IB illustrates the variations of $j_{21}^{+\prime}$ and $j_{21}^{-\prime}$ with ω_2/ω_1

TABLE IB

VARIATIONS OF $j_{21}^{+\prime}$ AND $j_{21}^{-\prime}$ WITH ω_2/ω_1
FOR $v_0/\omega_0 = 0.1$, $\omega_0 = \omega_1$

ω_2/ω_1	$j_{21}^{+\prime}$	$j_{21}^{-\prime}$
3	4.550×10^{-3}	2.200×10^{-2}
5	1.950×10^{-3}	4.545×10^{-3}
7	1.082×10^{-3}	1.949×10^{-3}
9	6.957×10^{-4}	1.085×10^{-3}
11	4.787×10^{-4}	6.897×10^{-4}

for $v_0/\omega_0 = 0.10$ and $\omega_0 = \omega_1$. It is seen that as the ratio (ω_2/ω_1) is increased beyond the value 3, both $j_{21}^{+\prime}$ and j_{21}^{-} decrease in magnitude, first quickly and then slowly. In the neighborhood of $\omega_2/\omega_1 \sim 2$, one expects a resonant increase in j_{21}^{-}, but since the resonance $\omega_2 = 2\omega_1$ corresponds to a zero difference frequency (perhaps a dc component of the current density), it is hardly of any interest; it has therefore not been investigated in detail.

Visvanathan (*40*) has also considered the generation of the third harmonic in the current density in a plasma; he considers the modulation of the electron temperature and treats the problem of third-harmonic generation by an elementary theory. His work has already been incorporated in their review by Wetzel and Tang (*1*). Silin (*11*) has investigated the time-dependent nature of the conductivity of a strongly ionized plasma due to collisions between electrons and ions and has shown that odd harmonics may be generated because of this; this case has not been included in the above analysis.

2. A dc Field and Two Microwave Fields

It was seen above that two microwave fields alone cannot give rise to the second harmonic or the sum-and-difference frequency components ($\omega_1 \pm \omega_2$) in the current density in a homogeneous plasma. A dc electric field applied simultaneously with the two microwave fields can, however, lead to the generation of both the even harmonics and the sum-and-difference frequencies. This was shown by Murphy (*26*) and by Sodha and Kaw (*23*), the former by using a quasimicroscopic approach involving the first two moments of the Boltzmann's equation, the Maxwell's equations, etc., and the latter by a thoroughly microscopic approach.

A very important feature of this type of generation is that the magnitudes of the second harmonic and the sum-and-difference frequencies generated are directly proportional to the amplitude of the dc electric field. Since the dc electric field can be readily adjusted to a large value, appreciable second harmonic and sum-and-difference frequencies may perhaps be generated for moderate values of the fundamental electric fields; one cannot be very certain about this conclusion because the present theories predicting this effect are limited to relatively low values of the dc field.

In the present case the plasma is under the influence of an electric field

$$\mathbf{E} = \mathbf{E}_d + \mathbf{E}_1 \exp(i\omega_1 t) + \mathbf{E}_2 \exp(i\omega_2 t) \tag{3.12}$$

directed along the x direction. For evaluating the various time-independent and time-dependent components of f^0, f^1, and f^2, one can still use Eqs. (3.3a–c), the acceleration a being given by

$$a = a_d + a_1 \exp(i\omega_1 t) + a_2 \exp(i\omega_2 t) \tag{3.13}$$

and the expected time dependence of the components of the distribution function, in accordance with Eq. (2.14), neglecting the terms with frequencies $2\omega_1, 3\omega_2, (\omega_1 \pm 2\omega_2)$ and $(\omega_2 \pm 2\omega_1)$ in f^0 and f^2. This last approximation is applicable in the range of validity of the present treatment only (discussed in detail in Appendix B).

For further analysis, the following approximations have been used:

a. $f_{12}^1, f_{13}^1, f_{22}^1, f_{23}^1, f_{12}^{1\pm}, f_{12}^{1\pm'}, f_{21}^{1\pm'} \ll f_{11}^1$, and f_{21}^1

$f_{11}^0, f_{21}^0, f_{12}^0, f_{22}^0, f_{12}^{0\pm}, f_{11}^2, f_{12}^2, f_{21}^2, f_{22}^2, f_{12}^{2\pm} \ll f_0^{\,0}$

which, as in the earlier case, correspond to the assumption that the magnitudes of all harmonic and combination frequency components of the current density are much smaller than those of the fundamentals.

b. $(m/M)\alpha\mathscr{E}_d^2 \ll (v/\omega_1)$ or (v/ω_2), where $\mathscr{E}_d = E_d/E_{00}$. Though this assumption limits the maximum value of the dc field for which the theory is applicable, yet it has been used because it permits explicit evaluation of the relevant components of the distribution function from the system of coupled equations, Eqs. (3.3a–c).

Proceeding as before [for details see Sodha and Kaw (23)], the second harmonic and the sum-and-difference frequency components of the current density can be readily shown to be given by

$$\mathbf{J}_{12} = (m/M)\alpha A_{12}\,\mathscr{E}_1^{\,2}\mathbf{E}_d \exp(2i\omega_1 t) \tag{3.14a}$$

$$\mathbf{J}_{22} = (m/M)\alpha A_{22}\,\mathscr{E}_2^{\,2}\mathbf{E}_d \exp(2i\omega_2 t) \tag{3.14b}$$

$$\mathbf{J}_{12}^+ = (m/M)\alpha A_{12}^+\,\mathscr{E}_1\mathscr{E}_2\,\mathbf{E}_d \exp\{i(\omega_1 + \omega_2)t\} \tag{3.14c}$$

$$\mathbf{J}_{12}^- = (m/M)\alpha A_{12}^-\,\mathscr{E}_1\tilde{\mathscr{E}}_2\,\mathbf{E}_d \exp\{i(\omega_1 - \omega_2)t\} \tag{3.14d}$$

(expressions for the other high frequency components remain unaltered under the approximation (b) which we have taken), where A_{12}, A_{22}, A_{12}^+, and A_{12}^- contain integrals similar to those expressed in Eqs. (3.7a, b); these can be evaluated as before, under similar assumptions, and have been presented in Appendix B. It is noted that for $n = 0$, all the A vanish in this case, also showing thereby that the second harmonic and the sum-and-difference frequencies vanish for an electron collision frequency independent of electron velocity, even in the presence of an external dc electric field. For $n = 1$, the expressions for the A are

$$A_{12} = \frac{\omega_p^{\,2}\omega_0^{\,2}}{\omega_1^{\,3}\pi} \left[\frac{1}{\pi^{1/2}} \left(\frac{3}{4}\frac{v_0^{\,3}}{\omega_1^{\,3}} - \frac{17}{60}\frac{v_0}{\omega_1} \right) - i\left(\frac{69}{96}\frac{v_0^{\,2}}{\omega_1^{\,2}} + \frac{3}{20} \right) \right] \tag{3.15}$$

$$\begin{aligned}
A_{12}^+ = \frac{\omega_p^{\,2}\omega_0^{\,2}}{\pi(\omega_1 + \omega_2)^2} &\left[\frac{1}{\pi^{1/2}} \left\{ \frac{4}{3}v_0^{\,3}\left[2\left(\frac{1}{\omega_1^{\,4}} + \frac{1}{\omega_2^{\,4}} \right) + \frac{\omega_1^{\,2} + \omega_2^{\,2}}{\omega_1^{\,2}\omega_2^{\,2}(\omega_1 + \omega_2)^2} \right] \right.\right. \\
&\quad \left. - \frac{2}{5}v_0\left[\frac{4}{3}\left(\frac{1}{\omega_1^{\,2}} + \frac{1}{\omega_2^{\,2}} \right) + \frac{3}{\omega_1\omega_2} \right] \right\} \\
&\quad - \frac{3}{2}iv_0^{\,2}\left\{ \frac{23}{15}\left(\frac{1}{\omega_1^{\,3}} + \frac{1}{\omega_2^{\,3}} \right) + \frac{1}{3\omega_1\omega_2(\omega_1 + \omega_2)} \right. \\
&\quad \left.\left. + \frac{3}{5}\cdot\frac{\omega_1^{\,2} + \omega_2^{\,2}}{\omega_1^{\,2}\omega_2^{\,2}(\omega_1 + \omega_2)} \right\} \right]
\end{aligned} \tag{3.16}$$

One notes from Eqs. (3.14a–d) that the magnitudes of the second harmonic and the sum-and-difference frequency components in the current density are directly proportional to that of the dc electric field. Further, the proportionality of the A to ω_p^2 and v_0 is true for the present case as well, so that the second harmonic and sum-and-difference frequency components are directly proportional to the collision frequency and the electron density.

To illustrate the variation of the second harmonic and the sum-and-difference frequency components of the current density with the collision frequency and the ratio ω_2/ω_1, some numerical values have been tabulated. Table IIA shows the variation of

$$j_{12} = \frac{24m\pi\omega_0^3 kT}{\omega_p^2 e^2} \left| \frac{J_{12}}{E_1^2 E_d \exp(2i\omega_1 t)} \right|$$

$$j_{12}^+ = \frac{24m\pi\omega_0^3 kT}{\omega_p^2 e^2} \left| \frac{J_{12}^+}{E_1 E_2 E_d \exp\{i(\omega_1 + \omega_2)t\}} \right|$$

$$j_{12}^- = \frac{24m\pi\omega_0^3 kT}{\omega_p^2 e^2} \left| \frac{J_{12}^-}{E_1 \tilde{E}_2 E_d \exp\{i(\omega_1 - \omega_2)t\}} \right|$$

TABLE IIA

VARIATION OF j_{12}, j_{12}^+, AND j_{12}^- WITH v_0/ω_0 FOR $\omega_0 = \omega_1$, $\omega_2/\omega_1 = 0.4$

v_0/ω_0	j_{12}	j_{12}^+	j_{12}^-
0.01	6.005×10^{-1}	8.135×10^{-2}	6.015×10^{-2}
0.03	6.025×10^{-1}	2.466×10^{-1}	2.861×10^{-1}
0.05	6.070×10^{-1}	4.940×10^{-1}	7.095×10^{-1}
0.07	6.148×10^{-1}	6.717×10^{-1}	1.354

with v_0/ω_0 for $\omega_2/\omega_1 = 0.4$ and $\omega_0 = \omega_1$. While the variation of j_{12} with the collision frequency is very slight, the sum-and-difference frequency components suffer a large variation; further, j_{12}^- increases much more rapidly than does j_{12}^+. Table IIB illustrates the variation of $j_{12}^\pm$ with ω_2/ω_1 for $v_0/\omega_0 =$

TABLE IIB

VARIATION OF j_{12}^+ AND j_{12}^- WITH ω_2/ω_1 FOR $\omega_0 = \omega_1$,
$v_0/\omega_0 = 0.05$

ω_2/ω_1	j_{12}^+	j_{12}^-
0.2	1.892	4.096
0.4	4.940×10^{-1}	7.095×10^{-1}
0.6	1.845×10^{-1}	2.110×10^{-1}
0.8	1.022×10^{-1}	1.872

0.05 and $\omega_0 = \omega_1$. As ω_2/ω_1 increases, j_{12}^- decreases at first and then increases as we approach the value $\omega_2/\omega_1 \sim 1$. This anomalous behavior of j_{12}^- may be explained by the existence of a resonance in its magnitude at $\omega_1 \sim \omega_2$. This resonance has not been investigated in any detail because it corresponds to a zero difference frequency and hence a dc component of the current density.

3. Effect of a Magnetic Field

The generation of combination frequencies in a plasma in the presence of a magnetic field was considered for the first time by Vilenskii (6); he did not take into account the important contribution arising because of the f^2 component. The generation with the f^2 terms has been studied recently by Mittal and Kaw (16); it has been shown that neglect of f^2 in the presence of a magnetic field results not only in an appreciable error in the quantitative estimation of the magnitudes of these components in the current density, but also in a considerable modification of some of the qualitative conclusions. Sodha and Kaw (18) had attempted earlier the simpler problem of the generation of third harmonic in the presence of a magnetic field, taking f^2 into account; however, that treatment suffered from the use of a nonconvergent expansion of the distribution function. In the present corrected analysis, although the form of the expression for the third-harmonic component of current density remains unchanged, the numerical magnitudes are somewhat altered.

For the investigation of the present problem one considers an electric field having components

$$E_{x,y} = E_{1x,y} \exp(i\omega_1 t) + E_{2x,y} \exp(i\omega_2 t), \qquad E_z = 0$$

and a static magnetic field with the components

$$B_{x,y} = 0, \qquad B_z = B_0$$

so that the electric and magnetic fields are at right angles to each other. One has to use the complete set of Eqs. (2.4a–e) as such (of course, neglecting $\partial f/\partial x$ and $\partial f/\partial y$ because the plasma is homogeneous), where

$$a_{x,y} = a_{1x,y} \exp(i\omega_1 t) + a_{2x,y} \exp(i\omega_2 t)$$

and the expected time dependence of f^0, f^1, f^2 are given by

$$f^0 = f_0{}^0 + f_{12}^0 \exp(2i\omega_1 t) + f_{22}^0 \exp(2i\omega_2 t)$$
$$+ f_{12}^{0+} \exp\{i(\omega_1 + \omega_2)t\} + f_{12}^{0-} \exp\{i(\omega_1 - \omega_2)t\} + \cdots$$

$$(3.17a)$$

$$f^1_{x,y} = f^1_{11x,y}\exp(i\omega_1 t) + f^1_{13x,y}\exp(3i\omega_1 t) + f^1_{21x,y}\exp(i\omega_2 t)$$
$$+ f^1_{23x,y}\exp(3i\omega_2 t) + f^{1+\prime}_{12x,y}\exp\{i(\omega_1 + 2\omega_2)t\}$$
$$+ f^{1-\prime}_{12x,y}\exp\{i(\omega_1 - 2\omega_2)t\} + f^{1+\prime}_{21x,y}\exp\{i(\omega_2 + 2\omega_1)t\}$$
$$+ f^{1-\prime}_{21x,y}\exp\{i(\omega_2 - 2\omega_1)t\} + \cdots$$

$$(3.17\text{b})$$

$$f^2_{x,y} = f^2_{0x,y} + f^2_{12x,y}\exp(2i\omega_1 t) + f^2_{22x,y}\exp(2i\omega_2 t)$$
$$+ f^{2+}_{12x,y}\exp\{i(\omega_1 + \omega_2)t\} + f^{2-}_{12x,y}\exp\{i(\omega_1 - \omega_2)t\} + \cdots$$

$$(3.17\text{c})$$

The fact that f^2 is actually a tensor has been ignored in these calculations; the results of this section should therefore be treated with caution.

Proceeding as described in Appendix C, one obtains for the x components of the fundamental, third harmonic $3\omega_1$, and the sum frequency $\omega_1 + 2\omega_2$ terms in f^1, the expressions

$$[(v + i\omega_1)^2 + \omega_B{}^2]f^1_{11x} = [(v + i\omega_1)a_{1x} - \omega_B a_{1y}]\partial f_0{}^0/\partial v$$

$$[(v + 3i\omega_1)^2 + \omega_B{}^2]f^1_{13x}$$
$$= [(v + 3i\omega_1)a_{1x} - \omega_B a_{1y}]\frac{\partial}{\partial v}\left\{\frac{a^2_{1x} + a^2_{1y}}{24i\omega_1 v^2}\frac{\partial}{\partial v}\left[\frac{v + i\omega_1}{(v + i\omega_1)^2 + \omega_B{}^2}v^2\frac{\partial f_0{}^0}{\partial v}\right]\right\}$$
$$+ \frac{a^2_{1x}}{15}(v + 3i\omega_1)\frac{\partial}{\partial v}\left[\frac{v}{v + 2i\omega_1}\frac{\partial}{\partial v}\left\{\frac{(v + i\omega_1)a_{1x} - \omega_B a_{1y}}{(v + i\omega_1)^2 + \omega_B{}^2}\frac{1}{v}\frac{\partial f_0{}^0}{\partial v}\right\}\right]$$
$$- \frac{a^2_{1y}}{15}\omega_B\frac{\partial}{\partial v}\left[\frac{v}{v + 2i\omega_1}\frac{\partial}{\partial v}\left\{\frac{(v + i\omega_1)a_{1y} + \omega_B a_{1x}}{(v + i\omega_1)^2 + \omega_B{}^2}\frac{1}{v}\frac{\partial f_0{}^0}{\partial v}\right\}\right]$$
$$+ \frac{(v + 3i\omega_1)a^2_{1x}}{5(v + 2i\omega_1)}\frac{\partial}{\partial v}\left\{\frac{(v + i\omega_1)a_{1x} - \omega_B a_{1y}}{(v + i\omega_1)^2 + \omega_B{}^2}\frac{1}{v}\frac{\partial f_0{}^0}{\partial v}\right\}$$
$$- \frac{\omega_B a^2_{1y}}{5(v + 2i\omega_1)}\frac{\partial}{\partial v}\left\{\frac{(v + i\omega_1)a_{1y} + \omega_B a_{1x}}{(v + i\omega_1)^2 + \omega_B{}^2}\frac{1}{v}\frac{\partial f_0{}^0}{\partial v}\right\}$$

and

$$\left[\frac{\{v + i(\omega_1 + 2\omega_2)\}^2 + \omega_B{}^2}{v + i(\omega_1 + 2\omega_2)}\right]f^{1+\prime}_{12x}$$
$$= \frac{1}{24i\omega_2}\left\{\frac{a_{1x}[v + i(\omega_1 + 2\omega_2)] - a_y\omega_B}{v + i(\omega_1 + 2\omega_2)}\right\}$$
$$\times \frac{\partial}{\partial v}\left[\frac{1}{v^2}\frac{\partial}{\partial v}\left\{\frac{a^2_{2x} + a^2_{2y}}{(v + i\omega_2)^2 + \omega_B{}^2}v^2\frac{\partial f_0{}^0}{\partial v}\right\}\right] + \frac{a_{1x}a_{2x}}{15v^3}$$

$$\times \frac{\partial}{\partial v}\left[\frac{v^4}{v+2i\omega_2}\frac{\partial}{\partial v}\left\{\frac{(v+i\omega_2)a_{2x}-\omega_B a_{2y}}{(v+i\omega_2)^2+\omega_B^2}\frac{1}{v}\frac{\partial f_0^0}{\partial v}\right\}\right]$$

$$-\frac{a_{1y}a_{2y}\omega_B}{15v^3[v+i(\omega_1+2\omega_2)]}$$

$$\times \frac{\partial}{\partial v}\left[\frac{v^4}{v+2i\omega_2}\frac{\partial}{\partial v}\left\{\frac{(v+i\omega_2)a_{2y}+\omega_B a_{2x}}{(v+i\omega_2)^2+\omega_B^2}\frac{1}{v}\frac{\partial f_0^0}{\partial v}\right\}\right]$$

$$+\frac{1}{12i(\omega_1+\omega_2)}\left(a_{2x}-\frac{a_{2y}\omega_B}{v+i(\omega_1+2\omega_2)}\right)$$

$$\times \frac{\partial}{\partial v}\left\{\frac{1}{v^2}\frac{\partial}{\partial v}\left[(a_{1x}a_{2x}+a_{1y}a_{2y})v^2\frac{\partial f_0^0}{\partial v}\right.\right.$$

$$\times \left(\frac{v+i\omega_2}{(v+i\omega_2)^2+\omega_B^2}+\frac{v+i\omega_1}{(v+i\omega_1)^2+\omega_B^2}\right)$$

$$\left.\left.-\omega_B(a_{1y}a_{2x}-a_{2y}a_{1x})v^2\frac{\partial f_0^0}{\partial v}\left(\frac{1}{(v+i\omega_1)^2+\omega_B^2}-\frac{1}{(v+i\omega_2)^2+\omega_B^2}\right)\right\}\right]$$

$$+\frac{a_{2x}}{15v^3}\frac{\partial}{\partial v}\left[\frac{v^4}{v+i(\omega_1+\omega_2)}\frac{\partial}{\partial v}\left\{\left(\frac{a_{1x}a_{2x}(v+i\omega_2)-\omega_B a_{1x}a_{2y}}{(v+i\omega_2)^2+\omega_B^2}\right.\right.\right.$$

$$\left.\left.\left.+\frac{a_{2x}a_{1x}(v+i\omega_1)-\omega_B a_{2x}a_{1y}}{(v+i\omega_1)^2+\omega_B^2}\right)\frac{1}{v}\frac{\partial f_0^0}{\partial v}\right\}\right]$$

$$-\frac{a_{2y}\omega_B}{15v^3[v+i(\omega_1+2\omega_2)]}\frac{\partial}{\partial v}\left[\frac{v^4}{v+i(\omega_1+\omega_2)}\right.$$

$$\times \frac{\partial}{\partial v}\left\{\left(\frac{a_{1y}a_{2y}(v+i\omega_2)+\omega_B a_{1y}a_{2x}}{(v+i\omega_2)^2+\omega_B^2}\right.\right.$$

$$\left.\left.\left.+\frac{a_{2y}a_{1y}(v+i\omega_1)+a_{2y}a_{1x}\omega_B}{(v+i\omega_1)^2+\omega_B^2}\right)\frac{1}{v}\frac{\partial f_0^0}{\partial v}\right\}\right]$$

Expressions for the y components of f^1 may be obtained by interchanging the subscripts x and y and replacing ω_B by $(-\omega_B)$. The expressions for the other frequency components of f^1 may be obtained from those given above in a manner mentioned in Section III-A-1.

Proceeding as in Section III-A-1, the expressions for the x and y components of the third harmonic $3\omega_1$ and the sum frequency $\omega_1+2\omega_2$ terms of the current density are given by

$$J_{13x}=(m/M)\alpha[B_1\mathscr{E}_{1x}^3-B_2\mathscr{E}_{1x}^2\mathscr{E}_{1y}+B_3\mathscr{E}_{1x}\mathscr{E}_{1y}^2-B_4\mathscr{E}_{1y}^3]E_{00}\exp(3i\omega_1t)$$

$$(3.18a)$$

$$J_{13y} = (m/M)\alpha[B_1\mathscr{E}_{1y}^3 + B_2\mathscr{E}_{1y}^2\mathscr{E}_{1x} + B_3\mathscr{E}_{1y}\mathscr{E}_{1x}^2 + B_4\mathscr{E}_{1x}^3]E_{00}\exp(3i\omega_1 t)$$

$$(3.18b)$$

$$\begin{aligned}
J_{12x}^+ = (m/M)\alpha[&C_1\mathscr{E}_{1x}\mathscr{E}_{2x}^2 - C_2\mathscr{E}_{1y}\mathscr{E}_{2y}^2 + C_3\mathscr{E}_{1x}\mathscr{E}_{2y}^2 \\
&- C_4\mathscr{E}_{1y}\mathscr{E}_{2x}^2 + C_5\mathscr{E}_{1y}\mathscr{E}_{2x}\mathscr{E}_{2y} - C_6\mathscr{E}_{1x}\mathscr{E}_{2x}\mathscr{E}_{2y}] \\
&\times E_{00}\exp\{i(\omega_1 + 2\omega_2)t\}
\end{aligned}$$

$$(3.18c)$$

$$\begin{aligned}
J_{12y}^+ = (m/M)\alpha[&C_1\mathscr{E}_{1y}\mathscr{E}_{2y}^2 + C_2\mathscr{E}_{1x}\mathscr{E}_{2x}^2 + C_3\mathscr{E}_{1y}\mathscr{E}_{2x}^2 + C_4\mathscr{E}_{1x}\mathscr{E}_{2y}^2 \\
&+ C_5\mathscr{E}_{1x}\mathscr{E}_{2x}\mathscr{E}_{2y} + C_6\mathscr{E}_{1y}\mathscr{E}_{2x}\mathscr{E}_{2y}]E_{00}\exp\{i(\omega_1 + 2\omega_2)t\}
\end{aligned}$$

$$(3.18d)$$

where the expression for α is the same as before and all B and C are integrals involving the collision frequency. Explicit expressions can be obtained for the B and C under the approximations

$$\omega_1 \gg v, \qquad \omega_2 \gg v, \qquad |\omega_1 - \omega_2| \gg v, \qquad |\omega_1 - 2\omega_2| \gg v$$

and for some particular values of ω_B. Further, the dependence of the electron collision frequency on electron velocity is chosen as before, in the form

$$v = v_0 u^n$$

and $f_0{}^0$ is taken to be Maxwellian corresponding to the temperature of the gas, in accordance with the assumptions considered earlier. It is noted from these derived expressions that for $n = 0$, all B and C vanish; this means that for an electron collision frequency independent of electron velocity, the third harmonic and the combination frequencies vanish in a magnetoplasma also. For $n = 1$ the expressions for B and C are considerably simplified.

From Eq. (3.18a–d), the coupled components of the third harmonic $3\omega_1$ and the sum frequency $\omega_1 + 2\omega_2$ in the current density in a magnetoplasma are given by

$$\begin{aligned}
J_{13}' = J_{13x} + iJ_{13y} = (m/M)\alpha[&(B_1 + iB_4)(\mathscr{E}_{1x}^3 + i\mathscr{E}_{1y}^3) \\
&- (B_2 - iB_3)(\mathscr{E}_{1x} - i\mathscr{E}_{1y})\mathscr{E}_{1x}\mathscr{E}_{1y}]E_{00}\exp(3i\omega_1 t)
\end{aligned}$$

$$(3.19a)$$

$$\begin{aligned}
J_{13}'' = J_{13x} - iJ_{13y} = (m/M)\alpha[&(B_1 - iB_4)(\mathscr{E}_{1x}^3 - i\mathscr{E}_{1y}^3) \\
&- (B_2 + iB_3)(\mathscr{E}_{1x} + i\mathscr{E}_{1y})\mathscr{E}_{1x}\mathscr{E}_{1y}]E_{00}\exp(3i\omega_1 t)
\end{aligned}$$

$$(3.19b)$$

$$\begin{aligned}
J_{12}^{+\prime} = J_{12x}^+ + iJ_{12y}^+ = (m/M)\alpha[&(C_1 + iC_2)(\mathscr{E}_{1x}\mathscr{E}_{2x}^2 + i\mathscr{E}_{1y}\mathscr{E}_{2y}^2) \\
&+ (C_3 + iC_4)(\mathscr{E}_{1x}\mathscr{E}_{2y}^2 + i\mathscr{E}_{1y}\mathscr{E}_{2x}^2) \\
&+ (C_5 + iC_6)(\mathscr{E}_{1y}\mathscr{E}_{2y}\mathscr{E}_{2x} + i\mathscr{E}_{1x}\mathscr{E}_{2x}\mathscr{E}_{2y})] \\
&E_{00}\exp\{i(\omega_1 + 2\omega_2)t\}
\end{aligned}$$

$$(3.19c)$$

$$J_{12}^{+\prime\prime} = J_{12x}^{+} - iJ_{12y}^{+} = (m/M)\alpha[(C_1 - iC_2)(\mathcal{E}_{1x}\mathcal{E}_{2x}^2 - i\mathcal{E}_{1y}\mathcal{E}_{2y}^2)$$
$$+ (C_3 - iC_4)(\mathcal{E}_{1x}\mathcal{E}_{2y}^2 - i\mathcal{E}_{1y}\mathcal{E}_{2x}^2)$$
$$+ (C_5 - iC_6)(\mathcal{E}_{1y}\mathcal{E}_{2y}\mathcal{E}_{2x} - i\mathcal{E}_{1x}\mathcal{E}_{2x}\mathcal{E}_{2y})]$$
$$\times E_{00}\exp\{i(\omega_1 + 2\omega_2)t\} \qquad (3.19d)$$

It is noted that the amplitudes of the third harmonic and sum frequency components of the current density are of the order $(m/M)\alpha\mathcal{E}_p^2 E_q$ $(p, q = 1, 2)$, i.e., of the order $e^2/6m\omega_0^2 kT$ times the cube of the applied electrical vectors. Further, one notes that when

$$\mathcal{E}_{1x} - i\mathcal{E}_{1y} = \mathcal{E}_{2x} - i\mathcal{E}_{2y} = 0$$

J' vanishes whereas J'' remains of a finite magnitude, i.e., when only pure extraordinary modes of the fundamental electromagnetic waves are sent into the plasma; then only the ordinary components of the harmonic and combination frequencies are generated in the current density. It can be similarly shown that pure ordinary modes of the fundamental waves generate only extraordinary components of the harmonic and combination frequencies in the current density.

If one ignores f^2 in the expansion of the distribution function of electron velocities (as Vilenskii (6) did) Eqs. (3.19a–d) are reduced to the simple forms

$$J_{13}' = (m/M)\alpha B_{10}(\mathcal{E}_{1x} + i\mathcal{E}_{1y})(\mathcal{E}_{1x}^2 + \mathcal{E}_{1y}^2)E_{00}\exp(3i\omega_1 t) \qquad (3.20a)$$

$$J_{13}'' = (m/M)\alpha B_{20}(\mathcal{E}_{1x} - i\mathcal{E}_{1y})(\mathcal{E}_{1x}^2 + \mathcal{E}_{1y}^2)E_{00}\exp(3i\omega_1 t) \qquad (3.20b)$$

$$J_{12}^{+\prime} = (m/M)\alpha(\mathcal{E}_{2x} + i\mathcal{E}_{2y})\{C_{10}(\mathcal{E}_{1x}\mathcal{E}_{2x} + \mathcal{E}_{1y}\mathcal{E}_{2y})$$
$$+ iC_{20}(\mathcal{E}_{1y}\mathcal{E}_{2x} - \mathcal{E}_{1x}\mathcal{E}_{2y})\}E_{00}\exp\{i(\omega_1 + 2\omega_2)t\} \qquad (3.20c)$$

$$J_{12}^{+\prime\prime} = (m/M)\alpha(\mathcal{E}_{2x} - i\mathcal{E}_{2y})[C_{30}(\mathcal{E}_{1x}\mathcal{E}_{2x} + \mathcal{E}_{1y}\mathcal{E}_{2y})$$
$$- iC_{40}(\mathcal{E}_{1y}\mathcal{E}_{2x} - \mathcal{E}_{1x}\mathcal{E}_{2y})]E_{00}\exp\{i(\omega_1 + 2\omega_2)t\} \qquad (3.20d)$$

where the new B and C are again some integrals involving the velocity-dependent collision frequency. From this set of equations, it is noted that for

$$\mathcal{E}_{1x} + i\mathcal{E}_{1y} = \mathcal{E}_{2x} + i\mathcal{E}_{2y} = 0$$

as well as

$$\mathcal{E}_{2x} - i\mathcal{E}_{2y} = \mathcal{E}_{1x} - i\mathcal{E}_{1y} = 0$$

both J' and J'' vanish. This means that the approximate Eqs. (3.20a–d) do not predict any harmonic or combination frequency components in the current density if pure modes of the fundamental waves are sent into the plasma. Thus it is found that ignoring of f^2 in the expansion of f leads to certain conclusions, which are even qualitatively incorrect.

From Eqs. (3.18a–d) one also notes that the magnitudes of the third harmonic and sum frequency components of the current density are respectively proportional to the B and C. Since the B may have resonances (i.e., large values) around $\omega_B = \omega_1$ and $3\omega_1$, and the C may have resonances around $\omega_B = \omega_1$, ω_2 and $(\omega_1 + 2\omega_2)$, one also expects near these frequencies the occurrence of resonances in the third harmonic and sum frequency components of the current density. To have a numerical appreciation of the magnitude of these resonances and their dependence on the electron collision frequency, we have calculated the following results

$$j'_{13} = \frac{96m\pi kT\omega_0^{\,3}}{e^2\omega_p^{\,2}} \left| \frac{J'_{13}}{E_{1x}^3 \exp(3i\omega_1 t)} \right| \qquad \text{for} \quad \mathscr{E}_{1x} = -i\mathscr{E}_{1y}$$

(3.21a)

$$j''_{13} = \frac{96m\pi kT\omega_0^{\,3}}{e^2\omega_p^{\,2}} \left| \frac{J''_{13}}{E_{1x}^3 \exp(3i\omega_1 t)} \right| \qquad \text{for} \quad \mathscr{E}_{1x} = i\mathscr{E}_{1y}$$

(3.21b)

$$j^{+\prime}_{12} = \frac{6m\pi kT\omega_0^{\,3}}{e^2\omega_p^{\,2}} \left| \frac{J^{+\prime}_{12}}{E_{1x}E_{2x}^2 \exp\{i(\omega_1 + 2\omega_2)t\}} \right| \qquad \text{for} \quad \begin{aligned} \mathscr{E}_{1x} &= -i\mathscr{E}_{1y} \\ \mathscr{E}_{2x} &= -i\mathscr{E}_{2y} \end{aligned}$$

(3.21c)

$$j^{+\prime\prime}_{12} = \frac{6m\pi kT\omega_0^{\,3}}{e^2\omega_p^{\,2}} \left| \frac{J^{+\prime\prime}_{12}}{E_{1x}E_{2x}^2 \exp\{i(\omega_1 + 2\omega_2)t\}} \right| \qquad \text{for} \quad \begin{aligned} \mathscr{E}_{1x} &= i\mathscr{E}_{1y} \\ \mathscr{E}_{2x} &= i\mathscr{E}_{2y} \end{aligned}$$

(3.21d)

These represent the special case when $n = 1$, i.e., for an electron collision frequency directly proportional to a single power of electron velocity. Calculations similar to these have also been made for the difference frequency $\omega_1 - 2\omega_2$. (These are tabulated as $j^{-\prime}_{12}$ and $j^{-\prime\prime}_{12}$, respectively.) Table III(A)

TABLE III(A)

VARIATION OF j_3' AND j_3'' WITH ν_0/ω_0 FOR $\omega_B = 0, \omega$, AND 3ω

	j_3'		
ν_0/ω_0	$\omega_B = 0$	$\omega_B = \omega$	$\omega_B = 3\omega$
0.001	3.011×10^{-4}	4.167×10^{-2}	3.009×10^{2}
0.003	9.028×10^{-4}	4.167×10^{-2}	1.003×10^{2}
0.010	3.012×10^{-3}	4.152×10^{-2}	3.009×10^{1}
0.030	9.040×10^{-3}	4.122×10^{-2}	1.004×10^{1}
0.100	3.067×10^{-2}	3.909×10^{-2}	3.030

Table IIIa (*continued*)

	j_3''		
v_0/ω_0	$\omega_B = 0$	$\omega_B = \omega$	$\omega_B = 3\omega$
0.001	3.011×10^{-4}	9.167×10^{-2}	7.000×10^{-1}
0.003	9.028×10^{-4}	9.167×10^{-2}	7.000×10^{-1}
0.010	3.012×10^{-3}	9.158×10^{-2}	7.000×10^{-1}
0.030	9.040×10^{-3}	9.093×10^{-2}	7.000×10^{-1}
0.100	3.067×10^{-2}	8.366×10^{-2}	6.997×10^{-1}

illustrates the variations of j_{13}' and j_{13}'' with v_0/ω_0 for $\omega_B = 0$, ω_1, and $3\omega_1$; Table III(B) illustrates the variations of $j_{12}^{\pm\prime}$ and $j_{12}^{\pm\prime\prime}$ with v_0/ω_0 for $\omega_B = 0$ and $\omega_1 \pm 2\omega_2$. It is noted that the magnitudes of j_{13}'', $j_{12}^{+\prime\prime}$, and $j_{12}^{-\prime\prime}$ for $\omega_B = 3\omega_1$, $\omega_1 + 2\omega_2$ and $\omega_1 - 2\omega_2$, respectively, are much larger than those for $\omega_B = 0$; this is expected in view of resonances at these frequencies. For j_{13}'' the results for $\omega_B = \omega_1$ are also presented; there is, however, no appreci-

TABLE III(B)

VARIATION OF $j_{12}^{\pm\prime}$ AND $j_{12}^{\pm\prime\prime}$ WITH v_0/ω_0 FOR $\omega_1/\omega_2 = 7$

	$\omega_B = 0$	$\omega_B = \omega_1 + 2\omega_2$		$\omega_B = 0$	$\omega_B = \omega_1 - 2\omega_2$	
v_0/ω_0	$(j_{12}^{+\prime} = j_{12}^{+\prime\prime}) \times 10^4$	$j_{12}^{+\prime}$	$j_{12}^{+\prime\prime}$	$(j_{12}^{-\prime} = j_{12}^{-\prime\prime}) \times 10^4$	$j_{12}^{-\prime}$	$j_{12}^{-\prime\prime}$
0.100	14.430	0.171	0.017	11.799	0.151	0.450
0.050	7.103	0.342	0.033	5.945	0.304	0.899
0.010	1.413	1.708	0.167	1.614	1.514	4.496
0.005	0.706	3.419	0.331	0.580	3.027	8.995
0.001	0.141	17.090	1.670	0.116	15.140	44.960

able resonance in this case. Resonances are also expected in $j_{12}^{\pm\prime}$ and $j_{12}^{\pm\prime\prime}$ around $\omega_B = \omega_1$ and ω_2; these were also numerically studied, but have not been presented because of their weak nature. Tables III(A), (B) also show that for resonant frequencies the magnitudes of the generated harmonic and combination frequency components decrease as the collision frequency increases. Table III(C) illustrates the variations of $j_{12}^{\pm\prime}$ and $j_{12}^{\pm\prime\prime}$ with ω_1/ω_2; it is noted that as ω_1/ω_2 increases, the magnitudes of the combination frequency components generated decrease. The values of ω_1/ω_2 have been chosen in such a manner that $\omega_1 - 2\omega_2$ is always greater than ω_p, the plasma frequency, so that the regions of plasma resonances are excluded.

TABLE III(C)

VARIATION OF $j_{12}^{\pm\prime}$ AND $j_{12}^{\pm\prime\prime}$ WITH (ω_1/ω_2) FOR $v_0/\omega_0 = 0.01$

	$\omega_B = 0$	$\omega_B = \omega_1 + 2\omega_2$		$\omega_B = 0$	$\omega_B = \omega_1 - 2\omega_2$	
ω_1/ω_2	$(j_{12}^{+\prime} = j_{12}^{+\prime\prime}) \times 10^4$	$j_{12}^{+\prime}$	$j_{12}^{+\prime\prime}$	$(j_{12}^{-\prime} = j_{12}^{-\prime\prime}) \times 10^4$	$j_{12}^{-\prime}$	$j_{12}^{-\prime\prime}$
5	2.395	2.108	0.259	3.249	2.557	7.780
7	1.413	1.708	1.167	1.164	1.514	4.496
9	0.930	1.410	0.094	0.593	1.376	3.352
11	0.661	1.111	0.080	0.358	1.267	2.650

4. Generation around Plasma Resonance

When the microwave frequency is around the plasma frequency, plasma oscillations set in; if the amplitude of these oscillations is sufficiently large, then harmonics will be generated because of their consequent nonlinear nature. It seems that not many kinetic theory investigations of this phenomenon in homogeneous plasmas have been carried out; however, treatments based on the elementary theory have been considered by many workers. Here we shall outline Smerd's (38) analysis, which he carried out to explain the observed characteristics of bursts of solar radio emission. The three basic equations of Smerd are:

The equation of motion	$v'(\omega - kv) = -eE/m$	(3.22a)
The equation of continuity	$N'(\omega - kv) = kNv'$	(3.22b)
The Poisson's equation	$N - N_i = (k/4\pi e)E'$	(3.22c)

in which the prime denotes differentiation with respect to $(\omega t - kx)$, k is the propagation constant, N_i is the number density of ions, and the remaining symbols have the usual meanings.

Writing

$$u_0 = v - v_i$$

where v_i is the ion velocity, and providing the dc electron current is equal to that of the positive ions, the solutions to the above equations may be written as

$$\omega = \omega_p + kv_i$$

$$(\omega t - kx) = \left(l + \frac{1}{2}\right)\pi + (-1)^l \frac{u_{0\max}}{(\omega_p/k)}\left[1 - \left(\frac{u_0}{u_{0\max}}\right)^2\right]^{1/2}$$

$$+ (-1)^l \sin^{-1}(u_0/u_{0\max})$$

$$N/N_i = [1 - u_0/(\omega_p/k)]^{-1}$$

which together yield

$$\frac{E}{E_{\max}} = (-1)^{l+1} \frac{u_{0\max}}{(\omega_p/k)} \left[1 - \left(\frac{u_0}{u_{0\max}}\right)^2 \right]^{1/2}$$

where for $l = 0, \pm 2, \pm 4, \ldots, u_0$ is positive and for $l = \pm 1, \pm 3, \pm 5, \ldots, u_0$ is negative; $E_{\max} = (m/ek)\omega_p^2$ and ω_p is the plasma frequency. If we draw a graph between $E/E_{\max}$ and $\omega t - kx$ (with the origin $\omega t - kx = 0$ at $l = 0$ and $u = -u_{\max}$), we obtain a distorted sine curve, the distortion explaining the existence of harmonics. Smerd (38) shows a generation of 70% second harmonic for appropriate parameters. It need not be emphasized that this analysis would be far more improved if kinetic theory had been used; further, the effect of collisions has also not been considered at all in this treatment.

5. Generation Due to Nonuniform Electric Fields

Krenz and Kino (37) have shown by means of a simple elementary theory (which seems to agree quite well with their experiments) that the presence of a nonuniform electric field in an otherwise homogeneous plasma leads to the generation of even harmonics in the current density; The physical phenomena leading to such a generation are as follows: The motion of an electron in a uniform rf electric field is purely sinusoidal with time. However, in a nonuniform electric field the motion of the electron is no longer so, for if the field has a gradient along its direction, then the electron is exposed to a field of varying amplitude, and when the field has a gradient in a direction perpendicular to its own, it gives rise to a rf magnetic field. This rf magnetic field then interacts with the moving electron to give rise to $(\mathbf{v} \times \mathbf{B})$ forces, which have the lowest order terms at the second-harmonic frequency. Thus, any nonuniformity in the electric fields leads to the generation of harmonics.

Krenz and Kino (37) have analyzed the generation of the second harmonic in the current density by taking the nonuniform nature of the rf electric field as well as that of the stationary electron density into account. The latter mechanism has been discussed in detail by Wetzel and Tang (1) in their review; in the present treatment, we have therefore isolated the former mechanism. The treatment has been further extended [following Sodha and Kaw (24), and Paranjape (25)] to the case of two rf fields in which the sum-and-difference frequencies are generated in addition to the even harmonics.

The first moment of the collisionless Boltzmann equation may be written as

$$(\partial \mathbf{v}/\partial t) + (\mathbf{v} \cdot \nabla)\mathbf{v} = -(e/m)(\mathbf{E} + (\mathbf{v} \times \mathbf{B}/c)) \tag{3.23}$$

where $\mathbf{B}$ is the rf magnetic field. Assuming the electron velocity to be of the form

$$\mathbf{v} = \mathbf{v}_1 \exp(i\omega_1 t) + \mathbf{v}_{12} \exp(2i\omega_1 t) + \mathbf{v}_2 \exp(i\omega_2 t)$$
$$+ \mathbf{v}_{22} \exp(2i\omega_2 t) + \mathbf{v}_+ \exp\{i(\omega_1 + \omega_2)t\} + \mathbf{v}_- \exp\{i(\omega_1 - \omega_2)t\} \quad (3.24)$$

substituting in the above equation, and equating the coefficients of the various frequency components on both sides of the resulting equation, one obtains the equations

$$i\omega_1 \mathbf{v}_1 = -(e/m)\mathbf{E}_1, \qquad i\omega_2 \mathbf{v}_2 = -(e/m)\mathbf{E}_2 \quad (3.25a,b)$$

$$2i\omega_1 \mathbf{v}_{12} + \tfrac{1}{2}(\mathbf{v}_1 \cdot \nabla)\mathbf{v}_1 = -(e/2mc)(\mathbf{v}_1 \times \mathbf{B}_1) \quad (3.26)$$

$$2i\omega_2 \mathbf{v}_{22} + \tfrac{1}{2}(\mathbf{v}_2 \cdot \nabla)\mathbf{v}_2 = -(e/2mc)(\mathbf{v}_2 \times \mathbf{B}_2) \quad (3.27)$$

$$i(\omega_1 + \omega_2)\mathbf{v}_+ + \tfrac{1}{2}(\mathbf{v}_1 \cdot \nabla)\mathbf{v}_2 + \tfrac{1}{2}(\mathbf{v}_2 \cdot \nabla)\mathbf{v}_1$$
$$= -(e/2mc)[\mathbf{v}_1 \times \mathbf{B}_2 + \mathbf{v}_2 \times \mathbf{B}_1] \quad (3.28)$$

$$i(\omega_1 - \omega_2)\mathbf{v}_- + \tfrac{1}{2}(\mathbf{v}_1 \cdot \nabla)\tilde{\mathbf{v}}_2 + \tfrac{1}{2}(\tilde{\mathbf{v}}_2 \cdot \nabla)\mathbf{v}_1$$
$$= -(e/2mc)[\mathbf{v}_1 \times \tilde{\mathbf{B}}_2 + \tilde{\mathbf{v}}_2 \times \mathbf{B}_1] \quad (3.29)$$

where $\mathbf{B}_1$ and $\mathbf{B}_2$ are given by one of the Maxwell's equations, viz.,

$$\nabla \times \mathbf{E}_1 = -\frac{1}{c}\frac{\partial \mathbf{B}_1}{\partial t} = -\frac{i\omega_1}{c}\mathbf{B}_1 \quad (3.30a)$$

$$\nabla \times \mathbf{E}_2 = -\frac{1}{c}\frac{\partial \mathbf{B}_2}{\partial t} = -\frac{i\omega_2}{c}\mathbf{B}_2 \quad (3.30b)$$

Solving the above equations for $\mathbf{v}_{12}$, $\mathbf{v}_{22}$, $\mathbf{v}_+$ and $\mathbf{v}_-$, and using the vector identity

$$\nabla(\mathbf{A} \cdot \mathbf{B}) = (\mathbf{A} \cdot \nabla)\mathbf{B} + (\mathbf{B} \cdot \nabla)\mathbf{A} + \mathbf{A} \times (\nabla \times \mathbf{B}) + \mathbf{B} \times (\nabla \times \mathbf{A})$$

one obtains the following equations:

$$\mathbf{v}_{12} = (e^2/8im^2\omega_1{}^3)\nabla(E_1{}^2) \quad (3.31a)$$

$$\mathbf{v}_{22} = (e^2/8im^2\omega_2{}^3)\nabla(E_2{}^2) \quad (3.31b)$$

$$\mathbf{v}_+ = [e^2/2im^2\omega_1\omega_2(\omega_1 + \omega_2)]\nabla(\mathbf{E}_1 \ \mathbf{E}_2) \quad (3.31c)$$

$$\mathbf{v}_- = -[e^2/2im^2\omega_1\omega_2(\omega_1 - \omega_2)]\nabla(\mathbf{E}_1 \cdot \tilde{\mathbf{E}}_2) \quad (3.31d)$$

where $\tilde{E}_2$ denotes the complex conjugate of E_2. Thus, the second harmonic and the sum-and-difference frequency components of the electron velocity arise entirely because the electric field is nonuniform. No such components are generated for uniform microwave fields alone. The current density is given by

$$\mathbf{J} = -Ne\mathbf{v} \quad (3.32)$$

where N is the uniform stationary electron density; thus its second harmonic and sum-and-difference frequency components are given by

$$\mathbf{J}_{12} = (iNe^3/8m^2\omega_1{}^3)\nabla(E_1{}^2)\exp(2i\omega_1 t) \tag{3.33a}$$

$$\mathbf{J}_{22} = (iNe^3/8m^2\omega_2{}^3)\nabla(E_2{}^2)\exp(2i\omega_2 t) \tag{3.33b}$$

$$\mathbf{J}_{12}^{+} = [iNe^3/2m^2\omega_1\omega_2(\omega_1 + \omega_2)]\nabla(\mathbf{E}_1 \cdot \mathbf{E}_2)\exp\{i(\omega_1 + \omega_2)t\} \tag{3.33c}$$

$$\mathbf{J}_{12}^{-} = [iNe^3/2m^2\omega_1\omega_2(\omega_1 - \omega_2)]\nabla(\mathbf{E}_1 \cdot \tilde{\mathbf{E}}_2)\exp\{i(\omega_1 - \omega_2)t\} \tag{3.33d}$$

An interesting result worth mentioning is that the magnitudes of the sum-and-difference frequency components in the current density are proportional to the cosine of the angle between the two electric fields $\mathbf{E}_1$ and $\mathbf{E}_2$. Thus these components are a maximum when $\mathbf{E}_1$ and $\mathbf{E}_2$ are parallel, and vanish when they are perpendicular, to each other.

Paranjape (25) has pointed out that equating the time-independent components on both sides of Eq. (3.23) leads to the conditions

$$(\tilde{\mathbf{E}}_1 \cdot \nabla)\mathbf{E}_1 = 0 \tag{3.34}$$

$$(\tilde{\mathbf{E}}_2 \cdot \nabla)\mathbf{E}_2 = 0 \tag{3.35}$$

correct to a first order of approximation. These equations therefore put some limits on the type of electric field for which the above treatment is valid.

In the investigation described above the collisions have been assumed to be absent; the case of second harmonic generation in the presence of collisions and a static magnetic field has been analyzed by Krenz and Kino (36) and Krenz (35). The chief new result is the prediction of a resonance in the second harmonic output when the fundamental frequency equals the gyrofrequency; this appears because of the resonant increase in the input power at this frequency and disappears if the input power is held constant. Krenz and Kino (37) report appreciably large efficiencies of harmonic generation, which seem to be in reasonable agreement with their experiments.

B. Inhomogeneous Plasmas

The inhomogeneous nature of a plasma may arise because of the presence of either a finite electron density gradient or a finite temperature gradient. In the presence of these inhomogenities, the even harmonics and the sum-and-difference frequencies are generated in the current density by two microwave fields alone. This is not possible for a homogeneous plasma, where as we have seen earlier, the presence of two microwave fields alone leads only to the generation of odd harmonics and the second-order combination frequencies in the current density.

1. Two Microwave Fields

Consider first of all an inhomogeneous plasma under the influence of two microwave fields alone. Let us try to analyze the magnitudes of the harmonics and the combination frequencies in the current density. Let both the temperature and the electron density gradients be acting along the same direction (say, x axis) and let the electric field

$$\mathbf{E} = \mathbf{E}_1 \exp(i\omega_1 t) + \mathbf{E}_2 \exp(i\omega_2 t) \tag{3.36}$$

be also directed along the same direction.

For the present case, Eqs. (2.4a–c) take the forms

$$\frac{\partial f^0}{\partial t} + \frac{v}{3}\frac{\partial f^1}{\partial x} = \frac{a}{3v^2}\frac{\partial}{\partial v}(v^2 f^1) + \left(\frac{\partial f^0}{\partial t}\right)_c \tag{3.37a}$$

$$\frac{\partial f^1}{\partial t} + v\left(\frac{\partial f^0}{\partial x} + \frac{2}{5}\frac{\partial f^2}{\partial x}\right) + vf^1 = a\left[\frac{\partial f^0}{\partial v} + \frac{2}{5v^3}\frac{\partial}{\partial v}(v^3 f^2)\right] \tag{3.37b}$$

$$\frac{\partial f^2}{\partial t} + vf^2 + \frac{2v}{3}\frac{\partial f^1}{\partial x} = \frac{2v}{3}a\frac{\partial}{\partial v}\left(\frac{f^1}{v}\right) \tag{3.37c}$$

where

$$a = a_1 \exp(i\omega_1 t) + a_2 \exp(i\omega_2 t)$$

$$a_1 = eE_1/m, \qquad a_2 = eE_2/m$$

and the expected time dependence of f^0, f^1, and f^2, in accordance with the discussion following Eq. (2.14), are as follows:

$$
\begin{aligned}
f^0 = {}& f_0{}^0 + f_{11}^0 \exp(i\omega_1 t) + f_{12}^0 \exp(2i\omega_1 t) + f_{21}^0 \exp(i\omega_2 t) \\
& + f_{22}^0 \exp(2i\omega_2 t) + f_{12}^{0+} \exp\{i(\omega_1 + \omega_2)t\} + f_{12}^{0-} \exp\{i(\omega_1 - \omega_2)t\}
\end{aligned}
\tag{3.38a}
$$

$$
\begin{aligned}
f^1 = {}& f_0{}^1 + f_{11}^1 \exp(i\omega_1 t) + f_{12}^1 \exp(2i\omega_1 t) + f_{13}^1 \exp(3i\omega_1 t) \\
& + f_{21}^1 \exp(i\omega_2 t) + f_{22}^1 \exp(2i\omega_2 t) + f_{23}^1 \exp(3i\omega_2 t) \\
& + f_{12}^{1+} \exp\{i(\omega_1 + \omega_2)t\} + f_{12}^{1-} \exp\{i(\omega_1 - \omega_2)t\} \\
& + f_{12}^{1+\prime} \exp\{i(\omega_1 + 2\omega_2)t\} + f_{12}^{1-\prime} \exp\{i(\omega_1 - 2\omega_2)t\} \\
& + f_{21}^{1+\prime} \exp\{i(\omega_2 + 2\omega_1)t\} + f_{21}^{1-\prime} \exp\{i(\omega_2 - 2\omega_1)t\}
\end{aligned}
\tag{3.38b}
$$

$$
\begin{aligned}
f^2 = {}& f_0{}^2 + f_{11}^2 \exp(i\omega_1 t) + f_{12}^2 \exp(2i\omega_1 t) + f_{21}^2 \exp(i\omega_2 t) \\
& + f_{22}^2 \exp(2i\omega_2 t) + f_{12}^{2+} \exp\{i(\omega_1 + \omega_2)t\} + f_{12}^{2-} \exp\{i(\omega_1 - \omega_2)t\}
\end{aligned}
\tag{3.38c}
$$

One normally expects terms with frequencies $3\omega_1$, $3\omega_2$, $\omega_1 \pm 2\omega_2$, and $\omega_2 \pm 2\omega_1$ in f^0 and f^2 also, but since their contributions to the higher order components of the current density are negligible in the range of validity of the present treatment (discussed in detail below), they have not been considered.

The existence of the terms with frequencies ω_1 and ω_2 in f^0, and their relevance in determining the magnitudes of the sum-and-difference frequency components in the current density was first pointed out by Vilenskii (5, 6); however, a detailed analytical investigation of this phenomenon and the related mechanism of the generation of even harmonics was carried out by Wetzel (10, 29, 30). Most of Wetzel's work is incorporated in his recent review on harmonic generation [Wetzel and Tang (1)]. Chiyoda (12) and Sodha and Kaw (20–22) have discussed the generation of these higher order components due to the terms with frequencies $2\omega_1$, $2\omega_2$, and $\omega_1 \pm \omega_2$ in f^0 alone, without taking into account the terms pointed out by Vilenskii (5, 6). Gupta (14) has given a comprehensive treatment combining the contributions due to all these first-order terms and also taking into account the important contributions from the second-order terms f^2. In the following analysis, we shall closely follow Gupta (14).

Now, in a strictly rigorous analysis, one should seek a simultaneous solution of the Poisson equation and the Boltzmann transfer equation for electrons. However, since such an analysis leads very soon to serious mathematical difficulties, one may proceed with less rigorous assumptions, viz.:

a. The ionized gas is neutral at all points so that the electric fields due to the nonuniform charge distribution (not electron density) may be neglected. Such an assumption is justified in view of the rigorous calculations of Darwin (65), Ginzburg (66, 67), Kadomtsev (68), etc., which point toward the fact that the effective field and the average macroscopic field in the plasma are identical. A simple physical explanation of this fact cannot be readily given, as Ginzburg (69) has rightly pointed out.

b. The distribution function of electron velocities depends on the position coordinate x only through the spatial dependence of the electron density N and the gas temperature T. Thus,

$$\partial f/\partial x = \gamma N \, \partial f/\partial N + CT \, \partial f/\partial T$$

where

$$\gamma = (1/N) \, dN/dx \qquad \text{and} \qquad C = (1/T) \, dT/dx$$

Substituting for $a, f, (\partial f/\partial x), (\partial f/\partial t)$, and $(\partial f/\partial t)_c$ in Eqs. (3.37a–b), equating the time-independent terms and the coefficients of various frequency terms on both sides of each of the resulting equations, and using the approximations (1)

$$f^1_{12}, f^1_{22}, f^1_{13}, f^1_{23}, f^{1\pm}_{12}, f^{1\pm\prime}_{12}, f^{1\pm\prime}_{21} \ll f^1_{11} \quad \text{or} \quad f^1_{21}$$

and (2) the terms of the order of $\gamma^4 a_1$, $C^4 a_1$, $\gamma^2 C^2 a_1$, $\gamma^2 a_1{}^3$, etc., are negligible, one obtains explicit expressions for the various components of f^0, f^1, and f^2 (for details, see Appendix D). Approximation (1) means, as before, that the harmonic and combination frequency components of the current density have a much smaller magnitude than the fundamentals. Approximation (2) limits the range of applicability of our treatment to low values of the density and temperature gradients in that the terms of the order of a fifth power in γ, C, a_1, or a_2 should be negligible in comparison with those occurring in the third power. It is this latter assumption that permits us to neglect the terms with frequencies $3\omega_1$, $3\omega_2$, $\omega_1 \pm 2\omega_2$, and $\omega_2 \pm 2\omega_1$ in f^0 because their contributions to the harmonic and combination frequencies will be in the form of fifth and higher powers of γ, C, a_1, or a_2. The expressions for the dc second harmonic, $2\omega_1$, and the sum frequency, $(\omega_1 + \omega_2)$, components in f^1 are given by (for details see Appendix D)

$$f_0{}^1 = -(v/v)\Psi \tag{3.39a}$$

$$
\begin{aligned}
f_{12}^1 = & -\frac{a_1{}^2}{v + 2i\omega_1}\left[\frac{1}{12i\omega_1 v}\frac{\partial}{\partial v}\left\{\frac{v^2}{v + i\omega_1}\frac{\partial\Psi}{\partial v}\right\} + \frac{1}{6i\omega_1}\frac{\partial}{\partial v}\left\{\frac{v}{v + i\omega_1}\frac{\partial\Psi}{\partial v}\right\}\right. \\
& + \frac{2}{15}\frac{v^2}{(v + 2i\omega_1)}\frac{\partial}{\partial v}\left\{\frac{1}{v + i\omega_1}\frac{1}{v}\frac{\partial\Psi}{\partial v}\right\} + \frac{1}{6i\omega_1}\frac{\partial}{\partial v}\left\{\frac{1}{v^2}\frac{\partial}{\partial v}\left(\frac{v^3\Psi}{v}\right)\right\} \\
& \left. + \frac{2}{15v^3}\frac{\partial}{\partial v}\left\{\frac{v^4}{(v + i\omega_1)^2}\frac{\partial\Psi}{\partial v} + \frac{v^4}{v + i\omega_1}\frac{\partial}{\partial v}\left(\frac{\Psi}{v}\right)\right\}\right]
\end{aligned}
\tag{3.39b}
$$

$$
\begin{aligned}
f_{12}^{1+} = & -\frac{a_1 a_2}{v + i(\omega_1 + \omega_2)}\left[\frac{1}{6i(\omega_1 + \omega_2)v}\frac{\partial}{\partial v}\left\{\left(\frac{1}{v + i\omega_1} + \frac{1}{v + i\omega_2}\right)v^2\frac{\partial\Psi}{\partial v}\right\}\right. \\
& + \frac{2}{15}\cdot\frac{v^2}{[v + i(\omega_1 + \omega_2)]}\frac{\partial}{\partial v}\left\{\left(\frac{1}{v + i\omega_1} + \frac{1}{v + i\omega_2}\right)\frac{1}{v}\frac{\partial\Psi}{\partial v}\right\} \\
& + \frac{2}{15v^3}\frac{\partial}{\partial v}\left\{v^4\frac{\partial\Psi}{\partial v}\left[\frac{1}{(v + i\omega_1)^2} + \frac{1}{(v + i\omega_2)^2}\right]\right. \\
& \left. + v^4\frac{\partial}{\partial v}\left(\frac{\Psi}{v}\right)\left(\frac{1}{v + i\omega_1} + \frac{1}{v + i\omega_2}\right)\right\} \\
& + \frac{1}{6i}\frac{\partial}{\partial v}\left\{v\frac{\partial\Psi}{\partial v}\left[\frac{1}{\omega_2(v + i\omega_2)} + \frac{1}{\omega_1(v + i\omega_1)}\right]\right. \\
& \left.\left. + \frac{1}{v^2}\frac{\partial}{\partial v}\left(\frac{v^3}{v}\Psi\right)\left(\frac{1}{\omega_2} + \frac{1}{\omega_1}\right)\right\}\right]
\end{aligned}
\tag{3.39c}
$$

where

$$\Psi \equiv \gamma N\,\partial f_0{}^0/\partial N + CT\,\partial f_0{}^0/\partial T \tag{3.40}$$

It is noted that the expression for $f^1_{22}/a_2{}^2$ can be obtained from that of $f^1_{12}/a_1{}^2$ by replacing ω_1 with ω_2 ; the expression for $f^{1-}_{12}/a_1\tilde{a}_2$ can be obtained from that of $f^{1+}_{12}/a_1 a_2$ by replacing ω_2 with $-(\omega_2)$. The expressions for the remaining time-dependent components of f^1 are exactly the same as that for a homogeneous plasma; this result comes about because of the approximation (2) mentioned above. Using Eq. (3.1) and the expressions for the components of f^1 derived in Appendix D, one obtains for the harmonic and combination frequency components of the current density, the expressions

$$J_{12} = (m/M)\alpha[(\gamma - CP)A_{12d} + CA_{12t}]\mathscr{E}_1{}^2 \exp(2i\omega_1 t) \tag{3.41a}$$

$$J_{22} = (m/M)\alpha[(\gamma - CP)A_{22d} + CA_{22t}]\mathscr{E}_2{}^2 \exp(2i\omega_2 t) \tag{3.41b}$$

$$J^+_{12} = (m/M)\alpha[(\gamma - CP)A^+_{12d} + CA^+_{12t}]\mathscr{E}_1\mathscr{E}_2 \exp\{i(\omega_1 + \omega_2)t\} \tag{3.41c}$$

$$J^-_{12} = (m/M)\alpha[(\gamma - CP)A^-_{12d} + CA^-_{12t}]\mathscr{E}_1\tilde{\mathscr{E}}_2 \exp\{i(\omega_1 - \omega_2)t\} \tag{3.41d}$$

$$J_{13} = (m/M)\alpha A_{13}\,\mathscr{E}_1{}^2 E_1 \exp(3i\omega_1 t) \tag{3.41e}$$

$$J_{23} = (m/M)\alpha A_{23}\,\mathscr{E}_2{}^2 E_2 \exp(3i\omega_2 t) \tag{3.41f}$$

$$J^{+\prime}_{12} = (m/M)\alpha A^{+\prime}_{12}\mathscr{E}_2{}^2 E_1 \exp\{i(\omega_1 + 2\omega_2)t\} \tag{3.41g}$$

$$J^{-\prime}_{12} = (m/M)\alpha A^{-\prime}_{12}\tilde{\mathscr{E}}_2{}^2 E_1 \exp\{i(\omega_1 - 2\omega_2)t\} \tag{3.41h}$$

$$J^{+\prime}_{21} = (m/M)\alpha A^{+\prime}_{21}\mathscr{E}_1{}^2 E_2 \exp\{i(\omega_2 + 2\omega_1)t\} \tag{3.41i}$$

$$J^{-\prime}_{21} = (mM)\alpha A^{-\prime}_{21}\tilde{\mathscr{E}}_1{}^2 E_2 \exp\{i(\omega_2 - 2\omega_1)t\} \tag{3.41j}$$

where the A_d and A_t are integrals involving the collision frequency; expressions for some of them are given in Appendix D. P is a parameter defined by

$$P = \tfrac{3}{2} - (T/N)\,\partial N/\partial T \tag{3.42}$$

Thus its magnitude is determined by the dependence of the electron density on temperature. Gupta (*14*) has considered two special situations:

1. An equilibrium composition. This is the case of a thermal plasma in which the electron density is governed by Saha's equation, viz.,

$$N = AT^{3/4} \exp(-U/2kT)$$

where U is the ionization potential of the gas and A is a constant involving the number of neutral atoms, ions, etc.

In this case, obviously,

$$P = \tfrac{3}{4} - (U/2kT)$$

2. A frozen composition. In this case the electron density is assumed to be independent of temperature so that $P = \tfrac{3}{2}$.

In general it is very difficult to ascribe any practical situation to one of these two cases; however, since the two results are of the same order of magnitude, a large error is not expected if one of them is chosen. It may be mentioned that in Sodha and Kaw's (20–22) treatment, the dependence of N on T was such that $P = 0$; this is again justified only because it gives correct orders of magnitude in the results.

The integrals in the A_d and A_t may be put in a closed form when

Case I. $|\omega_1 - \omega_2| \gg v$, $|\omega_1 - 2\omega_2| \gg v$, $\omega_1 \gg v$, $\omega_2 \gg v$.

Case II. $v \gg 3\omega_1$, $v \gg 3\omega_2$,
$$v = v_0 u^n, \text{ and } f_0{}^0 \text{ is taken to be Maxwellian in form.}$$

This last assumption is justified in view of the fact that our treatment is correct only up to terms of the order of $(m/M)\alpha$ and that the contributions of the non-Maxwellian parts of $f_0{}^0$ to the harmonic and combination frequency components in the current density will be of the order of $(m/M)\alpha^2$, $m/M\alpha\gamma^2$, $(m/M)\alpha C^2$, etc., and hence negligible. The assumption of a Maxwellian form for $f_0{}^0$ further allows us to use the equation

$$\partial f_0{}^0/\partial N = f_0{}^0/N \tag{3.43}$$

The expressions for the A for $n = 1$ (and Case I) are given in Appendix D. When the electron collision frequency is independent of electron velocity (i.e., when $n = 0$), one obtains simpler expressions:

$$A_{12d} = \frac{kT}{e} \frac{\omega_p{}^2}{6i\pi\omega_2} \left[\frac{33}{16} \frac{v_0{}^2}{\omega_2{}^2} - \frac{15}{4} i \frac{v_0}{\omega_2} - \frac{3}{4} \right] \tag{3.44a}$$

$$\begin{aligned}
A_{12d}^+ = \frac{kT}{e} \frac{\omega_p{}^2\omega_0{}^2}{\pi i(\omega_1 + \omega_2)^2} &\left[v_0{}^2 \left\{ \frac{5}{2(\omega_1 + \omega_2)} \left(\frac{1}{\omega_2{}^2} + \frac{1}{\omega_1{}^2} \right) + \frac{3}{4} \left(\frac{1}{\omega_2{}^3} + \frac{1}{\omega_1{}^3} \right) \right\} \right. \\
&- iv_0 \left\{ \frac{5}{2\omega_1\omega_2} + \frac{9}{4} \left(\frac{1}{\omega_1{}^2} + \frac{1}{\omega_2{}^2} \right) + \frac{3}{4} (\omega_1 + \omega_2) \left(\frac{1}{\omega_1{}^3} + \frac{1}{\omega_2{}^3} \right) \right\} \\
&\left. - \frac{3}{2} \left\{ \left(\frac{1}{\omega_1} + \frac{1}{\omega_2} \right) + \frac{(\omega_1 + \omega_2)}{2} \left(\frac{1}{\omega_1{}^2} + \frac{1}{\omega_2{}^2} \right) \right\} \right]
\end{aligned} \tag{3.44b}$$

The expressions for the other A can be obtained from these in a manner discussed in Appendix D. One notes that the second harmonic and the sum-and-difference frequency components in an inhomogeneous plasma are finite in magnitude, even for a collision frequency independent of electron velocity. This result is in contrast to that obtained for a homogeneous plasma in the presence of two microwave fields and a dc field, where these components were seen to vanish. It may be pointed out further that in the present case, the second harmonic and the sum-and-difference frequency components are seen to possess a finite magnitude, even in the limiting case $v_0 \to 0$; this is not true for a homogeneous plasma, as has been seen earlier.

To study the variation of the second harmonic and the sum-and-difference frequency components of the current density with the various relevant parameters, let us first of all note that these components are affected about equally by the temperature and density gradients; thus, both γ and C appear with a unit power in the numerator in all cases. The variation of these components with both ω_p^2/ω_0^2 (hence the electron density) and v_0/ω_0 is found to be the same as that for a homogeneous plasma, viz., that a direct proportionality exists. To illustrate the nature and magnitude of the variation with the dependence of collision frequency on electron velocity, some calculations have been made for a frozen electron density and the cases when the electron collision frequency is independent of and proportional to the first power of electron velocity. Table IV illustrates the variations of

$$j_{12} = \left| \frac{24 i m \pi \omega_0^3}{\gamma e \omega_p^2} \frac{J_{12}}{E_1^2 \exp(2i\omega_1 t)} \right|$$

$$= 4i\pi \frac{\omega_0}{\omega_p^2} \frac{e}{kT} [(1 - P)A_{12d} + A_{12t}] \tag{3.45a}$$

$$j_{12}^+ = \left| \frac{24 i m \pi \omega_0^3}{\gamma e \omega_p^2} \frac{J_{12}^+}{E_1 E_2 \exp\{i(\omega_1 + \omega_2)t\}} \right|$$

$$= 4i\pi \frac{\omega_0}{\omega_p^2} \frac{e}{kT} [(1 - P)A_{12d}^+ + A_{12t}^+] \tag{3.45b}$$

and

$$j_{12}^- = \left| \frac{24 i m \pi \omega_0^3}{\gamma e \omega_p^2} \frac{J_{12}^-}{E_1 \tilde{E}_2 \exp\{i(\omega_1 - \omega_2)t\}} \right|$$

$$= 4i\pi \frac{\omega_0}{\omega_p^2} \frac{e}{kT} [(1 - P)A_{12d}^- + A_{12t}^-] \tag{3.45c}$$

with n and ω_1/ω_2 for $\omega_0 = \omega_2$, $\gamma = C$, $v_0/\omega = 0.01$, and a frozen composition of the plasma ($P = \frac{3}{2}$).

The variation of the second harmonic and the sum-and-difference frequency components in the current density with n is found to be appreciable—more so for lower values of ω_1/ω_2. The variation with the ratio (ω_1/ω_2) is also significantly large; the nature of variation is similar to that exhibited by these frequencies in a homogeneous plasma in the presence of a dc electric field. Thus a resonance is expected in the difference frequency at $\omega_1/\omega_2 \simeq 1$.

TABLE IV

VARIATION OF j_{12}, $j_{12}^{\pm}$ WITH ω_1/ω_2 FOR $\omega_0 = \omega_2$, $\nu_0/\omega_0 = 0.01$

	j_{12}		j_{12}^{+}		j_{12}^{-}	
ω_1/ω_2	$n=0$	$n=1$	$n=0$	$n=1$	$n=0$	$n=1$
2			2.250	2.145	0.750	5.510
4	0.500	0.227	0.936	0.817	0.562	1.029
6	no variation	no variation	0.583	0.527	0.416	0.565
8			0.422	0.351	0.328	0.428

2. Effect of a Magnetic Field

The generation of the sum-and-difference frequencies in an inhomogeneous plasma in the presence of a magnetic field does not seem to have been attempted so far. Sodha and Kaw (22) investigated the simpler problem of the generation of second harmonic in the current density in an inhomogeneous magnetoplasma. That treatment was, however, a direct extension of Chiyoda's (12) work and did not include the important contribution to the second harmonic arising because of the fundamental frequency term in f^0 and all f^2 terms. These are the terms whose relevance in determining the magnitude of the second harmonic was pointed out by Wetzel (30) and Gupta (15). In the present treatment, following Gupta (15), expressions for the second-harmonic current density have been derived, taking all the first and second-order contributions into account.

It will be assumed that the inhomogenities and the electric field in the plasma are in the xy plane and that the magnetic field is at right angles to them (i.e., along the z direction). One can write for the present case

$$E_{x,y} = E_{0x,y}\exp(i\omega t), \qquad E_z = 0$$

$$B_{x,y} = 0, \qquad B_z = B_0$$

and

$$\partial N/\partial z = \partial T/\partial z = 0$$

In the present case Eqs. (2.4a–e) may be used as they are, where

$$a_x = a_{1x}\exp(i\omega t), \qquad a_y = a_{1y}\exp(i\omega t)$$

and the expected time dependence of f^0, $f^1_{x,y}$, and $f^2_{x,y}$ are given by

$$f^0 = f_0^{\,0} + f_1^{\,0}\exp(i\omega t) + f_2^{\,0}\exp(2i\omega t) + \cdots$$

$$f^1_{x,y} = f^1_{0x,y} + f^1_{1x,y}\exp(i\omega t) + f^1_{2x,y}\exp(2i\omega t) + f^1_{3x,y}\exp(3i\omega t) + \cdots$$

$$f^2_{x,y} = f^2_{0x,y} + f^2_{1x,y}\exp(i\omega t) + f^2_{2x,y}\exp(2i\omega t) + \cdots$$

Proceeding as before, one obtains the following expression for the x component of the second-harmonic part of f^1:

$$-f^1_{2x}[(v + 2i\omega)^2 + \omega_B{}^2]$$

$$= \frac{a_x{}^2 + a_y{}^2}{12i\omega v}\left[(v + 2i\omega)\frac{\partial}{\partial v}\left\{v^2\,\frac{v + i\omega}{(v + i\omega)^2 + \omega_B{}^2}\,\frac{\partial\Psi_x}{\partial v}\right\}\right.$$

$$\left.- \omega_B\frac{\partial}{\partial v}\left\{v^2\,\frac{v + i\omega}{(v + i\omega)^2 + \omega_B{}^2}\,\frac{\partial\Psi_y}{\partial v}\right\}\right]$$

$$+ \frac{(v + 2i\omega)a_x - \omega_B a_y}{6i\omega}\frac{\partial}{\partial v}\left[\frac{(v + i\omega)a_x - \omega_B a_y}{(v + i\omega)^2 + \omega_B{}^2}\,v\,\frac{\partial\Psi_x}{\partial v}\right.$$

$$+ \frac{(v + i\omega)a_y + \omega_B a_x}{(v + i\omega)^2 + \omega_B{}^2}\,v\,\frac{\partial\Psi_y}{\partial v}$$

$$\left.+ \frac{1}{v^2}\frac{\partial}{\partial v}\left\{v^2\,\frac{(va_x - \omega_B a_y)\Psi_x + (va_y + \omega_B a_x)\Psi_y}{v^2 + \omega_B{}^2}\right\}\right]$$

$$+ \frac{2}{15}\left[a_x v^2\frac{\partial}{\partial v}\left\{\frac{(v + i\omega)a_x - \omega_B a_y}{(v + i\omega)^2 + \omega_B{}^2}\,\frac{1}{v}\,\frac{\partial\Psi_x}{\partial v}\right\}\right.$$

$$\left.- \frac{a_y \omega_B}{v + 2i\omega}\,v^2\frac{\partial}{\partial v}\left\{\frac{(v + i\omega)a_y + \omega_B a_x}{(v + i\omega)^2 + \omega_B{}^2}\,\frac{1}{v}\,\frac{\partial\Psi_y}{\partial v}\right\}\right]$$

$$+ \frac{2a_x}{15v^3}\,(v + 2i\omega)\frac{\partial}{\partial v}$$

$$\times\left[\frac{v^4}{v + i\omega}\left\{\frac{(v + i\omega)a_x - \omega_B a_y}{(v + i\omega)^2 + \omega_B{}^2}\,\frac{\partial\Psi_x}{\partial v} + \frac{\partial}{\partial v}\left(\frac{v\Psi_x - \omega_B\Psi_y}{v^2 + \omega_B{}^2}\right)\right\}\right]$$

$$- \frac{2a_y}{15v^3}\,\omega_B\frac{\partial}{\partial v}\left[\frac{v^4}{v + i\omega}\left\{\frac{(v + i\omega)a_y + \omega_B a_x}{(v + i\omega)^2 + \omega_B{}^2}\,\frac{\partial\Psi_y}{\partial v} + \frac{\partial}{\partial v}\left(\frac{v\Psi_y + \omega_B\Psi_x}{v^2 + \omega_B{}^2}\right)\right\}\right]$$

where

$$\Psi_{x,y} = \gamma_{x,y}N\,\partial f_0{}^0/\partial N + C_{x,y}T\,\partial f_0{}^0/\partial T$$

Expression for f^1_{2y} is identical to that for f^1_{2x} with the places of x and y interchanged and with ω_B replaced by $-\omega_B$. The x and y components of the second harmonic in the current density are then given by

$$\begin{aligned}
J_{x12} = (m/M)\alpha[&\{(\gamma_x - PC_x)A_{1d} + C_x A_{1t}\}\mathscr{E}_x{}^2\\
+ &\{(\gamma_y - PC_y)A_{2d} + C_y A_{2t}\}\mathscr{E}_x{}^2 + \{(\gamma_y - PC_x)A'_{2d} + C_x A'_{2t}\}\mathscr{E}_x{}^2\\
+ &\{(\gamma_x - PC_x)A_{3d} + C_x A_{3t}\}\mathscr{E}_y{}^2 + \{(\gamma_y - PC_y)A_{4d} + C_y A_{4t}\}\mathscr{E}_y{}^2\\
+ &\{(\gamma_y - PC_x)A'_{4d} + C_x A'_{4t}\}\mathscr{E}_y{}^2 + \{(\gamma_x - PC_x)A_{5d} + C_x A_{5t}\}\mathscr{E}_x\mathscr{E}_y\\
+ &\{(\gamma_y - PC_y)A_{6d} + C_y A_{6t}\}\mathscr{E}_x\mathscr{E}_y\\
+ &\{(\gamma_y - PC_x)A'_{6d} + C_x A'_{6t}\}\mathscr{E}_x\mathscr{E}_y]\exp(2i\omega t)
\end{aligned}$$

$$\tag{3.46a}$$

and

$$
\begin{aligned}
J_{y12} = (m/M)\alpha[&\{(\gamma_y - PC_y)A_{1d} + C_y A_{1t}\}\mathscr{E}_x{}^2 \\
&- \{(\gamma_x - PC_x)A_{2d} + C_x A_{2t}\}\mathscr{E}_y{}^2 - \{(\gamma_x - PC_y)A'_{2d} + C_y A'_{2t}\}\mathscr{E}_y{}^2 \\
&+ \{(\gamma_y - PC_y)A_{3d} + C_y A_{3t}\}\mathscr{E}_x{}^2 - \{(\gamma_x - PC_x)A_{4d} + C_x A_{4t}\}\mathscr{E}_x{}^2 \\
&- \{(\gamma_x - PC_y)A'_{4d} + C_y A'_{4t}\}\mathscr{E}_x{}^2 - \{(\gamma_y - PC_y)A_{5d} + C_y A_{5t}\}\mathscr{E}_x\mathscr{E}_y \\
&+ \{(\gamma_x - PC_x)A_{6d} + C_x A_{6t}\}\mathscr{E}_x\mathscr{E}_y \\
&+ \{(\gamma_x - PC_y)A'_{6d} + C_y A'_{6t}\}\mathscr{E}_x\mathscr{E}_y] \exp(2i\omega t)
\end{aligned}
\tag{3.46b}
$$

where the expressions for α and P are the same as in the preceding section (with $\omega_0 = \omega$) and the A are integrals involving the collision frequency. Under the assumption of small density and temperature gradients the expression for the third-harmonic component of the current density for the present case is the same as that for a homogeneous plasma. Explicit expressions for the A may be obtained for three cases:

1. $\omega_B = 0, \qquad \omega \gg \nu.$
2. $\omega_B = \omega, \qquad \omega \gg \nu.$
3. $\omega_B = 2\omega, \qquad \omega \gg \nu;$
   $\nu = \nu_0 u^n$, and when $f_0{}^0$ is taken to be Maxwellian corresponding to the temperature T of the gas (in accordance with the considerations given earlier).

In the Appendix E, however, only the expressions for $n = 0$ (i.e., for a velocity-independent collision frequency) have been derived. In this case the third harmonic remains of a finite magnitude; further, for a frozen composition plasma $P = \frac{3}{2}$, the above expressions for the components of the current density are reduced to the simpler forms:

$$
\begin{aligned}
J_{x12} = (m/M)\alpha[&\{\gamma_x A_{1d} + \gamma_y A_{2d}\}\mathscr{E}_x{}^2 + \{\gamma_x A_{3d} + \gamma_y A_{4d}\}\mathscr{E}_y{}^2 \\
&+ \{\gamma_x A_{5d} + \gamma_y A_{6d}\}\mathscr{E}_x\mathscr{E}_y] \exp(2i\omega t)
\end{aligned}
$$

and

$$
\begin{aligned}
J_{y12} = (m/M)\alpha[&\{\gamma_y A_{1d} - \gamma_x A_{2d}\}\mathscr{E}_y{}^2 + \{\gamma_y A_{3d} - \gamma_x A_{4d}\}\mathscr{E}_x{}^2 \\
&- \{\gamma_y A_{5d} - \gamma_x A_{6d}\}\mathscr{E}_x\mathscr{E}_y] \exp(2i\omega t)
\end{aligned}
$$

where A_{2d}, A_{4d}, and A_{6d} now include A'_{2d}, A'_{4d}, and A'_{6d}, respectively. It is noted that the contribution of the temperature gradient to the second-harmonic current density vanishes; it should be emphasized here that this conclusion is valid only in the special case of a frozen composition plasma with $n = 0$.

The coupled components of the second harmonic in the current density are then given by

$$
\begin{aligned}
J_{x12} + iJ_{y12} = (m/M)\alpha[&(A_{1d} - iA_{4d})(\gamma_x \mathscr{E}_x{}^2 + i\gamma_y \mathscr{E}_y{}^2) \\
&- i(A_{2d} + iA_{3d})(\gamma_x \mathscr{E}_y{}^2 + i\gamma_y \mathscr{E}_x{}^2) \\
&+ (A_{5d} + iA_{6d})(\gamma_x - i\gamma_y)\mathscr{E}_x \mathscr{E}_y]\,\exp(2i\omega t)
\end{aligned}
\tag{3.47a}
$$

and

$$
\begin{aligned}
J_{x12} - iJ_{y12} = (m/M)\alpha[&(A_{1d} + iA_{4d})(\gamma_x \mathscr{E}_x{}^2 - i\gamma_y \mathscr{E}_y{}^2) \\
&+ i(A_{2d} - iA_{3d})(\gamma_x \mathscr{E}_y{}^2 - i\gamma_y \mathscr{E}_x{}^2) \\
&+ (A_{5d} - iA_{6d})(\gamma_x + i\gamma_y)\mathscr{E}_x \mathscr{E}_y]\,\exp(2i\omega t)
\end{aligned}
\tag{3.47b}
$$

One notes from Eqs. (3.47a,b) that the amplitude of the second-harmonic component in the current density is of the order of $(m/M)\alpha\gamma\mathscr{E}^2$ or $e^2/6m\omega^2 kT$ times the product of γ and the square of the electric field. It is further noted that the magnitude of the second-harmonic component is directly proportional to the A, which may have resonances around $\omega_B = \omega$ and $\omega_B = 2\omega$; therefore, one can expect such resonances in the second harmonic also around these frequencies. To get an idea of the nature and magnitude of these resonances, the quantities

$$
j_2' = \left| \frac{48 i\pi m\omega^3}{e\omega_p{}^2} \frac{J_{x2} + iJ_{y2}}{(\gamma_x - i\gamma_y)E_x{}^2} \right|
$$

and

$$
j_2'' = \left| \frac{48 i\pi m\omega^3}{e\omega_p{}^2} \frac{J_{x2} - iJ_{y2}}{(\gamma_x + i\gamma_y)E_x{}^2} \right|
$$

have been calculated for the conditions $\mathscr{E}_{1x} - i\mathscr{E}_{1y} = 0$ and $\mathscr{E}_{1x} + i\mathscr{E}_{1y} = 0$, respectively, and for $n = 0$.

It was noted that j_2' has a very small magnitude for $\mathscr{E}_{1x} + i\mathscr{E}_{1y} = 0$ and and that j_2'' has a very small magnitude for $\mathscr{E}_{1x} - i\mathscr{E}_{1y} = 0$. This means that when a pure mode (ordinary or extraordinary) of the fundamental electromagnetic wave is sent into the inhomogeneous plasma, then primarily there is a generation of the same mode of the second harmonic. This conclusion is in contrast to that obtained in the case of a homogeneous magnetoplasma, where a pure mode of the fundamental generates only the other mode of the third harmonic. Table V illustrates the variations of j_2' and j_2'' with v_0/ω. It is noted from the table that for $\omega_B = \omega$, both j_2' and j_2'' exhibit resonances, whereas for $\omega_B = 2\omega$ only j_2' does so.

3. Around Plasma Frequency

Moriyama and Sumi (48, 49) have discussed the microwave generation of harmonics in a plasma around the plasma frequency in the absence and

TABLE V

VARIATION OF j_2' AND j_2'' WITH v_0/ω FOR $n = 0$

	j_2'			j_2''		
v_0/ω	$\omega_B = 0$	$\omega_B = \omega$	$\omega_B = 2\omega$	$\omega_B = 0$	$\omega_B = \omega$	$\omega_B = 2\omega$
0.01	8.00	1383.9	1600.0	8.00	930.0	1.00
0.03	8.00	460.3	533.4	8.00	313.6	0.05
0.05	8.01	275.9	320.1	8.01	189.5	0.03
0.07	8.02	196.7	228.7	8.02	136.8	0.02
0.10	8.04	138.42	160.3	8.04	97.8	0.02

presence of a magnetic field, respectively. They propose a mechanism in which the harmonics are strongly generated by a resonant coupling of an external field with an inner field in the presence of a nonuniform stationary plasma density. When a uniformly applied microwave field couples with the nonuniform electron density, an inner field of the plasma varying in space and time is established. The coupling of this inner field with the external one causes the second-harmonic field; similarly the second-harmonic field couples with the external field to give the third-harmonic field, and so on. Each harmonic field is therefore one order of magnitude smaller than that of the previous harmonic. When the microwave frequency is around the plasma frequency, there is a resonant increase in the magnitude of the harmonics. In the presence of a magnetic field however (49), the harmonic output has a resonant maximum around the electron cyclotron frequency ω_B. The results of Moriyama and Sumi (48, 49) are in good agreement with the experiments of Uenohara et al. (50) and Hill and Tetenbaum (58).

The two basic equations with which Moriyama and Sumi start are the Boltzmann equation

$$\partial f/\partial t + \mathbf{v} \cdot \partial f/\partial \mathbf{x} - (e\mathbf{E}/m) \cdot \partial f/\partial \mathbf{v} = (\partial f/\partial t)_c$$

and the Poisson's equation

$$\partial \mathbf{E}/\partial x = 4\pi e\left[N_0(x) - \int f\, d\mathbf{v}\right]$$

in which $-e$, m, and $\mathbf{v}$ represent respectively the charge, mass, and velocity of the electron; N_0 is the time-independent number density of electrons; and $\mathbf{E}$ is the total field impressed on the plasma (which consists of the externally applied microwave field and the inner field of the plasma). Fourier-analyzing f and $\mathbf{E}$ in space and time, one has

$$f(v, x, t) = \sum_k \sum_l f_{k,l}(v) \exp[i(kx - l\omega t)]$$

$$E(x, t) = \sum_k \sum_l E_{k,l} \exp[i(kx - l\omega t)]$$

where $k = \pi g/d$ and g, l are integers and d defines the length of the bounded dimension of the plasma. Substituting the above equation and an appropriate expression for the collision term $(\partial f/\partial t)_c$ in the Boltzmann equation, and keeping in mind that the external field is of the form $\sum_{l \neq 0} E_{0,l} \exp(-il\omega t)$, one obtains a nonlinear equation, the nonlinear terms being of a small magnitude. Assuming these small nonlinear terms to be of the order of a parameter λ, expanding $f_{k,l}$ in terms of λ as

$$f_{k,l}(v) = f_{k,l}^{(0)} + \lambda f_{k,l}^{(1)} + \lambda^2 f_{k,l}^{(2)} + \cdots$$

$$E_{k,l}(v) = E_{k,l}^{(0)} + \lambda E_{k,l}^{(1)} + \lambda^2 E_{k,l}^{(2)} + \cdots$$

substituting in the above nonlinear equation and also the corresponding equation obtained from the Poisson's equation and the above Fourier components, and equating the coefficients of the same powers of λ, one obtains the following set of equations:

Zero-order equations in λ:

$$(v - i\omega l + ikv)f_{k,l}^{(0)} = \frac{e}{m}\left(E_{0,l}\frac{\partial f_{k,0}}{\partial v} + \sum_{k' \neq 0} E_{k',l}^{(0)}\frac{\partial f_{k-k',0}}{\partial v}\right)$$

$$ikE_{k,l}^{(0)} = -4\pi e \int f_{k,l}^{(0)} \, dv$$

First-order equations in λ:

$$(v - i\omega l + ikv)f_{k,l}^{(1)} = \frac{e}{m}\left(\sum_{k' \neq 0} E_{k',l}^{(1)}\frac{\partial f_{k-k',0}}{\partial v} + \sum_{\substack{l' \neq 0, \\ l-l' \neq 0}} E_{0,l'}\frac{\partial f_{k,l-l'}^{(0)}}{\partial v}\right.$$

$$\left. + \sum_{\substack{k' \neq 0, l' \neq 0 \\ l-l' \neq 0}} E_{k',l'}^{(0)}\frac{\partial f_{k-k',l-l'}^{(0)}}{\partial v}\right)$$

$$ikE_{k,l}^{(1)} = -4\pi e \int f_{k,l}^{(1)} \, dv$$

Second-order equations in λ:

$$(v - i\omega l + ikv)f_{k,l}^{(2)} = \frac{e}{m}\left\{\sum_{k' \neq 0} E_{k',l}^{(2)}\frac{\partial f_{k-k',0}}{\partial v} + \sum_{\substack{l' \neq 0, \\ l-l' \neq 0}} E_{0,l'}\frac{\partial f_{k,l-l'}^{(1)}}{\partial v}\right.$$

$$\left. + \sum_{\substack{k' \neq 0, l' \neq 0 \\ l-l' \neq 0}}\left(E_{k',l'}^{(0)}\frac{\partial f_{k-k',l-l'}^{(1)}}{\partial v} + E_{k',l'}^{(1)}\frac{\partial f_{k-k',l-l'}^{(0)}}{\partial v}\right)\right\}$$

$$ikE_{k,l}^{(2)} = -4\pi e \int f_{k,l}^{(2)} \, dv$$

and so on. Since the explicit evaluation of $f_{0,0}$ and $f_{k,0}$ was difficult, solutions were therefore obtained for the case $k \neq 0$, $l \neq 0$, and when $f_{0,0}$ and $f_{k,0}$ are known. Thus one may assume both $f_{0,0}$ and $f_{k,0}$ to be Maxwellian:

$$f_{0,0} = N_{0,0}(m/2\pi k_0 T)^{3/2} \exp(-mv^2/2k_0 T)$$
$$f_{k,0} = N_{k,0}(m/2\pi k_0 T)^{3/2} \exp(-mv^2/2k_0 T)$$

where $N_{0,0}$ is the average number density of electrons, $N_{k,0}$ is the spatial variation of the electron density in the stationary state, and k_0 is the Boltzmann's constant. The ratio

$$f_{k,0}/f_{0,0} = N_{k,0}/N_{0,0} \equiv \sigma_k \qquad \sigma_k = \text{constant}$$

It is further assumed that the external field is given by

$$E_{0,l}\begin{cases} \neq 0 & \text{for} \quad |l| = 1 \\ = 0 & \text{for} \quad |l| \geq 2 \end{cases}$$

and also

$$f_{k',0}\begin{cases} \neq 0 & \text{for} \quad k' = k \\ = 0 & \text{for} \quad k' \neq k \end{cases}$$

The harmonic components of the field strength and the distribution function can now be expressed in terms of certain integrals that have been explicitly solved by Moriyama and Sumi *(44)* for a constant collision frequency $v(v) = v_0$ and a high frequency plasma at low pressure (i.e., for $\omega^2 \gg (kv)^2 \gg v_0^2$). It is also assumed that the inequalities

$$f_{k,l}^{(i)} \gg f_{2k,l}^{(i)} \gg f_{3k,l}^{(i)} \cdots$$
$$E_{k,l}^{(i)} \gg E_{2k,l}^{(i)} \gg E_{3k,l}^{(i)} \cdots$$

hold, which is true if $\sigma_k/(v/\omega) \ll 1$.

When v/ω is small, the condition means that the spatial variation of electron density is small. The components of the field strength may be written as

$$E_{k,1}^{(0)} = E_{0,1}\sigma_k\left(\frac{\omega_p}{\omega}\right)^2\left(1 - 2i\frac{v_0}{\omega}\right)\left[1 - \left(\frac{\omega_p}{\omega}\right)^2\left(1 - 2i\frac{v_0}{\omega}\right)\right]^{-1}$$

$$E_{k,2}^{(1)} = -\frac{1}{2}E_{0,1}\sigma_k\left(\frac{ieE_{0,1}}{m\omega}\right)\frac{k}{\omega}\left(\frac{\omega_p}{\omega}\right)^2\left(1 - \frac{11}{4}i\frac{v_0}{\omega}\right)$$
$$\times \left\{1 - \left(\frac{\omega_p}{\omega}\right)^2\left(1 - 2i\frac{v_0}{\omega}\right)\right\}^{-1}\left\{1 - \frac{1}{4}\left(\frac{\omega_p}{\omega}\right)^2\left(1 - i\frac{v_0}{\omega}\right)\right\}^{-1}$$

$$E_{k,3}^{(2)} = \frac{1}{6}E_{0,1}\sigma_k\left(\frac{ieE_{0,1}}{m\omega}\right)^2\left(\frac{k}{\omega}\right)^2\left(\frac{\omega_p}{\omega}\right)^2\left\{1 - \left(\frac{\omega_p}{\omega}\right)^2\left(1 - 2i\frac{v_0}{\omega}\right)\right\}^{-1}$$
$$\times \left\{1 - \frac{\omega_p^2}{9\omega^2}\left(1 - \frac{2i}{3}\frac{v_0}{\omega}\right)\right\}^{-1}$$
$$\times \left[1 - \frac{347}{108}i\frac{v_0}{\omega} + \frac{7}{36}\frac{\omega_p^2}{\omega^2}\left(1 - \frac{121}{28}i\frac{v_0}{\omega}\right)\left\{1 - \frac{\omega_p^2}{4\omega^2}\left(1 - i\frac{v_0}{\omega}\right)\right\}^{-1}\right]$$

all of which have resonances around $\omega \simeq \omega_p$, where $\omega_p = (4\pi N_{00}e^2/m)^{1/2}$. The resonant values for the additional condition $\omega \gg v_0$ may be written as

$$E_{k,1}^{(0)} = \frac{1}{2} E_{0,1}\left(\frac{\sigma_k \omega}{iv_0}\right)$$

$$E_{k,2}^{(1)} = -\frac{1}{3} E_{0,1}\sigma_k\left(\frac{ieE_{0,1}}{m\omega}\right)\frac{k}{iv_0}$$

$$E_{k,3}^{(2)} = \frac{17}{144} E_{0,1}\sigma_k\left(\frac{ieE_{0,1}}{m\omega}\right)^2 \frac{k^2}{iv_0\,\omega}$$

Similar evaluations could be made for the distribution function, but we are mainly concerned with the harmonic components of the current density, which may be defined by the relation

$$j_{k,l}^{(i)} = -e \int f_{k,l}^{(i)} v\, dv, \qquad i = 0, 1, 2, \ldots$$

Expressions similar to the ones obtained above for the electric field components may be obtained for the components of the current density also; however, we shall give only the resonant values here. Thus

$$j_{k,1}^{(0)} = \frac{1}{2} e\sigma_k N_{0,0}\left(\frac{eE_{0,1}}{mv_0}\right)$$

$$j_{k,2}^{(1)} = -\frac{2i}{3} e\sigma_k N_{0,0}\left(\frac{eE_{0,1}}{m\omega}\right)^2 \frac{k}{v_0}$$

$$j_{k,3}^{(2)} = -\frac{17}{48} e\sigma_k N_{0,0}\left(\frac{eE_{0,1}}{m\omega}\right)^3 \frac{k^2}{\omega v_0}$$

The longitudinal harmonic field in the plasma gives rise to a transversal field in the outer space. Defining R_i to be the radiation resistance for the ith harmonic, one can arrive at a parameter p_i (proportional to the power radiated at the ith harmonic) given by

$$p_i = \left|\sum_k \sum_l j_{k,l}^{i-1}\right|^2 R_i$$

Assuming the microwave field to be of a cosinusoidal form, using the equation of electrical neutrality (viz., $\sum_k E_{k,0} \exp(ikx) = 0$), one obtains the resonant values of p_1, p_2, and p_3 as

$$p_1 = \frac{e^2}{4}\left(\frac{eE_0}{mv_0}\right)^2 N_{00}^2(\sigma_k + \sigma_{-k})^2 R_1$$

$$p_2 = \frac{e^2}{9}\left(\frac{eE_0}{mv_0}\right)^2\left(\frac{ekE_0}{m\omega^2}\right)^2 N_{00}^2(\sigma_k - \sigma_{-k})^2 R_2$$

$$p_3 = \left(\frac{17}{192}\right)^2 e^2 \left(\frac{eE_0}{mv_0}\right)^2 \left(\frac{ekE_0}{m\omega^2}\right)^4 N_{00}^2 (\sigma_k + \sigma_{-k})^2 R_3$$

It is noted that if the plasma considered has a complete geometrical symmetry (i.e., if $\sigma_k = \sigma_{-k}$), the second harmonic and in fact all even harmonics vanish. On the other hand, if the plasma is antisymmetric, all odd harmonics disappear. Further, whereas the ratio p_2/p_1 depends upon the symmetry of the system, the ratio p_3/p_1 does not.

Moryama and Sumi (*48*) find that the third harmonic does not vanish for a constant collision frequency; this contradicts the results obtained by us in Section III-A, the reason being that we did not consider the inner field of the plasma in that section (this field is important in the vicinity of the plasma frequency; i.e., when $\omega \simeq \omega_p$). For $f = 3000$ MHz, $E_0 = 10^3$ volts/cm, $d = 10^{-1}$ cm [Uenohara *et al.* (*50*) experimental values], and $R_3/R_1 \simeq 70$ [extrapolated Baird and Coleman's (*43*) value], Moriyama and Sumi (*48*) find $p_3/p_1 \simeq 1.26 \times 10^{-3}$, which compares favorably with Uenohara's (*50*) value of 10^{-3} under these conditions. They have also considered briefly the effect of the velocity-dependent nature of the collision frequency on harmonic generation and conclude that the harmonic current density is reduced for this collision effect.

In the presence of a magnetic field, Moriyama and Sumi (*49*) have again carried out a similar analysis and the chief modifications are:

a. The second-harmonic current density has a maximum in the vicinity of $\omega_B/\omega \simeq 1$, which shifts a little to lower values of ω_B/ω as ω_p/ω increases. The height of the resonance increases with decreasing ω_p/ω and v_0/ω. The flatness of the resonance curve increases with increasing ω_p/ω. These results are in qualitative agreement with the experiments of Hill and Tetenbaum (*58*).

b. The second harmonic (and all even harmonics) has a finite magnitude even if the plasma considered is completely symmetrical geometrically; this result arises because there is always some asymmetry introduced by the magnetic field itself.

The generation of sum-and-difference frequencies in a resonant plasma when two frequencies $\omega_1 \simeq \omega_p$ and $\omega_2 \approx \omega_p$ are simultaneously propagating through it is a problem worth consideration; the effect has already been demonstrated by the experiments of Stern (*61*).

C. Plasma with Induced Inhomogeneities

In this section, we summarize the generation of harmonics and combination frequencies in a plasma arising because of some inhomogenities induced by an electromagnetic wave in its propagation through the plasma. This will incorporate the mechanisms put forward by Ginzburg (*42*), Baird and Coleman (*43*), and Moriyama and Sumi (*44*).

1. Excess Charge Density Due to an Electromagnetic Field

Ginzburg (42) has discussed a new type of nonlinear effect that is connected with the changes in the electron concentration, produced by the electrical vector of the incident electromagnetic wave. Since the conductivity and hence the propagation parameters depend on the electron density N, the properties of the plasma therefore depend upon the electrical vector, and the medium becomes nonlinear. Assuming the ions to be immobile so that they merely compensate for the equilibrium electron charge eN, the value of $e\,\Delta N$ is obviously equal to the density of the average microscopic charge $\bar{\rho}$ of the plasma. In turn

$$\bar{\rho} = e\,\Delta N = (1/4\pi)\mathrm{div}\,\mathbf{E} \tag{3.48}$$

Thus, whenever $\mathrm{div}\,\mathbf{E} \neq 0$, the nonlinear effect considered here should exist.

Ginzburg (42) treats two different cases in some detail:

a. *An inhomogeneous and isotropic plasma.* Here

$$\Delta N = -\frac{E}{4\pi e}\frac{\mathrm{grad}\,\varepsilon'}{\varepsilon'}$$

where

$$\varepsilon' = 1 - \frac{4\pi e^2 N}{m\omega(\omega - iv)}$$

the symbols having their usual meanings.

b. *A homogeneous magnetoactive plasma.* Here

$$\Delta N = -\frac{\omega n}{4\pi e c}\{E_{0a}\cos\theta_a\sin(\omega t - \mathbf{k}\cdot r) + E_{0b}\cos\theta_b\cos(\omega t - \mathbf{k}\cdot r)\}$$

where

$$\mathbf{E} = \mathbf{E}_{0a}\cos(\omega t - \mathbf{k}\cdot r) + \mathbf{E}_{0b}\sin(\omega t - \mathbf{k}\cdot r)$$

is the electrical vector of the incident wave, θ_a or θ_b, the angles between E_{0a} or E_{0b} and k, and finally n is the index of refraction.

It is seen that in either case the change ΔN is linearly dependent on the electrical vector and varies with the frequency of the incident wave. This type of nonlinear effect can therefore generate frequencies $\omega_1 \pm \omega_2$ in the current density when two electromagnetic waves of frequencies ω_1 and ω_2 are simultaneously allowed to propagate through the plasma. It seems, however, that a detailed analysis of such a generation (especially with a kinetic theory approach) has not been carried out.

It may be pointed out finally that Ginzburg and Gurevich (9) have shown that the magnitude of this type of nonlinear effect is much smaller (less than about 0.1 times or so for the usual ionospheric parameters) than that of the nonlinear effect associated with changes in the electron temperatures, electron collision frequency, etc.

Sodha and Sawhney (70) have explored the possibility of harmonic generation by a nonlinear effect arising because of the dependence of electron density on electron temperature (and hence the electric field). They conclude that this nonlinear effect cannot lead to harmonic generation at microwave frequencies because the ionization and recombination rates for typical plasmas are such that the electron density cannot follow the time-dependent fluctuations of the electron temperature.

2. Modulation of the Ionization Frequency

Baird and Coleman (43) have analyzed the modulation of electron density at the microwave frequency in a discharge located between two closely spaced parallel plates whose dimensions are small compared with the wavelength (so that the diffusion becomes important), the modulation arising through the ionization frequency which is assumed to be directly proportional to the ordered drift velocity. This modulated electron density can then be used in explaining the generation of harmonics and combination frequencies in the microwave discharge.

The current density in a plasma is given by

$$J(x, t) = -eN(x, t)v_d(x, t) \tag{3.49}$$

where N is the electron density and v_d the drift velocity. N and v_d are given by

$$\partial N/\partial t = v_i N + D\,\partial^2 N/\partial x^2 \tag{3.50a}$$

and

$$(\partial v_d/\partial t) + v_c v_d = -eE_T/m \tag{3.50b}$$

where v_i is the ionization frequency, v_c is the collision frequency for momentum transfer, D is the diffusion coefficient, and it has been assumed that diffusion is the dominant loss mechanism. Chiyoda and Tamaru (45) have shown by means of a rigorous Boltzmann equation analysis that the macrosopic equation, Eq. (3.50a), used by Baird and Coleman (43), tacitly neglects the terms $a\,\partial f^0/\partial v$, which are otherwise responsible for the generation of odd harmonics in the current density [Margenau and Hartman (2), Rosen (3), etc.]. It is for this reason that Baird and Coleman (43) have to introduce a new assumption regarding the proportionality of ionization frequency to the ordered drift velocity, in order to account for the generation of harmonics. In the above equations E_T is the total field, which from a linear analysis may be taken as

$$E_T = E_0 + E(\cos \omega t - (\omega/v_c)\sin \omega t)$$

Baird and Coleman (*43*) further assume that

$$v_i = \alpha \, |v_d| \tag{3.51}$$

an assumption whose validity is to be ultimately judged by the comparison of their results with the experiments. Thus

$$v_i = (e\alpha E / m v_c)(\beta + \cos \theta) \qquad \beta = E_0/E, \quad \theta = \omega t$$

where a linear expression for v_d has been used. To explain the generation of harmonics in the current density, one should expand v_d (hence v_i) and N in a Fourier-series form. Thus

$$v_i = \frac{e\alpha E}{m v_c}\left[\frac{a_0}{2} + \sum_{k=1}^{\infty} a_k \cos k\theta\right] \tag{3.52a}$$

and

$$N = N_0\left[1 + \sum_{s=1}^{\infty} \beta_s \sin(s\omega t)\right]\cos(\pi x/2l)$$

$$= N_0 \cos(\pi x/2l) \qquad \text{when} \quad \beta_s = 0 \tag{3.52b}$$

This particular dependence of N on x has been chosen because it is a linear solution of Eq. (3.50a) for N, with the boundary conditions $N = 0$ at $x = \pm l$ (i.e., at the metal plates). Moreover, since the modulation of N is not expected to be large, the β_s are small as compared with unity.

Substituting in Eq. (3.50b), equating the coefficients, and using Eq. (3.49), one obtains

$$J(x, t) = \frac{N_0 \, e^2 E}{m v_c}\left[(\beta + \cos \omega t) + \frac{\pi e \alpha a_0 E}{m\omega v_c} \sum_{s=2}^{\infty} (A_s \sin s\omega t)\right]\cos\left(\frac{\pi x}{2l}\right) \tag{3.53}$$

where

$$A_s = \frac{1}{2\pi a_0}\left[2\beta \frac{a_s}{s} + \frac{a_{s-1}}{s-1} + \frac{a_{s+1}}{s+1}\right] = A_s(\beta)$$

since the coefficients a_k are also functions of $\beta = E_0/E$ only. The ratio of the power radiated at the sth harmonic frequency to that radiated at the fundamental frequency is given by

$$\frac{P_s}{P_1} = \left[\frac{\pi \alpha e a_0 E}{m\omega v_c}\right]^2 \frac{R_s}{R_1} A_s^{\,2}$$

where R_s and R_1 are the radiation resistances for the sth harmonic and the fundamental frequency, respectively. The above may be expressed as

$$\frac{P_s}{P_1} = \frac{R_s}{R_1}\left(\frac{N_R}{N_T}\right)^2 A_s^{\,2} \tag{3.54}$$

where N_R is the number of electrons produced during one rf cycle and N_T is the total number of electrons between the two plates.

Baird and Coleman (*43*) plotted P_s/P_1 against $\beta = E_0/E$ for various values of s; their theoretical results fitted remarkably well into the experimental curves that they obtained. The chief result noted from the graphs is that all odd harmonics attain a maximum for $E_0 = 0$ (i.e., for no dc field), whereas all even harmonics vanish in this case.

An analysis similar to the one considered above was also made for the case of frequency mixing, and the power radiated at the sum-and-difference frequencies $(\omega_1 \pm \omega_2)$ was obtained in terms of the power radiated at one of the fundamental frequencies; plots similar to the one mentioned above were drawn and experimentally verified. It was noted that for $E_0 = 0$, the sum-and-difference frequencies vanish.

Wetzel and Tang (*1*) commented thus on Baird and Coleman (*43*) paper: "The experimental results obtained by Baird and Coleman remain a puzzle. They found that both odd and even harmonics radiated from a microwave discharge between two posts in a waveguide when an additional dc electric bias field was applied. Their measurements were remarkably consistent with the predictions of a simple theory based on the assumption that $v_i = \alpha|v_d|$. Unfortunately, there is no convincing justification for this assumption; our formulation would also lead to both odd and even harmonics in combined ac and dc electric fields, but in order to obtain anything resembling $v_i = \alpha|v_d|$, the condition $(2m/M)v_c \gg \omega$ ought to be satisfied, which is difficult at microwave frequencies." The authors do not add anything to this comment.

3. *Induced Time Dependence of Electron Density*

Moriyama and Sumi (*44*) have investigated the generation of even harmonics in the current density in a homogeneous plasma by taking into account the time dependence induced in the electron density by the applied electromagnetic field. The earlier treatments of Margenau and Hartman (*2*), Rosen (*3*) etc. (which have been outlined in Section III-A) were restricted to the case in which the electron density was time-independent, with the result that only odd harmonics were generated in the current density.

In accordance with the discussion following Eq. (2.12), one can assume that when the electron density is time-dependent, then the sum of the upper and the lower indices in f_m^k (i.e., $(k + m)$) may be odd. If one limits oneself to the generation of the second harmonic only, one can safely write

$$f_0^{\,1} = f_1^{\,2} = \cdots = 0$$

and thus

$$N = 4\pi \int_0^\infty [f_0^{\,0} + f_1^{\,0} \exp(i\omega t)]v^2 \, dv$$

$$= N_0 + N_1 \exp(i\omega t) \tag{3.55a}$$

The second-harmonic current density is given by

$$J_{2x} = -(4\pi e/3) \int_0^\infty v^3 f_2{}^1 \, dv \, \exp(2i\omega t) \tag{3.55b}$$

In the following analysis it has been assumed with Moriyama and Sumi (*44*) that the gas temperature is negligible, that the electrons make only elastic collisions with neutral molecules, and that the collision frequency depends on the electronic speed v in accordance with the relation $v = v_0 v^n$, where n is an integer. Following the recurrence relations of Margenau and Hartman (*2*), one obtains

$$f_2{}^1 = [a_1/2(v_0 v^n + 2i\omega)] \, df_1{}^0/dv \tag{3.56a}$$

where $f_1{}^0$ is given by the equation

$$\frac{d^2 f_1{}^0}{dv^2} + \left[\frac{2}{v} + \frac{v_0 v^{1+n}}{a} (v_0 v^n + 2i\omega) - \frac{n v_0 v^{n-1}}{v_0 v^n + 2i\omega} \right] \frac{df_1{}^0}{dv}$$

$$+ \left[\frac{v_0}{a} (3 + n)v^n - \frac{b}{a} \right] (v_0 v^n + 2i\omega) f_1{}^0 = 0 \tag{3.56b}$$

where $a = e^2 E^2 M/12 m^3$, $b = i\omega M/m$, and M and m are the masses of a molecule and an electron, respectively. When $n = 0$, i.e., when the electron collision frequency is independent of electron velocity, an approximate solution to the above equation is given by

$$f_1{}^0 = \overline{K} v^{-1/2} \exp\left(\frac{-\eta}{4} v^2 \right) J_{\pm\frac{1}{2}} \left(\pm i \left(\frac{\eta b}{v_0} \right)^{1/2} v \right) \tag{3.57}$$

where $\eta \equiv (v_0/a)(v_0 + 2i\omega)$ and $\overline{K}$ is a constant, which must be determined from Eq. (3.55a). This solution is different from that given by Moriyama and Sumi (*44*) because their equation (5) and the following analysis was slightly erroneous. The above correct solution was also given by Moriyama (*71*).

Substituting this value of $f_1{}^0$ in Eq. (3.55), one obtains the magnitude of N_1. From Eq. (3.55b), one has

$$J_{2x} = (e^2 E_1/2m(v_0 + 2i\omega))N_1 \exp(2i\omega t)$$

Using the obtained value of N_1, one obtains the magnitude of the second harmonic in the current density. The foregoing treatment is not very rigorous; some of the assumptions involved may be questioned. Thus there seems to be no valid reason for assuming the gas temperature to be so small that one of the two second-order components of the collision terms can be neglected. This assumption seems to have been made simply to get a soluble differential equation for $f_1{}^0$ and does not have much of a physical basis. Moreover, when N is time-dependent, one would expect $k + m$ to take both odd and even values; the reason for considering only odd values is obscure.

IV. Propagation of the Harmonic and Combination Frequency Electromagnetic Waves in a Plasma

In the preceding section, explicit expressions were derived for the harmonic and combination frequency components in the current density in a plasma due to strong alternating electric fields. Very often these electric fields are actually the electrical vectors of strong electromagnetic waves that are propagating through the plasma. In such cases one is interested not only in the harmonic and combination frequency components that are generated in the current density, but also in the corresponding electromagnetic fields that they establish in the plasma. One is further interested in the propagation characteristics of the harmonic and combination frequency electromagnetic waves that are thus generated. In order to study these, one usually starts with the wave equation (or its equivalent, the Maxwell equations*) and solves it for the components of the electrical vector.

The wave equation for the propagation of plane transverse electromagnetic waves is

$$\nabla^2 \mathbf{E} = \frac{\varepsilon\mu}{c^2}\frac{\partial^2 \mathbf{E}}{\partial t^2} + \frac{4\pi\mu}{c^2}\frac{\partial \mathbf{J}}{\partial t} \tag{4.1}$$

where $\mathbf{E}$ stands for the electrical vector, $\mathbf{J}$ is the corresponding expression for the current density, ε and μ are the dc dielectric constant and the permeability of the plasma, respectively, c is the velocity of light in vacuum; it has been assumed that div $\mathbf{E} = 0$ (since the waves are plane and transverse). One should note that $\mathbf{J}$ stands for the total current density in the plasma. This includes the contributions arising from the newly set up harmonic and combination frequency fields in the plasma.

In order to solve the wave Eq. (4.1), the expressions for $\mathbf{J}$ derived in Section III for uniform electric fields will be used, even though we know that the field of the wave is always inhomogeneous in space. This is equivalent to the assumption that the complex conductivity is local and that the current density at a given point is determined by the electric field at that very point. In a weak field, this condition is violated if the field amplitude changes substantially over a mean free path. In a strong field, on the other hand, a more

---

* One has to be careful while using the Maxwell equations to study the propagation of electromagnetic waves through a plasma, since these equations are rigorously applicable to continuous media only and a plasma is essentially discrete in nature. Under these conditions, one can use the Maxwell equations only if the distance between any two particles is smaller than the wavelength of the wave. This condition is violated only in some problems connected with the propagation in interstellar matter, where the plasma is so sparse that the interparticle distance is comparable to or even greater than the wavelength of the wave [Ginzburg (69)].

stringent condition is to be satisfied [Ginzburg and Gurevich (9)], viz., that the field amplitude must change little over the electron energy relaxation length $\langle v/\nu\rangle(M/2m)^{1/2}$. This is a much longer length than the mean free path $\langle v/\nu\rangle$ in view of the very small magnitude of the ratio m/M.

Most of the expressions for $\mathbf{J}$ (in Section III) have been derived only with alternating electric fields, and one may wonder why the effect of the alternating magnetic field of the wave is not usually taken into account. The reason is not far to seek. It is well known that, in cgs units, the electric and magnetic vectors of the wave are of the same order of magnitude; thus the ratio of the magnetic to the electrical force on the electrons in the plasmas is of the order of v/c. For nonrelativistic plasmas, therefore, the effect of the magnetic field of the wave may be safely neglected.

In what follows we have again divided the plasmas into two classes, viz., homogeneous and inhomogeneous. Two specific problems have been analyzed in each case: the nonlinear growth of the harmonic and combination frequency waves in a plasma; and the magnitude of these frequency components in the reflected wave from a plasma-free space interface. This part of the work is almost entirely due to Vilenskii ($5, 6$), Gurevich (8), Ginzburg and Gurevich (9), and Sodha and Kaw (18–23).

A. Propagation in an Homogeneous Infinite Plasma

1. In the Presence of a dc Electric Field

Sodha and Kaw (23) have discussed the generation and growth of harmonic and combination frequency waves in a homogeneous infinite plasma due to two strong plane-polarized electromagnetic waves, propagating with their electric vectors along the direction of an externally applied dc electric field. The corrresponding problem in the absence of an external dc electric field had earlier been discussed by Vilenskii (5), Gurevich (8) and Ginzburg and Gurevich (9). As has already been seen in Section III-A-2, in the presence of a dc electric field one obtains all harmonics and all combination frequencies in the current density. The total electric field along the x axis will therefore be made up of the dc electric field, the two fundamental fields, and the newly set up harmonic and combination frequency fields. Thus, one may write

$$
\begin{aligned}
\mathbf{E} = {} & \mathbf{E}_d + \mathbf{E}_1 \exp(i\omega_1 t) + \mathbf{E}_2 \exp(i\omega_2 t) + \mathbf{E}_{12} \exp(2i\omega_1 t) \\
& + \mathbf{E}_{22} \exp(2i\omega_2 t) + \mathbf{E}_{13} \exp(3i\omega_1 t) + \mathbf{E}_{23} \exp(3i\omega_2 t) \\
& + \mathbf{E}_{12}^{+} \exp\{i(\omega_1 + \omega_2)t\} + \mathbf{E}_{12}^{-} \exp\{i(\omega_1 - \omega_2)t\} \\
& + \mathbf{E}_{12}^{+\prime} \exp\{i(\omega_1 + 2\omega_2)t\} + \mathbf{E}_{12}^{-\prime} \exp\{i(\omega_1 - 2\omega_2)t\} \\
& + \mathbf{E}_{21}^{+\prime} \exp\{i(\omega_2 + 2\omega_1)t\} + \mathbf{E}_{21}^{-\prime} \exp\{i(\omega_2 - 2\omega_1)t\} \qquad (4.2a)
\end{aligned}
$$

with $\mathbf{E}_{12}$, $\mathbf{E}_{22}$, $\mathbf{E}_{13}$, $\mathbf{E}_{23}$, $\mathbf{E}_{12}^{\pm}$, $\mathbf{E}_{12}^{\pm\prime}$, $\mathbf{E}_{21}^{\pm\prime} \ll \mathbf{E}_1$ or $\mathbf{E}_2$. The alternating component of the corresponding expression for the current density may be written as

$$
\begin{aligned}
\mathbf{J}_a = {}& \sigma_1 \mathbf{E}_1 \exp(i\omega_1 t) + \sigma_2 \mathbf{E}_2 \exp(i\omega_2 t) + \sigma_{12} \mathbf{E}_{12} \exp(2i\omega_1 t) \\
& + \sigma_{22} \mathbf{E}_{22} \exp(2i\omega_2 t) + \sigma_{13} \mathbf{E}_{13} \exp(3i\omega_1 t) + \sigma_{23} \mathbf{E}_{23} \exp(3i\omega_2 t) \\
& + \sigma_{12}^+ \mathbf{E}_{12}^+ \exp\{i(\omega_1 + \omega_2)t\} + \sigma_{12}^- \mathbf{E}_{12}^- \exp\{i(\omega_1 - \omega_2)t\} \\
& + \sigma_{12}^{+\prime} \mathbf{E}_{12}^{\prime+} \exp\{i(\omega_1 + 2\omega_2)t\} + \sigma_{12}^{-\prime} \mathbf{E}_{12}^{-\prime} \exp\{i(\omega_1 - 2\omega_2)t\} \\
& + \sigma_{21}^{+\prime} \mathbf{E}_{21}^{+\prime} \exp\{i(\omega_2 + 2\omega_1)t\} + \sigma_{21}^{-\prime} \mathbf{E}_{21}^{-\prime} \exp\{i(\omega_2 - 2\omega_1)t\} \\
& + \frac{m}{M} \alpha [A_{12} \mathscr{E}_1^{\,2} \mathbf{E}_d \exp(2i\omega_1 t) + A_{22} \mathscr{E}_2^{\,2} \mathbf{E}_d \exp(2i\omega_2 t) \\
& + A_{13} \mathscr{E}_1^{\,2} \mathbf{E}_1 \exp(3i\omega_1 t) + A_{23} \mathscr{E}_2^{\,2} \mathbf{E}_2 \exp(3i\omega_2 t) \\
& + A_{12}^+ \mathscr{E}_1 \mathscr{E}_2 \mathbf{E}_d \exp\{i(\omega_1 + \omega_2)t\} + A_{12}^- \mathscr{E}_1 \tilde{\mathscr{E}}_2 \mathbf{E}_d \exp\{i(\omega_1 - \omega_2)t\} \\
& + A_{12}^{+\prime} \mathscr{E}_2^{\,2} \mathbf{E}_1 \exp\{i(\omega_1 + 2\omega_2)t\} + A_{12}^{-\prime} \tilde{\mathscr{E}}_2^{\,2} \mathbf{E}_1 \exp\{i(\omega_1 - 2\omega_2)t\} \\
& + A_{21}^{+\prime} \mathscr{E}_1^{\,2} \mathbf{E}_2 \exp\{i(\omega_2 + 2\omega_1)t\} + A_{21}^{-\prime} \tilde{\mathscr{E}}_1^{\,2} \mathbf{E}_2 \exp\{i(\omega_2 - 2\omega_1)t\}]
\end{aligned}
$$

$$\tag{4.2b}$$

where σ_1, σ_2, σ_{12}, σ_{22}, σ_{13}, σ_{23}, $\sigma_{12}^\pm$, $\sigma_{12}^{\pm\prime}$ and $\sigma_{21}^{\pm\prime}$ denote the complex conductivities corresponding to frequencies ω_1, ω_2, $2\omega_1$, $2\omega_2$, $3\omega_1$, $3\omega_2$, $\omega_1 \pm \omega_2$, $\omega_1 \pm 2\omega_2$, $\omega_2 \pm 2\omega_1$, respectively; the usual expression for $\sigma(\omega)$, viz.,

$$
\sigma = \frac{e^2 N}{3m} \left\langle \frac{1}{v^2} \frac{d}{dv} \left(\frac{v^3}{v + i\omega} \right) \right\rangle
$$

[Sodha and Palumbo (72)] where $\langle \ \rangle$ denotes averaging over the velocity distribution, is valid, of course, in the present case also. One should use a Maxwellian form of $f_0^{\,0}$, the time-independent part of the isotropic component of the distribution function, in evaluating the averages because our expression for the current density is accurate only under this assumption (see Section III-A-2). An expression for only the alternating component of the current density has been written above; this is so because we are really interested in the quantity $\partial \mathbf{J}/\partial t$ and this will not contain any contribution because of the dc component of the current density.

Substituting for $\mathbf{J}_a$ and $\mathbf{E}$ in Eq. (4.1) from Eqs. (4.2), rearranging the terms, and equating the coefficients of various frequency components on both sides of the final equation, one obtains

$$
(\partial^2 \mathscr{E}_1 / \partial \xi^2) + \beta_1^{\,2} \mathscr{E}_1 = 0 \tag{4.3a}
$$

$$
(\partial^2 \mathscr{E}_2 / \partial \xi^2) + \beta_2^{\,2} \mathscr{E}_2 = 0 \tag{4.3b}
$$

$$
(\partial^2 \mathscr{E}_{12} / \partial \xi^2) + \beta_{12}^2 \mathscr{E}_{12} = (m/M)\alpha a_{12} \mathscr{E}_d \mathscr{E}_1^{\,2} \tag{4.3c}
$$

$$(\partial^2 \mathscr{E}_{22}/\partial\xi^2) + \beta_{22}^2 \mathscr{E}_{22} = (m/M)\alpha a_{22} \mathscr{E}_d \mathscr{E}_2{}^2 \tag{4.3d}$$

$$(\partial^2 \mathscr{E}_{13}/\partial\xi^2) + \beta_{13}^2 \mathscr{E}_{13} = (m/M)\alpha a_{13} \mathscr{E}_1{}^3 \tag{4.3e}$$

$$(\partial^2 \mathscr{E}_{23}/\partial\xi^2) + \beta_{23}^2 \mathscr{E}_{23} = (m/M)\alpha a_{23} \mathscr{E}_2{}^3 \tag{4.3f}$$

$$(\partial^2 \mathscr{E}_{12}^+/\partial\xi^2) + \beta_{12}^{+2} \mathscr{E}_{12}^+ = (m/M)\alpha a_{12}^+ \mathscr{E}_1 \mathscr{E}_2 \mathscr{E}_d \tag{4.3g}$$

$$(\partial^2 \mathscr{E}_{12}^-/\partial\xi^2) + \beta_{12}^{-2} \mathscr{E}_{12}^- = (m/M)\alpha a_{12}^- \mathscr{E}_1 \tilde{\mathscr{E}}_2 \mathscr{E}_d \tag{4.3h}$$

$$(\partial^2 \mathscr{E}_{12}^{+\prime}/\partial\xi^2) + \beta_{12}^{+\prime 2} \mathscr{E}_{12}^{+\prime} = (m/M)\alpha a_{12}^{+\prime} \mathscr{E}_1 \mathscr{E}_2{}^2 \tag{4.3i}$$

$$(\partial^2 \mathscr{E}_{12}^{-\prime}/\partial\xi^2) + \beta_{12}^{-\prime 2} \mathscr{E}_{12}^{-\prime} = (m/M)\alpha a_{12}^{-\prime} \mathscr{E}_1 \tilde{\mathscr{E}}_2{}^2 \tag{4.3j}$$

$$(\partial^2 \mathscr{E}_{21}^{+\prime}/\partial\xi^2) + \beta_{21}^{+\prime 2} \mathscr{E}_{21}^{+\prime} = (m/M)\alpha a_{21}^{+\prime} \mathscr{E}_2 \mathscr{E}_1{}^2 \tag{4.3k}$$

$$(\partial^2 \mathscr{E}_{21}^{-\prime}/\partial\xi^2) + \beta_{21}^{-\prime 2} \mathscr{E}_{21}^{-\prime} = (m/M)\alpha a_{21}^{-\prime} \mathscr{E}_2 \tilde{\mathscr{E}}_1{}^2 \tag{4.3l}$$

where it has been assumed that the propagation is along the z axis,

$$\xi = (\varepsilon\mu)^{1/2}\omega_0 z/c \tag{4.4a}$$

is a dimensionless distance parameter, and for a general frequency ω we have

$$\beta^2 = \frac{\omega^2}{\omega_0{}^2}\left[1 - \frac{4\pi i\sigma}{\varepsilon\omega}\right] = (-n + ik)^2 \tag{4.4b}$$

and

$$a = (4\pi i\omega/\varepsilon\omega_0{}^2)A$$

Any particular β or a is obtained by substituting the appropriate values of ω, σ, and A on the right sides.

a. Nonlinear growth of the harmonics and combination frequencies. Equations (4.3a) and (4.3b), which describe the propagation of the fundamental components of the frequencies ω_1 and ω_2, are linear in nature because, by applying the Maxwellian form of $f_0{}^0$, we have been using a linear expression for the conductivity. Each solution of these equations will involve two arbitrary constants; these are evaluated by using the boundary conditions

$$\mathscr{E}_1 = \mathscr{E}_{10} \quad \text{and} \quad \mathscr{E}_2 = \mathscr{E}_{20} \quad \text{at} \quad \xi = 0$$

and the appropriate radiation conditions

$$\mathscr{E}_1 = \mathscr{E}_2 = 0 \quad \text{at} \quad \xi = \infty$$

Thus one obtains

$$\mathscr{E}_1 = \mathscr{E}_{10} \exp(i\beta_1\xi) \tag{4.5a}$$

$$\mathscr{E}_2 = \mathscr{E}_{20} \exp(i\beta_2\xi) \tag{4.5b}$$

(note the fact that $\beta_1 = (-n_1 + ik_1)$ and so

$$\exp(i\beta_1\xi) = \exp(-k_1\xi)\exp(-in_1\xi)$$

which would mean a wave traveling along the positive z direction).

To solve the remaining Eqs. (4.3), which are nonlinear in nature, one may use a successive approximation technique similar to that of Epstein (*73*). The procedure has been outlined here by solving Eq. (4.3c). Expanding $\mathscr{E}_{12}$ as

$$\mathscr{E}_{12} = \mathscr{E}'_{12} + \alpha\mathscr{E}''_{12}$$

substituting in Eq. (4.3c), and equating the coefficients of the similar powers of α, one obtains

$$(\partial^2\mathscr{E}'_{12}/\partial\xi^2) + \beta_{12}^2\,\mathscr{E}'_{12} = 0 \tag{4.6a}$$

$$(\partial^2\mathscr{E}''_{12}/\partial\xi^2) + \beta_{12}^2\,\mathscr{E}''_{12} = (m/M)a_{12}\mathscr{E}_1{}^2\mathscr{E}_d \tag{4.6b}$$

Using the value of $\mathscr{E}_1$ given by Eq. (4.5a) on the right-hand side of Eq. (4.6b), one can solve both equations and obtain

$$\mathscr{E}_{12} = (K_{12} + \alpha K'_{12})\exp(i\beta_{12}\,\xi)$$
$$+ (m/M)\alpha\mathscr{E}_{10}^2\,\mathscr{E}_d a_{12}\exp(2i\beta_1\xi)/(\beta_{12}^2 - 4\beta_1{}^2) \tag{4.6c}$$

Using the boundary condition $\mathscr{E}_{12} = 0$ at $\xi = 0$ (which means only that no second harmonic is present before the wave propagates some distance in the plasma) and retaining terms only up to the order of $(m/M)\alpha$, one obtains

$$\mathscr{E}_{12} = \frac{m}{M}\,\alpha a_{12}\,\mathscr{E}_{10}^2\,\mathscr{E}_d\,\frac{\exp(i\beta_{12}\,\xi) - \exp(2i\beta_1\xi)}{(2\beta_1 + \beta_{12})(2\beta_1 - \beta_{12})} \tag{4.7a}$$

Proceeding in a similar manner and using the boundary conditions

$$\mathscr{E}_{22} = \mathscr{E}_{13} = \mathscr{E}_{23} = \mathscr{E}_{12}^{\pm} = \mathscr{E}_{12}^{\pm\prime} = \mathscr{E}_{21}^{\pm\prime} = 0 \qquad \text{at}\quad \xi = 0$$

one may solve the remaining Eqs. (4.3). The solutions are

$$\mathscr{E}_{22} = \frac{m}{M}\,\alpha a_{22}\,\mathscr{E}_{20}^2\,\mathscr{E}_d\,\frac{\exp(i\beta_{22}\,\xi) - \exp(2i\beta_2\,\xi)}{(2\beta_2 + \beta_{22})(2\beta_2 - \beta_{22})} \tag{4.7b}$$

$$\mathscr{E}_{13} = \frac{m}{M}\,\alpha a_{13}\,\mathscr{E}_{10}^3\,\frac{\exp(i\beta_{13}\,\xi) - \exp(3i\beta_1\xi)}{(3\beta_1 + \beta_{13})(3\beta_1 - \beta_{13})} \tag{4.7c}$$

$$\mathscr{E}_{23} = \frac{m}{M}\,\alpha a_{23}\,\mathscr{E}_{20}^3\,\frac{\exp(i\beta_{23}\,\xi) - \exp(3i\beta_2\,\xi)}{(3\beta_2 + \beta_{23})(3\beta_2 - \beta_{23})} \tag{4.7d}$$

$$\mathscr{E}_{12}^{+} = \frac{m}{M}\,\alpha a_{12}^{+}\,\mathscr{E}_{10}\,\mathscr{E}_{20}\,\mathscr{E}_d\,\frac{\exp(i\beta_{12}^{+}\,\xi) - \exp\{i(\beta_1 + \beta_2)\xi\}}{(\beta_1 + \beta_2 + \beta_{12}^{+})(\beta_1 + \beta_2 - \beta_{12}^{+})} \tag{4.7e}$$

$$\mathscr{E}_{12}^{-} = \frac{m}{M}\,\alpha a_{12}^{-}\,\mathscr{E}_{10}\,\tilde{\mathscr{E}}_{20}\,\mathscr{E}_d\,\frac{\exp(i\beta_{12}^{-}\,\xi) - \exp\{i(\beta_1 - \tilde{\beta}_2)\xi\}}{(\beta_1 - \tilde{\beta}_2 + \beta_{12}^{-})(\beta_1 - \tilde{\beta}_2 - \beta_{12}^{-})} \tag{4.7f}$$

$$\mathscr{E}_{12}^{+\prime} = \frac{m}{M}\,\alpha a_{12}^{+\prime}\,\mathscr{E}_{10}\,\mathscr{E}_{20}^2\,\frac{\exp(i\beta_{12}^{+\prime}\,\xi) - \exp\{i(\beta_1 + 2\beta_2)\xi\}}{(\beta_1 + 2\beta_2 + \beta_{12}^{+\prime})(\beta_1 + 2\beta_2 - \beta_{12}^{+\prime})} \tag{4.7g}$$

$$\mathcal{E}_{12}^{-\prime} = \frac{m}{M} \alpha a_{12}^{-\prime} \mathcal{E}_{10} \tilde{\mathcal{E}}_{20}^{2} \frac{\exp(i\beta_{12}^{-\prime}\xi) - \exp\{i(\beta_1 - 2\tilde{\beta}_2)\xi\}}{(\beta_1 - 2\tilde{\beta}_2 + \beta_{12}^{-\prime})(\beta_1 - 2\tilde{\beta}_2 - \beta_{12}^{-\prime})} \tag{4.7h}$$

$$\mathcal{E}_{21}^{+\prime} = \frac{m}{M} \alpha a_{21}^{+\prime} \mathcal{E}_{20} \mathcal{E}_{10}^{2} \frac{\exp(i\beta_{21}^{+\prime}\xi) - \exp\{i(\beta_2 + 2\beta_1)\xi\}}{(\beta_2 + 2\beta_1 + \beta_{21}^{+\prime})(\beta_2 + 2\beta_1 - \beta_{21}^{+\prime})} \tag{4.7i}$$

and

$$\mathcal{E}_{21}^{-\prime} = \frac{m}{M} \alpha a_{21}^{-\prime} \mathcal{E}_{20} \tilde{\mathcal{E}}_{10}^{2} \frac{\exp(i\beta_{21}^{-\prime}\xi) - \exp\{i(\beta_2 - 2\tilde{\beta}_1)\xi\}}{(\beta_2 - 2\tilde{\beta}_1 + \beta_{21}^{-\prime})(\beta_2 - 2\tilde{\beta}_1 - \beta_{21}^{-\prime})} \tag{4.7j}$$

Equations (4.7a–j) give us the magnitudes of the electrical vectors of the harmonics and the combination frequencies in a plasma when the two incident waves are traveling with their electrical vectors along the direction of the externally applied dc electric field. These can be used to investigate the non-linear growth of these frequencies in a plasma in the following manner [Sodha and Palumbo (4)]. Let us take the case of the electrical vector $\mathcal{E}_{13}$ corresponding to the third-harmonic frequency $3\omega_1$. One can write Eq. (4.7c) as

$$\mathcal{E}_{13} = \frac{m}{M} \alpha a_{13} \mathcal{E}_{10}^{3} \frac{\mathcal{F}_{13}(\xi)}{(3\beta_1 + \beta_{13})(3\beta_1 - \beta_{13})}$$

where $\mathcal{F}_{13}(\xi) = \exp(i\beta_{13}\xi) - \exp(3i\beta_1\xi)$ incorporates the entire space dependence of $\mathcal{E}_{13}$. To investigate the form of the dependence of $\mathcal{E}_{13}$ on ξ in detail, one must study the characteristics of $\mathcal{F}_{13}(\xi)$ as ξ is varied. Writing $\beta = -n + ik$, one obtains the following expression for the amplitude of $\mathcal{F}_{13}(\xi)$:

$$\begin{aligned}
|\mathcal{F}_{13}(\xi)| = [&\exp(-2k_{13}\xi) + \exp(-6k_1\xi) \\
&+ 2\exp\{-i(k_{13} + 3k_1)\xi\}\cos\{(n_{13} - 3n_1)\xi\}]^{1/2}
\end{aligned} \tag{4.8}$$

Consider a numerical example in which $\omega_p^2/\omega_0^2 = 0.36$, $\omega_0 = \omega_1$, the collision frequency dependence on the electron velocity is $v = v_0 u$ and $(v_0/\omega_0) = 0.10$; using the expressions

$$\beta_1^2 = (-n_1 + ik_1)^2$$

$$= 1 - \frac{\omega_p^2}{\omega^2} - i\frac{\omega_p^2}{\omega^2}\frac{8}{3\pi^{1/2}}\frac{v_0}{\omega}$$

and

$$\beta_3^2 = (-n_3 + ik_3)^2$$

$$= 9\left[1 - \frac{\omega_p^2}{9\omega^2} - \frac{i}{27}\frac{\omega_p^2}{\omega^2}\frac{8}{3\pi^{1/2}}\frac{v_0}{\omega}\right]$$

(which are readily derivable from the definitions of β^2, $\omega_p{}^2$ and v_0), one obtains

$$n_1 = 0.8, \qquad k_1 = 0.034, \qquad n_{13} = 2.939, \qquad k_{13} = 0.003$$

Figure 1 illustrates the variation of the relative values of the amplitude of the third harmonic $|\mathscr{F}_{13}(\xi)|$ with ξ/ξ_1, where $\xi_1 = (\pi/2)/(n_{13} - 3n_1) = 2.914$.

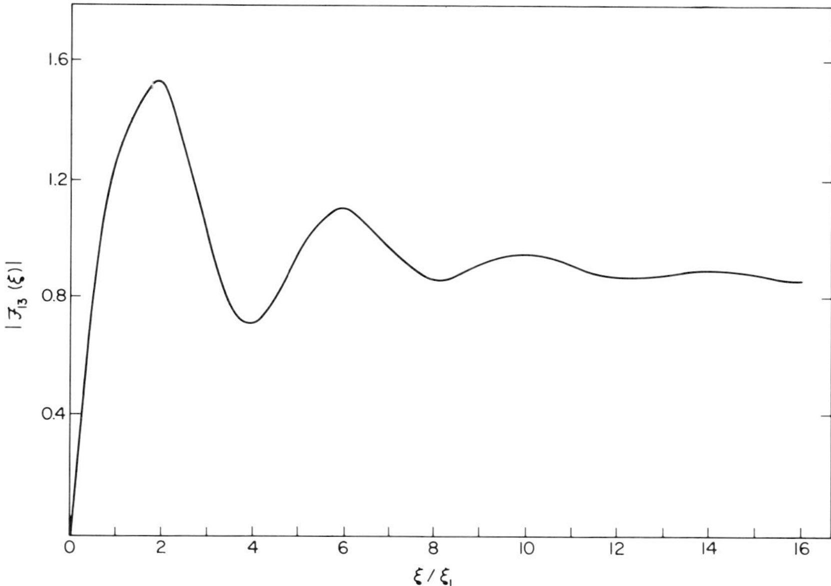

FIG. 1. Nonlinear growth of the third harmonic of an electromagnetic wave in a homogeneous plasma.

The maxima and minima in the amplitude of the third harmonic arise from the interference of two waves propagating with the propagation parameters $3\beta_1$ and β_{13} as expressed in Eq. (4.7c). As ξ increases, the wave corresponding to the propagation parameter $3\beta_1$ gets relatively weaker and the maxima and minima tend to flatten out.

Exactly similar variations are predicted for the other harmonic and combination frequency components also; they have, however, not been presented here.

b. Magnitude of the harmonic and combination frequency components in the reflected wave from a plasma and free space interface. Suppose that two plane-polarized electromagnetic waves of frequencies ω_1 and ω_2, plane-polarized along the x direction (which is also the direction of the externally applied dc electric field), are incident on the plasma free space interface (viz., the plane

$\xi = 0$) from the free space side. Let the region $-\infty \le \xi \le 0$ correspond to free space and the region $0 \le \xi \le \infty$ correspond to the plasma. The incident electrical vector is then given by

$$\frac{\mathbf{E}_i}{E_{00}} = \mathbf{B}_{i1} \exp\left\{i\left(\omega_1 t - \frac{\omega_1}{\omega_0}\xi\right)\right\} + \mathbf{B}_{i2} \exp\left\{i\left(\omega_2 t - \frac{\omega_2}{\omega_0}\xi\right)\right\} \qquad (4.9)$$

The electrical vector in free space will be given by

$$\begin{aligned}
\mathscr{E} = {} & \mathscr{E}_{i1} \exp(i\omega_1 t) + \mathscr{E}_{i2} \exp(i\omega_2 t) + \mathscr{E}_{\gamma 1} \exp(i\omega_1 t) \\
& + \mathscr{E}_{\gamma 2} \exp(i\omega_2 t) + \mathscr{E}_{\gamma 12} \exp(2i\omega_1 t) + \mathscr{E}_{\gamma 22} \exp(2i\omega_2 t) \\
& + \mathscr{E}_{\gamma 13} \exp(3i\omega_1 t) + \mathscr{E}_{\gamma 23} \exp(3i\omega_2 t) + \mathscr{E}_{\gamma 12}^{+} \exp i(\omega_1 + \omega_2)t \\
& + \mathscr{E}_{\gamma 12}^{-} \exp\{i(\omega_1 - \omega_2)t\} + \mathscr{E}_{\gamma 12}^{+\prime} \exp\{i(\omega_1 + 2\omega_2)t\} \\
& + \mathscr{E}_{\gamma 12}^{-\prime} \exp\{i(\omega_1 - 2\omega_2)t\} + \mathscr{E}_{\gamma 21}^{+\prime} \exp\{i(\omega_2 + 2\omega_1)t\} \\
& + \mathscr{E}_{\gamma 21}^{-\prime} \exp\{i(\omega_2 - 2\omega_1)t\}
\end{aligned}$$

where

$$\mathscr{E}_{i1} = B_{i1} \exp\left(-i\frac{\omega_1}{\omega_0}\xi\right), \qquad\qquad \mathscr{E}_{i2} = B_{i2} \exp\left(-i\frac{\omega_2}{\omega_0}\xi\right)$$

$$\mathscr{E}_{\gamma 1} = B_{\gamma 1} \exp\left(i\frac{\omega_1}{\omega_0}\xi\right), \qquad\qquad \mathscr{E}_{\gamma 2} = B_{\gamma 2} \exp\left(i\frac{\omega_2}{\omega_0}\xi\right)$$

$$\mathscr{E}_{\gamma 12} = B_{\gamma 12} \exp\left(2i\frac{\omega_1}{\omega_0}\xi\right), \qquad\qquad \mathscr{E}_{\gamma 22} = B_{\gamma 22} \exp\left(2i\frac{\omega_2}{\omega_0}\xi\right)$$

$$\mathscr{E}_{\gamma 13} = B_{\gamma 13} \exp\left(3i\frac{\omega_1}{\omega_0}\xi\right), \qquad\qquad \mathscr{E}_{\gamma 23} = B_{\gamma 23} \exp\left(3i\frac{\omega_2}{\omega_0}\xi\right)$$

$$\mathscr{E}_{\gamma 12}^{+} = B_{\gamma 12}^{+} \exp\left\{\frac{i(\omega_1 + \omega_2)}{\omega_0}\xi\right\}, \qquad \mathscr{E}_{\gamma 12}^{-} = B_{\gamma 12}^{-} \exp\left\{\frac{i(\omega_1 - \omega_2)}{\omega_0}\xi\right\}$$

$$\mathscr{E}_{\gamma 12}^{+\prime} = B_{\gamma 12}^{+\prime} \exp\left\{\frac{i(\omega_1 + 2\omega_2)}{\omega_0}\xi\right\}, \qquad \mathscr{E}_{\gamma 12}^{-\prime} = B_{\gamma 12}^{-\prime} \exp\left\{\frac{i(\omega_1 - 2\omega_2)}{\omega_0}\xi\right\}$$

$$\mathscr{E}_{\gamma 21}^{+\prime} = B_{\gamma 21}^{+\prime} \exp\left\{\frac{i(\omega_2 + 2\omega_1)}{\omega_0}\xi\right\}, \qquad \mathscr{E}_{\gamma 21}^{-\prime} = B_{\gamma 21}^{-\prime} \exp\left\{\frac{i(\omega_2 - 2\omega_1)}{\omega_0}\xi\right\}$$

the terms with subscript γ denote the components of the reflected wave and the B are the magnitudes of the various electrical vectors at $\xi = 0$. To a second order of approximation the electrical vector inside the plasma is given by the general solution of Eqs. (4.3a–b), viz.,

$$\begin{aligned}
\mathscr{E} = {} & \mathscr{E}_1 \exp(i\omega_1 t) + \mathscr{E}_2 \exp(i\omega_2 t) + \mathscr{E}_{12} \exp(2i\omega_1 t) \\
& + \mathscr{E}_{22} \exp(2i\omega_2 t) + \mathscr{E}_{13} \exp(3i\omega_1 t) + \mathscr{E}_{23} \exp(3i\omega_2 t)
\end{aligned}$$

$$+ \mathscr{E}_{12}^{+} \exp\{i(\omega_1 + \omega_2)t\} + \mathscr{E}_{12}^{-} \exp\{i(\omega_1 - \omega_2)t\}$$
$$+ \mathscr{E}_{12}^{+\prime} \exp\{i(\omega_1 + 2\omega_2)t\} + \mathscr{E}_{12}^{-\prime} \exp\{i(\omega_1 - 2\omega_2)t\}$$
$$+ \mathscr{E}_{21}^{+\prime} \exp\{i(\omega_2 + 2\omega_1)t\} + \mathscr{E}_{21}^{-\prime} \exp\{i(\omega_2 - 2\omega_1)t\}$$

where

$$\mathscr{E}_1 = K_1 \exp(i\beta_1 \xi); \qquad \mathscr{E}_2 = K_2 \exp(i\beta_2 \xi)$$

$$\mathscr{E}_{12} = K_{12} \exp(i\beta_{12}\xi) + \frac{m}{M} \alpha a_{12} \mathscr{E}_d \frac{K_1{}^2 \exp(2i\beta_1 \xi)}{\beta_{12}^2 - 4\beta_1{}^2}$$

$$\mathscr{E}_{22} = K_{22} \exp(i\beta_{22}\xi) + \frac{m}{M} \alpha a_{22} \mathscr{E}_d \frac{K_2{}^2 \exp(2i\beta_2 \xi)}{\beta_{22}^2 - 4\beta_2{}^2}$$

$$\mathscr{E}_{13} = K_{13} \exp(i\beta_{13}\xi) + \frac{m}{M} \alpha a_{13} \frac{K_1{}^3 \exp(3i\beta_1 \xi)}{\beta_{13}^2 - 9\beta_1{}^2}$$

$$\mathscr{E}_{23} = K_{23} \exp(i\beta_{23}\xi) + \frac{m}{M} \alpha a_{23} \frac{K_2{}^3 \exp(3i\beta_2 \xi)}{\beta_{23}^2 - 9\beta_2{}^2}$$

$$\mathscr{E}_{12}^{+} = K_{12}^{+} \exp(i\beta_{12}^{+}\xi) + \frac{m}{M} \alpha a_{12}^{+} \mathscr{E}_d \frac{K_1 K_2 \exp\{i(\beta_1 + \beta_2)\xi\}}{\beta_{12}^{+2} - (\beta_1 + \beta_2)^2}$$

$$\mathscr{E}_{12}^{-} = K_{12}^{-} \exp(i\beta_{12}^{-}\xi) + \frac{m}{M} \alpha a_{12}^{-} \mathscr{E}_d \frac{K_1 \tilde{K}_2 \exp\{i(\beta_1 - \tilde{\beta}_2)\xi\}}{\beta_{12}^{-2} - (\beta_1 - \tilde{\beta}_2)^2}$$

$$\mathscr{E}_{12}^{+\prime} = K_{12}^{+\prime} \exp(i\beta_{12}^{+\prime}\xi) + \frac{m}{M} \alpha a_{12}^{+\prime} \frac{K_1 K_2{}^2 \exp\{i(\beta_1 + 2\beta_2)\xi\}}{\beta_{12}^{+\prime 2} - (\beta_1 + 2\beta_2)^2}$$

$$\mathscr{E}_{12}^{-\prime} = K_{12}^{-\prime} \exp(i\beta_{12}^{-\prime}\xi) + \frac{m}{M} \alpha a_{12}^{-\prime} \frac{K_1 \tilde{K}_2{}^2 \exp\{i(\beta_1 - 2\tilde{\beta}_2)\xi\}}{\beta_{12}^{-\prime 2} - (\beta_1 - 2\tilde{\beta}_2)^2}$$

$$\mathscr{E}_{21}^{+\prime} = K_{21}^{+\prime} \exp(i\beta_{21}^{+\prime}\xi) + \frac{m}{M} \alpha a_{21}^{+\prime} \frac{K_2 K_1{}^2 \exp\{i(\beta_2 + 2\beta_1)\xi\}}{\beta_{21}^{+\prime 2} - (\beta_2 + 2\beta_1)^2}$$

$$\mathscr{E}_{21}^{-\prime} = K_{21}^{-\prime} \exp(i\beta_{21}^{-\prime}\xi) + \frac{m}{M} \alpha a_{21}^{-\prime} \frac{K_2 \tilde{K}_1{}^2 \exp\{i(\beta_2 - 2\tilde{\beta}_1)\xi\}}{\beta_{21}^{-\prime 2} - (\beta_2 - 2\tilde{\beta}_1)^2}$$

where the K are arbitrary constants. Since $\mathscr{E}$ and $(1/\mu)(d\mathscr{E}/d\xi)$ are continuous across $\xi = 0$, one may equate the expressions for the various components of the electrical vector and their differential coefficients with respect to ξ divided by the permeability of the relevant medium in free space and in the plasma at $\xi = 0$. All the unknown constants can then be evaluated, and we get the following expressions for the components of the reflected wave:

$$\frac{B_{y1}}{B_{i1}} = \frac{1 + (\omega_0/\omega_1)(\beta_1/\mu)}{1 - (\omega_0/\omega_1)(\beta_1/\mu)}; \qquad \frac{B_{y2}}{B_{i2}} = \frac{1 + (\omega_0/\omega_2)(\beta_2/\mu)}{1 - (\omega_0/\omega_2)(\beta_2/\mu)} \qquad (4.10\text{a, b})$$

$$\frac{B_{\gamma 12}}{B_{i1}^2} = \frac{4}{\mu}\frac{m}{M}\,\alpha\mathscr{E}_d\,a_{12}\bigg/\left[(2\beta_1 + \beta_{12})\left\{\frac{\beta_{12}}{\mu} - \frac{2\omega_1}{\omega_0}\right\}\left\{1 - \frac{\omega_0}{\omega_1}\frac{\beta_1}{\mu}\right\}^2\right] \qquad (4.10c)$$

$$\frac{B_{\gamma 22}}{B_{i2}^2} = \frac{4}{\mu}\frac{m}{M}\,\alpha\mathscr{E}_d\,a_{22}\bigg/\left[(2\beta_2 + \beta_{22})\left\{\frac{\beta_{22}}{\mu} - \frac{2\omega_2}{\omega_0}\right\}\left\{1 - \left(\frac{\omega_0}{\omega_2}\frac{\beta_2}{\mu}\right)\right\}^2\right] \qquad (4.10d)$$

$$\frac{B_{\gamma 13}}{B_{i1}^3} = \frac{8}{\mu}\frac{m}{M}\,\alpha a_{13}\bigg/\left[(3\beta_1 + \beta_{13})\left\{\frac{\beta_{13}}{\mu} - \frac{3\omega_1}{\omega_0}\right\}\left\{1 - \left(\frac{\omega_0}{\omega_1}\frac{\beta_1}{\mu}\right)\right\}^3\right] \qquad (4.10e)$$

$$\frac{B_{\gamma 23}}{B_{i1}^3} = \frac{8}{\mu}\frac{m}{M}\,\alpha a_{23}\bigg/\left[(3\beta_2 + \beta_{23})\left\{\frac{\beta_{23}}{\mu} - \frac{3\omega_2}{\omega_0}\right\}\left\{1 - \left(\frac{\omega_0}{\omega_2}\frac{\beta_2}{\mu}\right)\right\}^3\right] \qquad (4.10f)$$

$$\frac{B_{\gamma 12}^+}{B_{i1}B_{i2}} = \frac{4}{\mu}\frac{m}{M}\,\alpha$$

$$\times\, a_{12}^+\,\mathscr{E}_d\bigg/\left[(\beta_1 + \beta_2 + \beta_{12}^+)\left\{\frac{\beta_{12}^+}{\mu} - \frac{\omega_1 + \omega_2}{\omega_0}\right\}\left\{1 - \left(\frac{\omega_0}{\omega_1}\frac{\beta_1}{\mu}\right)\right\}\left\{1 - \left(\frac{\omega_0}{\omega_2}\frac{\beta_2}{\mu}\right)\right\}\right]$$

$$(4.10g)$$

$$\frac{B_{\gamma 12}^-}{B_{i1}\tilde{B}_{i2}} = \frac{4}{\mu}\frac{m}{M}\,\alpha$$

$$\times\, a_{12}^-\,\mathscr{E}_d\bigg/\left[(\beta_1 - \tilde{\beta}_2 + \beta_{12}^-)\left\{\frac{\beta_{12}^-}{\mu} - \frac{\omega_1 - \omega_2}{\omega_0}\right\}\left\{1 - \left(\frac{\omega_0}{\omega_1}\frac{\beta_1}{\mu}\right)\right\}\left\{1 - \left(\frac{\omega_0}{\omega_2}\frac{\tilde{\beta}_2}{\mu}\right)\right\}\right]$$

$$(4.10h)$$

$$\frac{B_{\gamma 12}^{+\prime}}{B_{i1}B_{i2}^2} = \frac{8}{\mu}\frac{m}{M}\,\alpha$$

$$\times\, a_{12}^{+\prime}\bigg/\left[(\beta_1 + 2\beta_2 + \beta_{12}^{+\prime})\left\{\frac{\beta_{12}^{+\prime}}{\mu} - \frac{\omega_1 + 2\omega_2}{\omega_0}\right\}\left\{1 - \left(\frac{\omega_0}{\omega_1}\frac{\beta_1}{\mu}\right)\right\}\left\{1 - \left(\frac{\omega_0}{\omega_2}\frac{\beta_2}{\mu}\right)\right\}^2\right]$$

$$(4.10i)$$

$$\frac{B_{\gamma 12}^{-\prime}}{B_{i1}\tilde{B}_{i2}^2} = \frac{8}{\mu}\frac{m}{M}\,\alpha$$

$$\times\, a_{12}^{-\prime}\bigg/\left[(\beta_1 - 2\tilde{\beta}_2 + \beta_{12}^{-\prime})\left(\frac{\beta_{12}^{-\prime}}{\mu} - \frac{\omega_1 - 2\omega_2}{\omega_0}\right)\left\{1 - \left(\frac{\omega_0}{\omega_1}\frac{\beta_1}{\mu}\right)\right\}\left\{1 - \left(\frac{\omega_0}{\omega_2}\frac{\tilde{\beta}_2}{\mu}\right)\right\}^2\right]$$

$$(4.10j)$$

$$\frac{B_{\gamma 21}^{+\prime}}{B_{i2}B_{i1}^2} = \frac{8}{\mu}\frac{m}{M}\,\alpha$$

$$\times\, a_{21}^{+\prime}\bigg/\left[(\beta_2 + 2\beta_1 + \beta_{21}^{+\prime})\left(\frac{\beta_{21}^{+\prime}}{\mu} - \frac{\omega_2 + 2\omega_1}{\omega_0}\right)\left\{1 - \left(\frac{\omega_0}{\omega_2}\frac{\beta_2}{\mu}\right)\right\}\left\{1 - \left(\frac{\omega_0}{\omega_1}\frac{\beta_1}{\mu}\right)\right\}^2\right]$$

$$(4.10k)$$

$$\frac{B_{\gamma 21}^{-\prime}}{B_{i2}\,\tilde{B}_{i1}^2} = \frac{8}{\mu}\frac{m}{M}\,\alpha$$

$$\times a_{21}^{-\prime}\left/\left[(\beta_2 - 2\tilde{\beta}_1 + \beta_{21}^{-\prime})\left\{1 - \left(\frac{\omega_0}{\omega_2}\right)\frac{\beta_2}{\mu}\right\}\left\{1 - \left(\frac{\omega_0}{\omega_1}\frac{\tilde{\beta}_1}{\mu}\right)\right\}^2\left\{\frac{\beta_{21}^{-\prime}}{\mu} - \frac{\omega_2 - 2\omega_1}{\omega_0}\right\}\right]\right.$$

$$(4.101)$$

From Eqs. (4.10) one notes that the second harmonic and the sum-and-difference frequencies in the reflected wave are of the order of $(m/M)\alpha$ or $e^2/6m\omega_0^2 kT$ times $\mathscr{E}_{ip}\mathscr{E}_{iq}\mathscr{E}_d$ $(p, q = 1, 2)$. The second harmonic and the sum-and-difference frequency components in the reflected wave are proportional only to the second powers of the incident electrical vectors (which are usually small). In contrast, the other harmonic and combination frequency components considered are proportional to the third powers of the incident electrical vectors. Therefore the former have a larger magnitude than the latter. One further notes the important fact that the dc field $\mathscr{E}_d$ gives a free parameter that can be readily adjusted to a large value to give appreciable second-harmonic and sum-and-difference frequency components.

To study the variation of the various harmonic and combination frequency components in the reflected wave with the collision frequency and the ratio ω_2/ω_1, some calculations have been made for the case $n = 1$, i.e., when the electron collision frequency is proportional to the electron velocity. Let us first of all discuss the variation of the components that are generated even in the absence of the dc field; the components that are generated only because of the presence of the dc field will be treated separately in a later discussion.

TABLE VI(A)

VARIATIONS OF $e_{\gamma 13}$, $e_{\gamma 21}^{+\prime}$ AND $e_{\gamma 21}^{-\prime}$ WITH ν_0/ω_0 FOR $\omega_2/\omega_1 = 3$

ν_0/ω_0	$e_{\gamma 13}$	$e_{\gamma 21}^{+\prime}$	$e_{\gamma 21}^{-\prime}$
0.001	3.52×10^{-6}	1.230×10^{-6}	2.680×10^{-5}
0.003	1.054×10^{-5}	3.700×10^{-6}	8.080×10^{-5}
0.010	3.517×10^{-5}	1.231×10^{-5}	2.692×10^{-4}
0.030	1.055×10^{-4}	3.699×10^{-5}	8.054×10^{-4}
0.100	3.582×10^{-4}	1.243×10^{-4}	2.623×10^{-3}

Table VI(A) illustrates the variation of

$$e_{\gamma 13} = \frac{6m\omega_0^2 kT}{e^2}\left|\frac{E_{\gamma 13}}{E_{i1}^3}\right| \quad \text{for} \quad \frac{\omega_p^2}{\omega_0^2} = 0.36$$

$$e_{\gamma 21}^{+\prime} = \frac{6m\omega_0^2 kT}{e^2}\left|\frac{E_{\gamma 21}^{+\prime}}{E_{i2}\,E_{i1}^2}\right|$$

$$\text{for} \quad \frac{\omega_p^2}{\omega_0^2} = 0.40$$

$$e_{\gamma 21}^{-\prime} = \frac{6m\omega_0^2 kT}{e^2}\left|\frac{E_{\gamma 21}^{-\prime}}{E_{i2}\,\tilde{E}_{i1}^2}\right|$$

with v_0/ω_0 when $\omega_0 = \omega_1$ and the wave with frequency ω_2 has been assumed to be weak in intensity. Table VI(B) illustrates the variations of $e_{\gamma 21}^{+\prime}$ and $e_{\gamma 21}^{-\prime}$

TABLE VI(B)

VARIATIONS OF $e_{\gamma 21}^{+\prime}$ AND $e_{\gamma 21}^{-\prime}$ WITH ω_2/ω_1 FOR $v_0/\omega_0 = 0.100$

ω_2/ω_1	$e_{\gamma 21}^{+\prime}$	$e_{\gamma 21}^{-\prime}$
3	1.243×10^{-4}	2.623×10^{-3}
5	3.696×10^{-5}	1.848×10^{-4}
7	1.575×10^{-5}	4.790×10^{-5}
9	8.230×10^{-6}	1.920×10^{-5}
11	4.770×10^{-6}	1.510×10^{-5}

with ω_2/ω_1 for $v_0/\omega_0 = 0.100$ and $\omega_0 = \omega_1$. The minimum value of ω_2/ω_1 has been chosen in such a manner that $\omega_2 - 2\omega_1$ is always greater than ω_p so that the wave with this frequency is able to propagate through the plasma. The variations of $e_{\gamma 13}$, $e_{\gamma 21}^{+\prime}$ and $e_{\gamma 21}^{-\prime}$ with v_0/ω_0 and ω_2/ω_1 are seen to be similar in form to those of j_{13}, $j_{\gamma 21}^{+\prime}$, and $j_{\gamma 21}^{-\prime}$, which were considered earlier. However, one important point is noted; the fall in the magnitude of $e_{\gamma 21}^{-\prime}$ with increasing ω_2/ω_1 (beyond the value 3) is much steeper than that of $e_{\gamma 21}^{+\prime}$, and moreover, for higher values of ω_2/ω_1, the steepness of the fall is considerably reduced. This is the nature of variation that one expects for $e_{\gamma 21}^{-\prime}$ in view of the existence of a resonance at $\omega_2/\omega_1 = 2$.

Now, in order to get an idea of the absolute magnitudes of the third harmonic and the second-order combination frequency components in the reflected waves let us consider some typical cases, where $T = 400°$ K, $\omega_0 = \omega_1 = 10^5$ rad/sec, and $v_0/\omega_0 = 0.100$:

1. $E_{i1} = 1.5 \times 10^{-2}$ volts/cm; one obtains $E_{\gamma 13} = 8.565 \times 10^{-4}$ volts/cm, which is about 5.71% of the fundamental.
2. $E_{i1} = 3 \times 10^{-2}$ volts/cm, $E_{i2} = 3 \times 10^{-3}$ volts/cm, $\omega_2/\omega_1 = 3$; one obtains $E_{\gamma 21}^{+\prime} = 3.801 \times 10^{-4}$ volts/cm, which is about 1.27% of E_{i1}.
3. $E_{i1} = 9.486 \times 10^{-3}$ volts/cm, $E_{i2} = 3 \times 10^{-3}$ volts/cm; one obtains $E_{\gamma 21}^{-\prime} = 6 \times 10^{-4}$ volts/cm, which is about 6.31% of E_{i1}.

The same example has not been chosen for all three cases because of the large disparity in the magnitudes of these three components.

TABLE VII(A)

VARIATIONS OF $e_{\gamma 12}$, $e_{\gamma 12}^{+}$, AND $e_{\gamma 12}^{-}$ WITH v_0/ω_0 FOR $(\omega_2/\omega_1) = 0.4$

v_0/ω_0	$e_{\gamma 12}$	$e_{\gamma 12}^{+}$	$e_{\gamma 12}^{-}$
0.01	1.524×10^{-3}	3.072×10^{-4}	5.293×10^{-4}
0.03	1.530×10^{-3}	9.315×10^{-4}	2.518×10^{-3}
0.05	1.540×10^{-3}	1.866×10^{-3}	6.240×10^{-3}
0.07	1.561×10^{-3}	2.537×10^{-3}	1.191×10^{-2}

Table VII(A) illustrates the variations of

$$e_{\gamma 12} = \frac{6m\omega_0{}^2 kT}{e^2} \left| \frac{E_{\gamma 12}}{E_{i1}^2 E_d} \right|$$

$$= 2 \frac{\omega_p{}^2}{\omega_0{}^2} j_{12} |F_{12}(\beta)|$$

$$e_{\gamma 12}^+ = \frac{6m\omega_0{}^2 kT}{e^2} \left| \frac{E_{\gamma 12}^+}{E_d E_{i1} E_{i2}} \right|$$

$$= \left(1 + \frac{\omega_2}{\omega_0}\right) \frac{\omega_p{}^2}{\omega_0{}^2} j_{12}^+ |F_{12}^+(\beta)|$$

$$e_{\gamma 12}^- = \frac{6m\omega_0{}^2 kT}{e^2} \left| \frac{E_{\gamma 12}^-}{E_d E_{i1} \tilde{E}_{i2}} \right|$$

$$= \left(1 - \frac{\omega_2}{\omega_0}\right) \frac{\omega_p{}^2}{\omega_0{}^2} j_{12}^- |F_{12}^-(\beta)|$$

with v_0/ω_0 for $\omega_p{}^2/\omega_0{}^2 = 0.02$, $\omega_0 = \omega_1$ and $\omega_2/\omega_1 = 0.4$, where

$$F_{12}(\beta) = \frac{4}{\mu} \bigg/ \left[(2\beta_1 + \beta_{12}) \left\{ \frac{\beta_{12}}{\mu} - \frac{2\omega_1}{\omega_0} \right\} \left\{ 1 - \frac{\omega_0}{\omega_1} \frac{\beta_1}{\mu} \right\}^2 \right]$$

$$F_{12}^+(\beta) = \frac{4}{\mu} \bigg/ \left[(\beta_1 + \beta_2 + \beta_{12}^+) \left(\frac{\beta_{12}^+}{\mu} - \frac{\omega_1 + \omega_2}{\omega_0} \right) \left(1 - \frac{\omega_0}{\omega_1} \frac{\beta_1}{\mu} \right) \left(1 - \frac{\omega_0}{\omega_2} \frac{\beta_2}{\mu} \right) \right]$$

$$F_{12}^-(\beta) = \frac{4}{\mu} \bigg/ \left[(\beta_1 - \tilde{\beta}_2 + \beta_{12}^-) \left(\frac{\beta_{12}^-}{\mu} - \frac{\omega_1 - \omega_2}{\omega_0} \right) \left(1 - \frac{\omega_0}{\omega_1} \frac{\beta_1}{\mu} \right) \left(1 - \frac{\omega_0}{\omega_2} \frac{\tilde{\beta}_2}{\mu} \right) \right]$$

Similarly, Table VII(B) illustrates the variations of $e_{\gamma 12}^+$ and $e_{\gamma 12}^-$ with ω_2/ω_1 for $v_0/\omega_0 = 0.05$ (the rest of the parameters being the same as before). It is

TABLE VII(B)

VARIATIONS OF $e_{\gamma 12}^+$ AND $e_{\gamma 12}^-$ WITH ω_2/ω_1 FOR $v_0/\omega_0 = 0.05$

ω_2/ω_1	$e_{\gamma 12}^+$	$e_{\gamma 12}^-$
0.2	1.116×10^{-2}	2.216×10^{-2}
0.4	1.866×10^{-3}	6.240×10^{-3}
0.6	5.948×10^{-4}	2.846×10^{-3}
0.8	2.902×10^{-4}	6.466×10^{-2}

noted from Table VII(A) that the magnitudes of the second harmonic and the sum-and-difference frequency components increase with the collision frequency. From Table VII(B), one notes that as ω_2/ω_1 is increased, the magnitude of the sum frequency decreases continuously and that of the difference frequency decreases first and then increases. This anomolous behavior of the difference frequency arises because one expects a resonance in the magnitude of this component at $\omega_2 \sim \omega_1$. However, since at that point the difference frequency will be zero, one expects a large dc component, which is hardly of any interest.

To get an idea of the absolute order of magnitude of the second harmonic and the sum-and-difference frequencies generated in the reflected wave, consider a typical case with $\omega_p^2/\omega_0^2 = 0.02$, $v_0/\omega_0 = 0.05$, $\omega_0 = \omega_1 = 10^7$ rad/sec, $\omega_2/\omega_1 = 0.8$, $T = 300°$ K; for $E_{i1} = E_{i2} = 10^{-3}$ esu/cm and $E_d = 3 \times 10^{-5}$ esu/cm, one obtains $E_{y12}^- = 1.975 \times 10^{-5}$ esu/cm, which is about 2% of the fundamental waves. Thus we see that appreciable difference frequencies are generated for low values of the dc field and moderate values of the fundamental electrical vectors; the same conclusion can be drawn about the second harmonic and the sum frequency also. In the above example, we had to choose the magnitude of the dc electric field to be very low in view of the limitations of the analysis (discussed in detail in Appendix B); it need hardly be emphasized that such a limitation need not be imposed in any experiments that may be conducted in this direction.

2. In the Presence of a Magnetic Field

The propagation of the combination frequency waves in a homogeneous magnetoplasma has been studied in some detail by Vilenskii (6). However, as has already been mentioned, the expressions for the combination frequency components of the current density obtained by him were not complete because he had neglected some important terms while writing down the expansion of the distribution function. No investigation of the propagation of these components with the more accurate expressions of the current density obtained in Section III has been attempted so far. Sodha and Kaw (18) have, however, discussed the propagation of the third harmonic of an electromagnetic wave in a magnetoplasma; as we have seen earlier, the form of the third-harmonic current density expression used by them is correct, though the magnitudes are in error. Here, we have outlined Sodha and Kaw's (18) treatment (using the correct expressions derived in Section III-A-3) in order to illustrate the procedure involved in tackling such problems.

Consider a plane-polarized electromagnetic wave of frequency ω to be propagating in the z direction, which is also the direction of the externally applied dc magnetic field. The plasma then finds itself under the influence of an electric field having the components

$$E_x = E_{1x} \exp(i\omega t) + E_{3x} \exp(3i\omega t) \tag{4.11a}$$

$$E_y = E_{1y} \exp(i\omega t) + E_{3y} \exp(3i\omega t) \tag{4.11b}$$

$$E_z = 0 \tag{4.11c}$$

and a magnetic field **B** with components

$$B_x = B_y = 0, \; B_z = 0 \tag{4.12a,b}$$

where $E_3 \ll E_1$ is the third-harmonic field set up by the wave itself. The x and y components of the current density are then given by

$$
\begin{aligned}
J_x = {}& (\sigma_1' E_{1x} - \sigma_1'' E_{1y})\exp(i\omega t) \\
& + (\sigma_3' E_{3x} - \sigma_3'' E_{3y})\exp(3i\omega t) \\
& + \frac{m}{M}\,\alpha[B_1 \mathscr{E}_{1x}^2 E_{1x} - B_2 \mathscr{E}_{1x}^2 E_{1y} \\
& + B_3 \mathscr{E}_{1y}^2 E_{1x} - B_4 \mathscr{E}_{1y}^2 E_{1y}]\exp(3i\omega t)
\end{aligned}
\tag{4.13a}
$$

and

$$
\begin{aligned}
J_y = {}& (\sigma_1'' E_{1x} + \sigma_1' E_{1y})\exp(i\omega t) \\
& + (\sigma_3'' E_{3x} + \sigma_3' E_{3y})\exp(3i\omega t) \\
& + \frac{m}{M}\,\alpha\{B_1 \mathscr{E}_{1y}^2 E_{1y} + B_2 \mathscr{E}_{1y}^2 E_{1x} \\
& + B_3 \mathscr{E}_{1x}^2 E_{1y} + B_4 \mathscr{E}_{1x}^2 E_{1x}\}\exp(3i\omega t)
\end{aligned}
\tag{4.13b}
$$

respectively, where σ'_1 and σ''_1 are the components of the conductivity tensor corresponding to the frequency ω and are given by [Sodha and Palumbo (74)] as

$$\sigma_1' = \frac{e^2 N}{3m}\left\langle \frac{1}{v^2}\frac{d}{dv}\left[\frac{v^3(v + i\omega)}{(v + i\omega)^2 + \omega_B^{\,2}}\right]\right\rangle$$

and

$$\sigma_1'' = \frac{e^2 N}{3m}\left\langle \frac{1}{v^2}\frac{d}{dv}\left[\frac{v^3 \omega_B}{(v + i\omega)^2 + \omega_B^{\,2}}\right]\right\rangle$$

One can use a Maxwellian form for $f_0^{\,0}$ in evaluating these averages because our expressions for the third-harmonic current density are correct only up to the order of $(m/M)\alpha$ (the non-Maxwellian terms in $f_0^{\,0}$ would contribute terms of the order of $(m/M)\alpha^2$, which have been neglected).

One notes from the above expressions for the components of the current density in a plasma that the x component is determined not only by the x component of the applied electric field, but also by its y component; in other words the x and y components are inextricably coupled inside the plasma. This is a well-known result and appears because of the anisotropy introduced in the plasma by the external magnetic field. Because of this coupling, the

x and y components of the electrical vectors of the fundamental as well as the harmonic components do not propagate independently but in the form of elliptically polarized modes ($\mathscr{E}_x + \alpha'\mathscr{E}_y$). When the direction of propagation coincides with that of the magnetic field, Sodha and Palumbo (75) have shown that $\alpha' = \pm i$; thus, in our case also, the coupled modes of the fundamental and the third harmonic will be ($\mathscr{E}_{1x} \pm i\mathscr{E}_{1y}$) and ($\mathscr{E}_{3x} \pm i\mathscr{E}_{3y}$), respectively. The two modes in each case are known as the extraordinary and ordinary modes of propagation.

a. Nonlinear growth of the third harmonic. Writing the x and y components of Eq. (4.1) and substituting the values of the components of **E** and **J** from Eqs. (4.12) and (4.13), one obtains

$$\frac{\partial^2 \mathscr{E}_{1x}}{\partial \xi^2}\exp(i\omega t) + \frac{\partial^2 \mathscr{E}_{3x}}{\partial \xi^2}\exp(3i\omega t) = -[\mathscr{E}_{1x}\exp(i\omega t) + 9\mathscr{E}_{3x}\exp(3i\omega t)]$$
$$+ \frac{4\pi i}{\varepsilon\omega}\left[(\sigma_1'\mathscr{E}_{1x} - \sigma_1''\mathscr{E}_{1y})\exp(i\omega t)\right.$$
$$+ 3(\sigma_3'\mathscr{E}_{3x} - \sigma_3''\mathscr{E}_{3y})\exp(3i\omega t)$$
$$+ \frac{3m}{M}\alpha(B_1\mathscr{E}_{1x}^3 - B_2\mathscr{E}_{1x}^2\mathscr{E}_{1y}$$
$$\left.+ B_3\mathscr{E}_{1x}\mathscr{E}_{1y}^2 - B_4\mathscr{E}_{1y}^3)\exp(3i\omega t)\right]$$

$$\frac{\partial^2 \mathscr{E}_{1y}}{\partial \xi^2}\exp(i\omega t) + \frac{\partial^2 \mathscr{E}_{3y}}{\partial \xi^2}\exp(3i\omega t)$$
$$= -[\mathscr{E}_{1y}\exp(i\omega t) + 9\mathscr{E}_{3y}\exp(3i\omega t)]$$
$$+ \frac{4\pi i}{\varepsilon\omega}\left[(\sigma_1''\mathscr{E}_{1x} + \sigma_1'\mathscr{E}_{1y})\exp(i\omega t)\right.$$
$$+ 3(\sigma_3''\mathscr{E}_{3x} + \sigma_3'\mathscr{E}_{3y})\exp(3i\omega t)$$
$$+ \frac{3m}{M}\alpha(B_1\mathscr{E}_{1y}^3 + B_2\mathscr{E}_{1y}^2\mathscr{E}_{1x} + B_3\mathscr{E}_{1y}\mathscr{E}_{1x}^2$$
$$\left.+ B_4\mathscr{E}_{1x}^3)\exp(3i\omega t)\right]$$

where $\xi = (\varepsilon\mu)^{1/2}(\omega z/C)$.

Multiplying the latter equation by $\pm i$, adding to the former, and equating the coefficients of $\exp(i\omega t)$ and $\exp(3i\omega t)$ on both sides of the final equation, the two different modes of propagation of each of the waves of frequency ω

and 3ω are given by

$$\frac{\partial^2}{\partial \xi^2}(\mathscr{E}_{1x} + i\mathscr{E}_{1y}) + \beta_1'^2(\mathscr{E}_{1x} + i\mathscr{E}_{1y}) = 0 \tag{4.14a}$$

$$\frac{\partial^2}{\partial \xi^2}(\mathscr{E}_{1x} - i\mathscr{E}_{1y}) + \beta_1''^2(\mathscr{E}_{1x} - i\mathscr{E}_{1y}) = 0 \tag{4.14b}$$

$$\frac{\partial^2}{\partial \xi^2}(\mathscr{E}_{3x} + i\mathscr{E}_{3y}) + \beta_3'^2(\mathscr{E}_{3x} + i\mathscr{E}_{3y})$$
$$= \frac{m}{M}\,\alpha[b_{14}'(\mathscr{E}_{1x}^3 + i\mathscr{E}_{1y}^3) - b_{23}''\,\mathscr{E}_{1x}\mathscr{E}_{1y}(\mathscr{E}_{1x} - i\mathscr{E}_{1y})] \tag{4.14c}$$

$$\frac{\partial^2}{\partial \xi^2}(\mathscr{E}_{3x} - i\mathscr{E}_{3y}) + \beta_3''^2(\mathscr{E}_{3x} - i\mathscr{E}_{3y})$$
$$= \frac{m}{M}\,\alpha[b_{14}''(\mathscr{E}_{1x}^3 - i\mathscr{E}_{1y}^3) - b_{23}'\,\mathscr{E}_{1x}\mathscr{E}_{1y}(\mathscr{E}_{1x} + i\mathscr{E}_{1y})] \tag{4.14d}$$

where

$$\beta_1'^2 = 1 - \frac{4\pi i}{\varepsilon\omega}(\sigma_1' + i\sigma_1'') \qquad\qquad \beta_1''^2 = 1 - \frac{4\pi i}{\varepsilon\omega}(\sigma_1' - i\sigma_1'')$$

$$\beta_3'^2 = 9\left[1 - \frac{4\pi i}{3\varepsilon\omega}(\sigma_3' + i\sigma_3'')\right] \qquad\qquad \beta_3''^2 = 9\left[1 - \frac{4\pi i}{3\varepsilon\omega}(\sigma_3' - i\sigma_3'')\right]$$

$$b_{14}' = \frac{12\pi i}{\varepsilon\omega}(B_1 + iB_4) \qquad\qquad b_{14}'' = \frac{12\pi i}{\varepsilon\omega}(B_1 - iB_4)$$

$$b_{23}' = \frac{12\pi i}{\varepsilon\omega}(B_2 + iB_3) \qquad \text{and} \qquad b_{23}'' = \frac{12\pi i}{\varepsilon\omega}(B_2 - iB_3)$$

Following a successive approximation technique similar to that outlined in the preceding subsection, the solutions of the above equations with the appropriate radiation condition and the boundary conditions

$$\mathscr{E}_{1x} = \mathscr{E}_{1x}^0, \qquad \mathscr{E}_{1y} = \mathscr{E}_{1y}^0 \qquad \text{and} \qquad \mathscr{E}_{3x} = \mathscr{E}_{3y} = 0 \qquad \text{at} \quad \xi = 0$$

are

$$\mathscr{E}_{1x} + i\mathscr{E}_{1y} = K_1'\exp(i\beta_1'\xi) \tag{4.15a}$$

$$\mathscr{E}_{1x} - i\mathscr{E}_{1y} = K_1''\exp(i\beta_1''\xi) \tag{4.15b}$$

$$\mathscr{E}_{3x} + i\mathscr{E}_{3y}$$
$$= \frac{m}{4M}\,\alpha K_1''\left[K_1''^2(b_{14}' - ib_{23}'')\frac{\exp(i\beta_3'\xi) - \exp(3i\beta_1''\xi)}{(3\beta_1'' - \beta_3')(3\beta_1'' + \beta_3')}\right.$$
$$\left. + K_1'^2(3b_{14}' + ib_{23}'')\frac{\exp(i\beta_3'\xi) - \exp[i(2\beta_1' + \beta_1'')\xi]}{(2\beta_1' + \beta_1'' - \beta_3')(2\beta_1' + \beta_1'' + \beta_3')}\right] \tag{4.15c}$$

$$\mathscr{E}_{3x} - i\mathscr{E}_{3y}$$

$$= \frac{m}{4M} \alpha K_1' \left[K_1'^2 (b_{14}'' + ib_{23}') \frac{\exp(i\beta_3'' \xi) - \exp(3i\beta_1'\xi)}{(3\beta_1' - \beta_3'')(3\beta_1' + \beta_3'')} \right.$$

$$\left. + K_1''^2 (3b_{14}'' - ib_{23}') \frac{\exp(i\beta_3''\xi) - \exp[i(\beta_1' + 2\beta_1'')\xi]}{(\beta_1' + 2\beta_1'' + \beta_3')(\beta_1' + 2\beta_1'' - \beta_3'')} \right] \quad (4.15d)$$

where

$$K_1' = \mathscr{E}_{1x}^0 + i\mathscr{E}_{1y}^0 \quad (4.16a)$$

$$K_1'' = \mathscr{E}_{1x}^0 - i\mathscr{E}_{1y}^0 \quad (4.16b)$$

and only terms up to the order of $(m/M)\alpha$ have been retained in the third-harmonic components.

From Eqs. (4.15) it is seen that the amplitude of the electrical vector of the third harmonic of an electromagnetic wave is of the order of $(m/M)\alpha\mathscr{E}_1^3 E_{00}$ or $e^2/6m\omega^2 kT$ times the cube of the amplitude of the fundamental. It is also seen from Eqs. (4.15) that the amplitudes of the ordinary and extraordinary modes of the third harmonic of an electromagnetic wave are respectively proportional to the amplitudes of the extraordinary and ordinary modes of the fundamental. Thus, if we send a pure mode of the fundamental in a plasma, only the other mode of the third harmonic will be generated.

The growth of the two modes of propagation of the third harmonic of an electromagnetic wave in a magnetoplasma can be studied with the help of Eqs. (4.15). The results are fairly similar to those obtained earlier for the growth of the third harmonic in a plasma in the absence of a magnetic field.

b. Third harmonic in the reflected wave. Let a plane-polarized electromagnetic wave whose electrical vector is given by $\mathbf{E}_i/E_{00} = \mathbf{B}_{i1} \exp[i(\omega t - \xi)]$ be incident normally on the magnetoplasma-free space interface, viz., the plane $\xi = 0$, from the free space side. The electrical vector in free space is then given by

$$\mathscr{E} = \mathscr{E}_{i1} \exp(i\omega t) + \mathscr{E}_{y1} \exp(i\omega t) + \mathscr{E}_{y3} \exp(3i\omega t) \quad (4.17)$$

where

$$\mathscr{E}_{i1} = A_{i1} \exp(-i\xi) \qquad \mathscr{E}_{y1} = A_{y1} \exp(i\xi) \qquad \mathscr{E}_{y3} = A_{y3} \exp(3i\xi)$$

and the last two terms in Eq. (4.17) represent the reflected wave. To a second order of approximation the electrical vector inside the plasma is given by the general solution of Eqs. (4.14), viz.,

$$\mathscr{E} = \mathscr{E}_1 \exp(i\omega t) + \mathscr{E}_3 \exp(3i\omega t)$$

where

$$\mathscr{E}_{1x} + i\mathscr{E}_{1y} = K_1' \exp(i\beta_1'\xi)$$

$$\mathscr{E}_{1x} - i\mathscr{E}_{1y} = K_1'' \exp(i\beta_1''\xi)$$

$$\mathscr{E}_{3x} + i\mathscr{E}_{3y} = K_3' \exp(i\beta_3'\xi) + \frac{m\alpha}{4M} K_1'' \left[K_1''^2 \frac{(b_{14}' - ib_{23}'')}{\beta_3'^2 - 9\beta_1''^2} \exp(3i\beta_1''\xi) \right.$$

$$\left. + K_1'^2 \frac{(3b_{14}' + ib_{23}'')\exp\{i(\beta_1'' + 2\beta_1')\xi\}}{\beta_3'^2 - (\beta_1'' + 2\beta_1')^2} \right]$$

$$\mathscr{E}_{3x} - i\mathscr{E}_{3y} = K_3'' \exp(i\beta_3''\xi) + \frac{m\alpha}{4M} K_1' \left[K_1'^2 \frac{(b_{14}'' + ib_{23}')\exp(3i\beta_1'\xi)}{\beta_3''^2 - 9\beta_1'^2} \right.$$

$$\left. + K_1''^2(3b_{14}'' - ib_{23}') \frac{\exp\{i(\beta_1' + 2\beta_1'')\xi\}}{\beta_3''^2 - (\beta_1' + 2\beta_1'')^2} \right]$$

Since E_x, E_y, $(1/\mu)(dE_x/dz)$ and $(1/\mu)(dE_y/dz)$ are continuous across $z = 0$, one may equate the expressions for $\mathscr{E}_{1x} \pm i\mathscr{E}_{1y}$, $\mathscr{E}_{3x} \pm i\mathscr{E}_{3y}$, $(1/\mu)(d/d\xi)(\mathscr{E}_{1x} \pm i\mathscr{E}_{1y})$ and $(1/\mu)(d/d\xi)(\mathscr{E}_{3x} \pm i\mathscr{E}_{3y})$ in the plasma and in the free space at $\xi = 0$. From these eight equations, the eight unknown constants can be obtained. The expressions of interest are

$$A_{y1x} = \frac{K_1'}{4}\left(1 + \frac{\beta_1'}{\mu}\right) + \frac{K_1''}{4}\left(1 + \frac{\beta_1''}{\mu}\right) \tag{4.18a}$$

$$A_{y1y} = \frac{K_1'}{4i}\left(1 + \frac{\beta_1'}{\mu}\right) - \frac{K_1''}{4i}\left(1 + \frac{\beta_1''}{\mu}\right) \tag{4.18b}$$

$$A_{y3x} = \frac{K_3' + K_3''}{2} + \frac{m\alpha}{8M}\left[\frac{K_1''^3(b_{14}' - ib_{23}'')}{\beta_3'^2 - 9\beta_1''^2} + \frac{K_1''K_1'^2(3b_{14}' + ib_{23}'')}{\beta_3'^2 - (\beta_1'' + 2\beta_1')^2}\right.$$

$$\left. + \frac{K_1'^3(b_{14}'' + ib_{23}')}{\beta_3''^2 - 9\beta_1'^2} + \frac{K_1'K_1''^2(3b_{14}'' - ib_{23}')}{\beta_3''^2 - (\beta_1' + 2\beta_1'')^2}\right] \tag{4.18c}$$

$$A_{y3y} = \frac{K_3' - K_3''}{2i} + \frac{m\alpha}{8Mi}\left[\frac{K_1''^3(b_{14}' - ib_{23}'')}{\beta_3'^2 - 9\beta_1''^2} + \frac{K_1''K_1'^2(3b_{14}' + ib_{23}'')}{\beta_3'^2 - (\beta_1'' + 2\beta_1')^2}\right.$$

$$\left. - \frac{K_1'^3(b_{14}'' + ib_{23}')}{\beta_3''^2 - 9\beta_1'^2} - \frac{K_1'K_1''^2(3b_{14}'' - ib_{23}')}{\beta_3''^2 - (\beta_1' + 2\beta_1'')^2}\right] \tag{4.18d}$$

where

$$K_1' = \frac{2(A_{i1x} + iA_{i1y})}{1 - (\beta_1'/\mu)} \tag{4.19a}$$

$$K_1'' = \frac{2(A_{i1x} - iA_{i1y})}{1 - (\beta_1''/\mu)} \tag{4.19b}$$

$$K_3' = \frac{m\alpha}{4M} K_1'' \left[\frac{(3\beta_1''/\mu - 3)K_1''^2(b_{14}' - ib_{23}'')}{(3\beta_1'' + \beta_3')(3\beta_1'' - \beta_3')} \right.$$

$$\left. + \frac{\{(\beta_1'' + 2\beta_1')/\mu - 3\}K_1'^2(3b_{14}' + ib_{23}'')}{(2\beta_1' + \beta_1'' + \beta_3')(2\beta_1' + \beta_1'' - \beta_3')} \right] \frac{1}{(\beta_3'/\mu) - 3} \tag{4.19c}$$

$$K_3'' = \frac{m\alpha}{4M} K_1' \left[\frac{(3\beta_1'/\mu - 3)K_1'^2(b_{14}'' + ib_{23}')}{(3\beta_1' + \beta_3'')(3\beta_1' - \beta_3'')} \right.$$

$$\left. + \frac{\{(\beta_1' + 2\beta_1'')/\mu - 3\}K_1''^2(3b_{14}'' - ib_{23}')}{(2\beta_1'' + \beta_1' + \beta_3'')(2\beta_1'' + \beta_1' - \beta_3'')} \right] \frac{1}{(\beta_3''/\mu) - 3} \tag{4.19d}$$

It is seen from Eqs. (4.15) and (4.19) that the amplitude of the third harmonic of an electromagnetic wave in a magnetoplasma and on reflection from a magnetoplasma-free space interface is proportional to b_{14}', b_{14}'', b_{23}', and b_{23}'', which have maxima around $\omega = \omega_B$ and $\omega = \omega_B/3$ (because B_1, B_2, B_3, and B_4 have maxima around these frequencies; see Section

TABLE VIII

VARIATIONS OF e_3' AND e_3'' WITH ν_0/ω FOR $\omega_p{}^2/\omega^2 = 0.36$

	e_3'		
ν_0/ω	$\omega_B = 0$	$\omega_B = \omega$	$\omega_B = 3\omega$
0.001	3.517×10^{-6}	3.916×10^{-4}	9.272×10^{-2}
0.003	1.054×10^{-5}	3.916×10^{-4}	8.335×10^{-2}
0.010	3.519×10^{-5}	3.901×10^{-4}	6.713×10^{-2}
0.030	1.055×10^{-4}	3.875×10^{-4}	4.451×10^{-2}
0.100	3.582×10^{-4}	3.675×10^{-4}	2.028×10^{-2}

	e_3''		
ν_0/ω	$\omega_B = 0$	$\omega_B = \omega$	$\omega_B = 3\omega$
0.001	3.517×10^{-6}	1.275×10^{-7}	4.481×10^{-3}
0.003	1.054×10^{-5}	1.013×10^{-6}	4.481×10^{-3}
0.010	3.519×10^{-5}	8.758×10^{-6}	4.481×10^{-3}
0.030	1.055×10^{-4}	5.212×10^{-5}	4.481×10^{-3}
0.100	3.582×10^{-4}	2.288×10^{-4}	4.479×10^{-3}

III-A-3;) hence, one may expect resonances in the amplitudes of the third harmonic in the vicinity of these frequencies. The sharpness of these resonances is, however, completely flattened by the increase in the relevant propagation parameters around these frequencies.

Table VIII illustrates the variations of

$$e_3' = \frac{6m\omega^2 kT}{e^2} \left| \frac{E_{y3x} + iE_{y3y}}{E_{i1x}^3} \right| \quad \text{for} \quad E_{i1x} = iE_{i1y}$$

$$e_3'' = \frac{6m\omega^2 kT}{e^2} \left| \frac{E_{y3x} - iE_{y3y}}{E_{i1x}^3} \right| \quad \text{for} \quad E_{i1x} = -iE_{i1y}$$

with v_0/ω for the three cases $\omega_B = 0$, ω and 3ω. It is seen that only e_3' exhibits a resonance at $\omega_B = 3\omega$; however, as was expected, this resonance is not so sharp as the corresponding resonance in the current density (because of the resonant increase also in the relevant propagation parameters around these frequencies). The variation with the collision frequency is in general small; it is, however, significant for e_3'' for the resonant case $\omega_B = \omega$.

In order to obtain the absolute magnitude of the third harmonic in the reflected wave, consider a typical example in which

$$|E_{i1x}| = |E_{i1y}| = 1.2 \times 10^{-2} \text{ volts/cm}, \quad v_0/\omega = 0.01,$$

$$T = 300° \text{ K}, \quad \omega_B = \omega = 10^5 \text{ sec}^{-1}$$

Then the extraordinary third-harmonic component is about 7.2×10^{-4} volts/cm. Thus, about 6% of the third harmonic is generated in the reflected wave for moderate values of the fundamental.

B. Propagation in an Inhomogeneous Infinite Plasma

1. In the Absence of a Magnetic Field

The problem of the generation and propagation of harmonic and combination frequency waves in an inhomogeneous plasma (in the absence of a magnetic field) when two plane-polarized electromagnetic waves of frequencies ω_1 and ω_2 are propagating with their electrical vectors along the direction of the density and temperature gradients has been investigated in detail by Sodha and Kaw (20, 21). In this case all the harmonics and combination frequencies are generated in the current density (as has been shown in Section III-A-1). These components in the current density give rise to the corresponding microwave fields so that the total electric field in the plasma is again given by Eq. (4.2a). The alternating component of the total current density is then [using Eq. (3.35)]

$$
\begin{aligned}
\mathbf{J}_a ={}& \sigma_1 \mathbf{E}_1 \exp(i\omega_1 t) + \sigma_2 \mathbf{E}_2 \exp(i\omega_2 t) + \sigma_{12}\mathbf{E}_{12} \exp(2i\omega_1 t) \\
&+ \sigma_{13}\mathbf{E}_{13}\exp(3i\omega_1 t) + \sigma_{23}\mathbf{E}_{23}\exp(3i\omega_2 t) + \sigma_{22}\mathbf{E}_{22}\exp(2i\omega_2 t) \\
&+ \sigma_{12}^{+}\mathbf{E}_{12}^{+}\exp\{i(\omega_1+\omega_2)t\} + \sigma_{12}^{-}\mathbf{E}_{12}^{-}\exp\{i(\omega_1-\omega_2)t\} \\
&+ \sigma_{21}^{+\prime}\mathbf{E}_{21}^{+\prime}\exp\{i(\omega_2+2\omega_1)t\} + \sigma_{21}^{-\prime}\mathbf{E}_{21}^{-\prime}\exp\{i(\omega_2-2\omega_1)t\} \\
&+ \sigma_{12}^{+\prime}\mathbf{E}_{12}^{+\prime}\exp\{i(\omega_1+2\omega_2)t\} + \sigma_{12}^{-\prime}\mathbf{E}_{12}^{-\prime}\exp\{i(\omega_1-2\omega_2)t\} \\
&+ (m/M)\alpha[\{(\gamma - CP)A_{12d} + CA_{12t}\}\mathscr{E}_1{}^2 \exp(2i\omega_1 t) \\
&+ \{(\gamma - CP)A_{22d} + CA_{22t}\}\mathscr{E}_2{}^2 \exp(2i\omega_2 t) + A_{13}\mathscr{E}_1{}^2\mathbf{E}_1 \exp(3i\omega_1 t) \\
&+ A_{23}\mathscr{E}_2{}^2\mathbf{E}_2 \exp(3i\omega_2 t) + \{(\gamma - CP)A_{12d}^{+} + CA_{12t}^{+}\} \\
&\times \mathscr{E}_1\mathscr{E}_2 \exp\{i(\omega_1+\omega_2)t\} + \{(\gamma - CP)A_{12d}^{-} + CA_{12t}^{-}\} \\
&\times \mathscr{E}_1\tilde{\mathscr{E}}_2 \exp\{i(\omega_1-\omega_2)t\} + A_{12}^{+\prime}\mathscr{E}_2{}^2\mathbf{E}_1 \exp\{i(\omega_1+2\omega_2)t\} \\
&+ A_{12}^{-\prime}\tilde{\mathscr{E}}_2{}^2\mathbf{E}_1 \exp\{i(\omega_1-2\omega_2)t\} + A_{21}^{+\prime}\mathscr{E}_1{}^2\mathbf{E}_2 \exp\{i(\omega_2+2\omega_1)t\} \\
&+ A_{21}^{+\prime}\tilde{\mathscr{E}}_1{}^2\mathbf{E}_2 \exp\{i(\omega_2-2\omega_1)t\}]
\end{aligned}
\tag{4.20}
$$

for which the various symbols have been already defined. Proceeding as in Section A, one obtains the following equations for the propagation of the fundamental, harmonic, and combination frequency waves:

$$
(\partial^2\mathscr{E}_1/\partial\xi^2) + \beta_1{}^2\mathscr{E}_1 = 0 \qquad (\partial^2\mathscr{E}_2/\partial\xi^2) + \beta_2{}^2\mathscr{E}_2 = 0 \tag{4.21a, b}
$$

$$
(\partial^2\mathscr{E}_{12}/\partial\xi^2) + \beta_{12}^2\mathscr{E}_{12} = (m/M)\alpha[(\gamma - CP)a_{12d} + Ca_{12t}]\mathscr{E}_1{}^2 \tag{4.21c}
$$

$$
(\partial^2\mathscr{E}_{22}/\partial\xi^2) + \beta_{22}^2\mathscr{E}_{22} = (m/M)\alpha[(\gamma - CP)a_{22d} + Ca_{22t}]\mathscr{E}_2{}^2 \tag{4.21d}
$$

$$
(\partial^2\mathscr{E}_{12}^{+}/\partial\xi^2) + \beta_{12}^{+2}\mathscr{E}_{12}^{+} = (m/M)\alpha[(\gamma - CP)a_{12d}^{+} + Ca_{12t}^{+}]\mathscr{E}_1\mathscr{E}_2 \tag{4.21e}
$$

$$
(\partial^2\mathscr{E}_{12}^{-}/\partial\xi^2) + \beta_{12}^{-2}\mathscr{E}_{12}^{-} = (m/M)\alpha[(\gamma - CP)a_{12d}^{-} + Ca_{12t}^{-}]\mathscr{E}_1\tilde{\mathscr{E}}_2 \tag{4.21f}
$$

$$
(\partial^2\mathscr{E}_{13}/\partial\xi^2) + \beta_{13}^2\mathscr{E}_{13} = (m/M)\alpha a_{13}\mathscr{E}_1{}^3 \tag{4.21g}
$$

$$
(\partial^2\mathscr{E}_{23}/\partial\xi^2) + \beta_{23}^2\mathscr{E}_{23} = (m/M)\alpha a_{23}\mathscr{E}_2{}^3 \tag{4.21h}
$$

$$
(\partial^2\mathscr{E}_{12}^{+\prime}/\partial\xi^2) + \beta_{12}^{+\prime 2}\mathscr{E}_{12}^{+\prime} = (m/M)\alpha a_{12}^{+\prime}\mathscr{E}_1\mathscr{E}_2{}^2 \tag{4.21i}
$$

$$
(\partial^2\mathscr{E}_{12}^{-\prime}/\partial\xi^2) + \beta_{12}^{-\prime 2}\mathscr{E}_{12}^{-\prime} = (m/M)\alpha a_{12}^{-\prime}\mathscr{E}_1\tilde{\mathscr{E}}_2{}^2 \tag{4.21j}
$$

$$
(\partial^2\mathscr{E}_{21}^{+\prime}/\partial\xi^2) + \beta_{21}^{+\prime 2}\mathscr{E}_{21}^{+\prime} = (m/M)\alpha a_{21}^{+\prime}\mathscr{E}_2\mathscr{E}_1{}^2 \tag{4.21k}
$$

$$
(\partial^2\mathscr{E}_{21}^{-\prime}/\partial\xi^2) + \beta_{21}^{-\prime 2}\mathscr{E}_{21}^{-\prime} = (m/M)\alpha a_{21}^{-\prime}\mathscr{E}_2\tilde{\mathscr{E}}_1{}^2 \tag{4.21l}
$$

for which all the β have already been defined in Section A and

$$
a = (4\pi i\omega/\varepsilon\omega_0{}^2)(A/E_{00})
$$

a. Nonlinear growth of the harmonic and combination frequency waves. It has been assumed in deriving Eqs. (4.21) that the electromagnetic waves are propagating in the z direction and that their electrical vectors and the density and temperature gradients are all along the z direction. Thus, in the investigation of the present problem, one is interested in the case in which the electromagnetic wave propagates at right angles to the density and temperature gradients.

Now, all the β in the foregoing equations are functions of x (because they depend through σ on N, which is a function of x), and so these equations cannot be easily solved, in general. However, if one assumes that dN/dx is sufficiently small, then N (and hence β) may be taken to be practically independent of x over a small section of the incident wave front. One can then state that the various sections of the incident wave front are propagating independently in regions of different electron densities (the electron density in each region being a constant), and thus the propagation of the harmonic and combination frequency waves in the inhomogeneous plasma may be studied in a simple manner. Further, this assumption will lead to a tilting of the wave front as it propagates in the inhomogeneous plasma.

Taking the β to be independent of x, employing a technique similar to that of Epstein (73), and using the appropriate radiation condition (viz., $\mathscr{E} = 0$ at $\xi = \infty$) and the boundary conditions

$$\mathscr{E}_1 = \mathscr{E}_{10}, \qquad \mathscr{E}_2 = \mathscr{E}_{20}$$

$$\mathscr{E}_{12} = \mathscr{E}_{22} = \mathscr{E}_{13} = \mathscr{E}_{23} = \mathscr{E}_{12}^{\pm} = \mathscr{E}_{12}^{\pm\prime} = \mathscr{E}_{21}^{\pm\prime} = 0 \qquad \text{at} \quad \xi = 0$$

one gets

$$\mathscr{E}_{12} = \frac{m}{M}\, \alpha\, \frac{[(\gamma - CP)a_{12d} + Ca_{12t}]\mathscr{E}_{10}^2}{(2\beta_1 + \beta_{12})(2\beta_1 - \beta_{12})}\, [\exp(i\beta_{12}\xi) - \exp(2i\beta_1\xi)] \qquad (4.22a)$$

$$\mathscr{E}_{22} = \frac{m}{M}\, \alpha\, \frac{[(\gamma - CP)a_{22d} + Ca_{22t}]\mathscr{E}_{20}^2}{(2\beta_2 + \beta_{22})(2\beta_2 - \beta_{22})}\, [\exp(i\beta_{22}\xi) - \exp(2i\beta_2\xi)] \qquad (4.22b)$$

$$\mathscr{E}_{12}^+ = \frac{m}{M}\, \alpha\, \frac{[(\gamma - CP)a_{12d}^+ + Ca_{12t}^+]\mathscr{E}_{10}\mathscr{E}_{20}}{(\beta_1 + \beta_2 + \beta_{12}^+)(\beta_1 + \beta_2 - \beta_{12}^+)}\, [\exp(i\beta_{12}^+\xi) - \exp\{i(\beta_1 + \beta_2)\xi\}]$$

$$(4.22c)$$

$$\mathscr{E}_{12}^- = \frac{m}{M}\, \alpha\, \frac{[(\gamma - CP)a_{12d}^- + Ca_{12t}^-]\mathscr{E}_{10}\tilde{\mathscr{E}}_{20}}{(\beta_1 - \tilde{\beta}_2 + \beta_{12}^-)(\beta_1 - \tilde{\beta}_2 - \beta_{12}^-)}\, [\exp(i\beta_{12}^-\xi) - \exp\{i(\beta_1 - \tilde{\beta}_2)\xi\}]$$

$$(4.22d)$$

The remaining equations are identical to those obtained for the case of a homogeneous plasma in Section IV-A-1. This identity arises because of the nature of the approximations used in either case (in the former the higher order terms in the dc electric field are neglected, and in the latter the higher order terms of the inhomogeneity are neglected).

Equations (4.22a–d) give us the magnitudes of the electrical vectors of the second harmonic and the sum-and-difference frequency electromagnetic waves in a plasma when the two incident waves are traveling in a direction at right angles to that of the density and temperature gradients. These can be used for studying the nonlinear growth of these frequencies in a plasma, in the manner outlined in Section IV-A-1.

b. Magnitude of the reflected components. Following Sodha and Kaw (*21*), if one restricts one's analysis to the reflection of a small section of the incident plane wave front (in which the electron density may be taken to be independent of x), then the β may be taken as constants in this case also. Proceeding in a manner similar to that of Section IV-A-1, one can then evaluate the magnitude of the electrical vectors of the reflected harmonic and combination frequency components for the present case, in which the electromagnetic waves are propagating at right angles to the density and temperature gradients. The expressions for the second harmonic and sum-and-difference frequency components are

$$\frac{A_{y12}}{A_{i1}^2} = \frac{4}{\mu}\frac{m}{M}\,\alpha\,\frac{(\gamma - CP)a_{12d} + Ca_{12t}}{(2\beta_1 + \beta_{12})\left\{\dfrac{\beta_{12}}{\mu} - \dfrac{2\omega_1}{\omega_0}\right\}\left\{1 - \left(\dfrac{\omega_0}{\omega_1}\dfrac{\beta_1}{\mu}\right)\right\}^2} \tag{4.23a}$$

$$\frac{A_{y22}}{A_{i2}^2} = \frac{4}{\mu}\frac{m}{M}\,\alpha\,\frac{(\gamma - CP)a_{22d} + Ca_{22t}}{(2\beta_2 + \beta_{22})\left\{\dfrac{\beta_{22}}{\mu} - \dfrac{2\omega_2}{\omega_0}\right\}\left\{1 - \left(\dfrac{\omega_0}{\omega_2}\dfrac{\beta_2}{\mu}\right)\right\}^2} \tag{4.23b}$$

$$\frac{A_{y12}^+}{A_{i1}A_{i2}} = \frac{4}{\mu}\left(\frac{m}{M}\,\alpha\right)$$

$$\times\,\frac{(\gamma - CP)a_{12d}^+ + Ca_{12t}^+}{(\beta_1 + \beta_2 + \beta_{12}^+)\left[\dfrac{\beta_{12}^+}{\mu} - \dfrac{(\omega_1 + \omega_2)}{\omega_0}\right]\left(1 - \dfrac{\omega_0}{\omega_1}\dfrac{\beta_1}{\mu}\right)\left(1 - \dfrac{\omega_0}{\omega_2}\dfrac{\beta_2}{\mu}\right)} \tag{4.23c}$$

$$\frac{A_{y12}^-}{A_{i1}\tilde{A}_{i2}} = \frac{4}{\mu}\left(\frac{m}{M}\,\alpha\right)$$

$$\times\,\frac{(\gamma - CP)a_{12d}^- + Ca_{12t}^-}{(\beta_1 - \tilde{\beta}_2 + \beta_{12}^-)\left[\dfrac{\beta_{12}^-}{\mu} - \dfrac{(\omega_1 - \omega_2)}{\omega_0}\right]\left(1 - \dfrac{\omega_0}{\omega_1}\dfrac{\beta_1}{\mu}\right)\left(1 - \dfrac{\omega_0}{\omega_2}\dfrac{\tilde{\beta}_2}{\mu}\right)} \tag{4.23d}$$

The expressions for the odd harmonic and the second-order combination frequencies are identical to those for a homogeneous plasma.

Equations (4.23) give us the magnitudes of the second harmonic and the sum-and difference components in the reflected wave from an inhomogeneous plasma-free space interface when the incident waves of frequencies ω_1 and ω_2 are plane-polarized along the direction of the density and temperature gradients. To have an appreciation of the order of magnitude of these components in the reflected wave, some calculations have been made for the cases of a collision frequency independent of and proportional to the first power of electron velocity:

Taking $\mu = 1$, $\omega_0 = \omega_2$ and $\gamma = C$, Eqs. (4.23a–d) may be written as

$$e_{\gamma 12} = \omega_0^2 \left| \frac{E_{\gamma 12}}{\gamma E_{i1}^2} \right| = \frac{e}{m} \frac{\omega_p^2}{\omega_0^2} j_{12}\, G_{12}(\beta) \tag{4.24a}$$

$$e_{\gamma 12}^+ = \omega_0^2 \left| \frac{E_{\gamma 12}^+}{\gamma E_{i1} E_{i2}} \right| = \frac{e}{m} \frac{\omega_p^2}{\omega_0^2} \left(1 + \frac{\omega_1}{\omega_2} \right) j_{12}^+\, G_{12}^+(\beta) \tag{4.24b}$$

$$e_{\gamma 12}^- = \omega_0^2 \left| \frac{E_{\gamma 12}^-}{\gamma E_{i1} \tilde{E}_{i2}} \right| = \frac{e}{m} \frac{\omega_p^2}{\omega_0^2} \left(1 - \frac{\omega_1}{\omega_2} \right) j_{12}^-\, G_{12}^-(\beta) \tag{4.24c}$$

for which j_{12}, j_{12}^+, and j_{12}^- have been already defined in Section III-B-1 and

$$G_{12}(\beta) = \tfrac{4}{3}[(2\beta_1 + \beta_{12})(\beta_{12} - 2)(1 - \beta_1^2)]^{-1}$$

$$G_{12}^+(\beta) = \tfrac{4}{3}[(\beta_1 + \beta_2 + \beta_{12}^+)(\beta_{12}^+ - 1 - (\omega_2/\omega_1))(2 - 2\beta_1)(1 - (\omega_1/\omega_2)\beta_2)]^{-1}$$

$$G_{12}^-(\beta) = \tfrac{4}{3}[(\beta_1 - \tilde{\beta}_2 + \beta_{12}^-)(\beta_{12}^- - 1 + (\omega_2/\omega_1))(2 - 2\beta_1)(1 - (\omega_1/\omega_2)\tilde{\beta}_2]^{-1}$$

Table IX illustrates the variation of $e_{\gamma 12}^+$ and $e_{\gamma 12}^+$ with n and the ratio ω_1/ω_2 for $\omega_p^2/\omega_0^2 = 0.4$ and $v_0/\omega_0 = 0.01$. The variation with the collision

TABLE IX

VARIATION OF $e_{\gamma 12}$, $e_{\gamma 12}^\pm$ WITH ω_1/ω_2 AND n FOR $\omega_p^2/\omega_0^2 = 0.40$, $v_0/\omega_0 = 0.01$

ω_1/ω_2	$e_{\gamma 12} \times 10^{-15}$		$e_{\gamma 12}^+ \times 10^{-15}$		$e_{\gamma 12}^- \times 10^{-15}$	
	$n = 0$	$n = 1$	$n = 0$	$n = 1$	$n = 0$	$n = 1$
2			8.250	7.880	9.060	67.280
4	6.619	3.058	1.938	1.689	1.857	3.397
6	no variation	no variation	0.847	0.765	0.818	1.116
8			0.473	0.393	0.460	0.601

frequency has not been studied because it was found that this variation was negligibly small for small magnitudes of the collision frequency. The values of the ratio ω_1/ω_2 were chosen in such a manner that $\omega_1 - \omega_2 > \omega_p$ always; therefore the regions of plasma resonances are excluded.

It is seen from Table IX that the electrical vector of the difference frequency $\omega_1 - \omega_2$ has a larger magnitude than that of the sum frequency $(\omega_1 + \omega_2)$; the variation with ω_2/ω_1 is identical to the one obtained for these components of the current density. To have an idea of the absolute magnitudes of the electrical vectors of the second harmonic and the sum and difference frequencies in the inhomogeneous plasma, let us consider a typical case in which $|E_{i1}| = |E_{i2}| = 3 \times 10^{-4}$ volts/cm. Then, for $C = \gamma = 0.1$ cm^{-1}, $\omega_p^2/\omega_0^2 = 0.4$, $v_0/\omega_0 = 0.01$, $n = 0$, $\omega_1/\omega_2 = 2$, and $\omega_0 = \omega_2 = 10^5$ rad/sec, we have

$$|E_{\gamma 12}| = 1.986 \times 10^{-5} \quad \text{volts/cm}, \qquad |E_{\gamma 12}^-| = 2.718 \times 10^{-5} \quad \text{volt/cm}$$

$$|E_{\gamma 12}^+| = 2.475 \times 10^{-5} \quad \text{volts/cm}$$

Thus we see that sufficiently large magnitudes of the second harmonic and sum and difference frequency components may be obtained for usual values of the relevant parameters.

2. Effect of a Magnetic Field

The nonlinear generation of the sum and difference frequency components in an inhomogeneous magnetoplasma does not seem to have been attempted so far. However, Sodha and Kaw (22) have discussed the simpler phenomenon of the generation of second harmonic of an electromagnetic wave in an inhomogeneous plasma in the presence of a magnetic field. The expressions for the harmonic components derived in that paper are incomplete because the fundamental frequency term in f^0 and the f^2 terms was neglected when the expansion of the distribution function was written down. In the present analysis, the authors have modified that treatment (22), by using the more complete expressions derived in Section III-B-2 for the second harmonic components of current density in an inhomogeneous magnetoplasma.

a. Nonlinear growth of the second harmonic. Consider a plane polarized electromagnetic wave of frequency ω to be propagating in the Z direction (which is also the direction of the externally applied magnetic field). Let the inhomogenity in the plasma also exist only in the plane of the wave front, i.e., the xy plane. The plasma then finds itself under the influence of an electric field having components

$$E_x = E_{1x} \exp(i\omega t) + E_{2x} \exp(2i\omega t) + E_{3x} \exp(3i\omega t)$$

$$E_y = E_{1y} \exp(i\omega t) + E_{2y} \exp(2i\omega t) + E_{3y} \exp(3i\omega t)$$

$$E_z = 0$$

where E_2, $E_3 \ll E_1$, and a magnetic field having components $B_x = B_y = 0$ and $B_z = B_0$; $\partial T/\partial x$, $\partial T/\partial y$, $\partial N/\partial x$, $\partial N/\partial y$ = finite and $\partial T/\partial z = \partial N/\partial z = 0$.

We proceed in a manner similar to that in Section IV-A-2 for studying propagation of the third harmonic of an electromagnetic wave in a homogeneous magnetoplasma, using the relevant expressions for the current density from Section III-B-2. To avoid making the algebra unmanageable, we use the simpler Eqs. (3.47a,b), applicable only to frozen composition plasmas with an electron collision frequency independent of electron velocity, rather than the more general Eqs. (3.46a,b). One obtains the following equations, which describe the growth of the extraordinary and ordinary components of the fundamental and second harmonic waves:

$$\frac{\partial^2}{\partial \xi^2}(\mathscr{E}_{1x} + i\mathscr{E}_{1y}) + \beta_1'^2(\mathscr{E}_{1x} + i\mathscr{E}_{1y}) = 0$$

$$\frac{\partial^2}{\partial \xi^2}(\mathscr{E}_{1x} - i\mathscr{E}_{1y}) + \beta_1''^2(\mathscr{E}_{1x} - i\mathscr{E}_{1y}) = 0$$

$$\frac{\partial^2}{\partial \xi^2}(\mathscr{E}_{2x} + i\mathscr{E}_{2y}) + \beta_2'^2(\mathscr{E}_{2x} + i\mathscr{E}_{2y}) = \frac{m}{M}\alpha[a_{14d}''(\gamma_x \mathscr{E}_{1x}^2 + i\gamma_y \mathscr{E}_{1y}^2)$$
$$- ia_{23d}'(\gamma_x \mathscr{E}_{1y}^2 + i\gamma_y \mathscr{E}_{1x}^2) + a_{56d}'(\gamma_x - i\gamma_y)\mathscr{E}_{1x}\mathscr{E}_{1y}]$$

$$\frac{\partial^2}{\partial \xi^2}(\mathscr{E}_{2x} - i\mathscr{E}_{2y}) + \beta_2''^2(\mathscr{E}_{2x} - i\mathscr{E}_{2y}) = \frac{m}{M}\alpha[a_{14d}'(\gamma_x \mathscr{E}_{1x}^2 - i\gamma_y \mathscr{E}_{1y}^2)$$
$$+ ia_{23d}''(\gamma_x \mathscr{E}_{1y}^2 - i\gamma_y \mathscr{E}_{1x}^2) + a_{56d}''(\gamma_x + i\gamma_y)\mathscr{E}_{1x}\mathscr{E}_{1y}]$$

where

$$\beta_1'^2 = 1 - \frac{4\pi i}{\varepsilon\omega}(\sigma_1' + i\sigma_1''), \qquad \beta_1''^2 = 1 - \frac{4\pi i}{\varepsilon\omega}(\sigma_1' - i\sigma_1'')$$

$$\beta_2'^2 = 4\left[1 - \frac{2\pi i}{\varepsilon\omega}(\sigma_2' + i\sigma_2'')\right], \qquad \beta_2''^2 = 4\left[1 - \frac{2\pi i}{\varepsilon\omega}(\sigma_2' - i\sigma_2'')\right]$$

where σ' and σ'' have the same meaning as in Section IV-A-2, the meanings of the subscripts are obvious, and

$$a_{14d}' = \frac{8\pi i}{\varepsilon\omega}\frac{A_{1d} + iA_{4d}}{E_{00}} \qquad a_{14d}'' = \frac{8\pi i}{\varepsilon\omega}\frac{A_{1d} - iA_{4d}}{E_{00}}$$

$$a_{23d}' = \frac{8\pi i}{\varepsilon\omega}\frac{A_{2d} + iA_{3d}}{E_{00}} \qquad a_{23d}'' = \frac{8\pi i}{\varepsilon\omega}\frac{A_{2d} - iA_{3d}}{E_{00}}$$

$$a_{56d}' = \frac{8\pi i}{\varepsilon\omega}\frac{A_{5d} + iA_{6d}}{E_{00}} \qquad a_{56d}'' = \frac{8\pi i}{\varepsilon\omega}\frac{A_{5d} - iA_{6d}}{E_{00}}$$

Using the boundary conditions,

$$\mathscr{E}_{1x} = \mathscr{E}_{1x}^C, \qquad \mathscr{E}_{1y} = \mathscr{E}_{1y}^0, \qquad \mathscr{E}_{2x} = \mathscr{E}_{2y} = 0 \qquad \text{at} \quad \xi = 0$$

and the appropriate radiation condition ($\mathscr{E} = 0$ at $\xi = \infty$) and assuming the β to be independent of x, y for a very small portion of the wave front as in Section IV-B-1, the second-order solutions to the above equations may be written as

$$\mathscr{E}_{1x} + i\mathscr{E}_{1y} = K_1{}' \exp(i\beta_1{}'\xi)$$

$$\mathscr{E}_{1x} - i\mathscr{E}_{1y} = K_1'' \exp(i\beta_1''\xi)$$

$$
\begin{aligned}
\mathscr{E}_{2x} &+ i\mathscr{E}_{2y} \\
= \frac{m\alpha}{4M} \Bigg[&(\gamma_x - i\gamma_y)(a_{14d}'' + ia_{23d}' - ia_{56d}')K_1'^2 \frac{\{\exp(i\beta_2{}'\xi) - \exp(2i\beta_1{}'\xi)\}}{(2\beta_1{}' + \beta_2{}')(2\beta_1{}' - \beta_2{}')} \\
&+ (\gamma_x - i\gamma_y)(a_{14d}'' + ia_{23d}' + ia_{56d}')K_1''^2 \frac{\{\exp(i\beta_2{}'\xi) - \exp(2i\beta_1''\xi)\}}{(2\beta_1'' + \beta_2{}')(2\beta_1'' - \beta_2{}')} \\
&+ (\gamma_x + i\gamma_y)(a_{14d}'' - ia_{23d}')2K_1{}'K_1'' \frac{\{\exp(i\beta_2{}'\xi) - \exp i(\beta_1{}' + \beta_1'')\xi\}}{(\beta_1{}' + \beta_1'' + \beta_2{}')(\beta_1{}' + \beta_1'' - \beta_2{}')} \Bigg]
\end{aligned}
$$

$$
\begin{aligned}
\mathscr{E}_{2x} &- i\mathscr{E}_{2y} \\
= \frac{m\alpha}{4M} \Bigg[&(\gamma_x + i\gamma_y)(a_{14d}' - ia_{23d}'' - ia_{56d}'')K_1'^2 \frac{\{\exp(i\beta_2''\xi) - \exp(2i\beta_1{}'\xi)\}}{(\beta_2'' + 2\beta_1{}')(2\beta_1{}' - \beta_2'')} \\
&+ (\gamma_x + i\gamma_y)(a_{14d}' - ia_{23d}'' + ia_{56d}'')K_1''^2 \frac{\{\exp(i\beta_2''\xi) - \exp(2i\beta_1''\xi)\}}{(\beta_2'' + 2\beta_1'')(2\beta_1'' - \beta_2'')} \\
&+ (\gamma_x - i\gamma_y)(a_{14d}' + ia_{23d}'')2K_1{}'K_1'' \frac{\{\exp(i\beta_2''\xi) - \exp i(\beta_1{}' + \beta_1'')\xi\}}{(\beta_1{}' + \beta_1'' + \beta_2'')(\beta_1{}' + \beta_1'' - \beta_2'')} \Bigg]
\end{aligned}
$$

where

$$K_1{}' = \mathscr{E}_{1x}^0 + i\mathscr{E}_{1y}^0 \qquad \text{and} \qquad K_1'' = \mathscr{E}_{1x}^0 - i\mathscr{E}_{1y}^0$$

The equations above may be used to study the nonlinear growth of the second harmonic in an inhomogeneous magnetoplasma; not many new features arise in the results. One interesting point is noted however. When $\mathscr{E}_{1x} + i\mathscr{E}_{1y} = 0$, it is found that $\mathscr{E}_{2x} + i\mathscr{E}_{2y}$ nearly vanishes and only $\mathscr{E}_{2x} - i\mathscr{E}_{2y}$ is finite. This means that when a pure ordinary mode of the fundamental is sent into the inhomogeneous magnetoplasma, only the ordinary mode of the second harmonic is generated; the converse result may also be shown to be true. A similar but opposite result has already been pointed out for third-harmonic generation in a homogeneous magnetoplasma also. Thus one may generalize this result and say that when a pure mode of the fundamental is sent into a plasma, only the pure modes of all the harmonics will be generated.

b. Second harmonic in the reflected wave. Let a plane-polarized wave whose electrical vector is given by

$$\mathbf{E}_i/E_{00} = \mathbf{B}_{i1} \exp[i(\omega t - \xi)]$$

be normally incident on the magnetoplasma-free space interface, viz., the plane $\xi = 0$, from the free space side. Proceeding as before (of course, limiting the reflection to a very small portion of the incident wave front so that the β may be taken as independent of the coordinates), one obtains the following expressions for the extraordinary and ordinary components of the fundamental and the second harmonic in the reflected wave:

$$\frac{B_{y1x} + iB_{y1y}}{B_{i1x} + iB_{i1y}} = \frac{1 + (\beta_1'/\mu)}{1 - (\beta_1'/\mu)} \qquad \frac{B_{y1x} - iB_{y1y}}{B_{i1x} - iB_{i1y}} = \frac{1 + (\beta_1''/\mu)}{1 - (\beta_1''/\mu)}$$

$$
\begin{aligned}
B_{y2x} + iB_{y2y} = \frac{m\alpha}{4M\mu}\Bigg[& \frac{K_1'^2}{\beta_2' + 2\beta_1'}(\gamma_x - i\gamma_y)(a_{14d}'' + ia_{23d}' - ia_{56d}') \\
& + \frac{K_1''^2}{\beta_2' + 2\beta_1''}(\gamma_x - i\gamma_y)(a_{14d}'' + ia_{23d}' + ia_{56d}') \\
& + \frac{2K_1'K_1''}{\beta_2' + \beta_1' + \beta_1''}(\gamma_x + i\gamma_y)(a_{14d}'' - ia_{23d}')\Bigg]\frac{1}{(\beta_2'/\mu) - 2}
\end{aligned}
$$

$$
\begin{aligned}
B_{y2x} - iB_{y2y} = \frac{m\alpha}{4M\mu}\Bigg[& \frac{K_1'^2}{\beta_2'' + 2\beta_1'}(\gamma_x + i\gamma_y)(a_{14d}' - ia_{23d}'' - ia_{56d}'') \\
& + \frac{K_1''^2}{\beta_2'' + 2\beta_1''}(\gamma_x + i\gamma_y)(a_{14d}' - ia_{23d}'' + ia_{56d}'') \\
& + \frac{2K_1'K_1''}{\beta_2'' + \beta_1' + \beta_1''}(\gamma_x - i\gamma_y)(a_{14d}' + ia_{23d}'')\Bigg]\frac{1}{(\beta_2''/\mu) - 2}
\end{aligned}
$$

where

$$\frac{K_1'}{B_{i1x} + iB_{i1y}} = \frac{2}{1 - (\beta_1'/\mu)}, \qquad \frac{K_1''}{B_{i1x} - iB_{i1y}} = \frac{2}{1 - (\beta_1''/\mu)}$$

It is seen from the preceding equations that the amplitudes of the second-harmonic components in the reflected wave from an inhomogeneous magnetoplasma-free space interface are proportional to the a_d' and a_d''; as these may have resonances around $\omega_B = \omega$ and $\omega_B = 2\omega$, one also expects resonances in the magnitudes of the second harmonic around these frequencies. The sharpness of these resonances is, however, completely flattened out by the increase in the relevant propagation parameters around these frequencies.

TABLE X

VARIATION OF e_2' AND e_2'' WITH v_0/ω FOR $\omega_p^2/\omega^2 = 0.4$

	e_2'			e_2''		
v_0/ω	$\omega_B = 0$	$\omega_B = \omega$	$\omega_B = 2\omega$	$\omega_B = 0$	$\omega_B = \omega$	$\omega_B = 2\omega$
0.01	0.050	0.140	0.788	0.050	4.724	0.00488
0.03	0.050	0.189	0.654	0.050	1.593	0.00024
0.05	0.050	0.206	0.460	0.050	0.963	0.00014
0.07	0.050	0.213	0.373	0.050	0.695	0.00010
0.10	0.050	0.215	0.289	0.050	0.497	0.00010

Table X illustrates the variations of

$$e_2' = \frac{m\omega^2}{e} \left| \frac{E_{y2x} + iE_{y2y}}{(\gamma_x - i\gamma_y)E_{i1x}^2} \right| \quad \text{for} \quad E_{i1x} = iE_{i1y}$$

$$e_2'' = \frac{m\omega^2}{e} \left| \frac{E_{y2x} - iE_{y2y}}{(\gamma_x + i\gamma_y)E_{i1x}^2} \right| \quad \text{for} \quad E_{i1x} = -iE_{i1y}$$

with v_0/ω for a collision frequency independent of electron velocity for the three cases $\omega_B = 0$, ω, and 2ω. A strong resonance is exhibited only by e_2'' around $\omega_B = \omega$.

To have an idea of the absolute order of magnitude of the second harmonic generated in the reflected wave, consider a typical case in which $\omega = \omega_B = 10^7$ rad/sec^{-1}, $\gamma_y = 0$, $\gamma_x = 0.1$ cm^{-1}, $v_0/\omega = 0.05$, $\omega_p^2/\omega^2 = 0.4$; then, for $E_{i1y} = 0$, $E_{i1x} \sim 10^{-4}$ esu/cm, one obtains $|E_{y2x} - iE_{y2y}| \sim 8.5 \times 10^{-6}$ esu/cm, i.e., about 8.5 % of the fundamental amplitude.

V. CONCLUSION

In this review the authors have given a comprehensive account of the various mechanisms responsible for the generation of harmonics and combination frequencies in the current density in a plasma. The main effort has been to take all the important contributions into account, and wherever that is not done, it has been clearly pointed out. The authors have also pointed out at various stages that kinetic theory investigations of some mechanisms have yet to be carried out. In the future more stress should be laid on such theoretical investigations.

No correlation of the experimental and theoretical results has been attempted because most of the experimental results available at present are such that all the mechanisms are simultaneously responsible for the generation processes. Experiments under controlled conditions isolating individual mechanisms are required to check the results of the theory.

From a device point of view, it is believed that arcs and electric discharges are most suitable for giving reasonable efficiencies of harmonic generation, since in these one usually comes across situations where the dc electric fields, nonuniformities in the plasma, magnetic fields, etc., are all simultaneously acting and contributing to the generation of harmonics. Further, high mobility semiconductors at low temperatures (in which the present theoretical methods are applicable to a large extent) also hold great promise.

Appendix A. Evaluation of the Components of Distribution Function for an Homogeneous Plasma

Substituting the last of Eqs. (2.13) in Eq. (3.3c) and equating the time-independent terms and the coefficients of various frequency terms on both sides of the final equation, one gets

$$v f_0^2 = \frac{v}{3} \frac{\partial}{\partial v} \left[\frac{1}{v} (\tilde{a}_1 f_{11}^1 + \tilde{a}_2 f_{21}^1) \right] \tag{A.1a}$$

$$(v + 2i\omega_1) f_{12}^2 = \frac{v}{3} \frac{\partial}{\partial v} \left[\frac{1}{v} (a_1 f_{11}^1 + \tilde{a}_1 f_{13}^1 + \tilde{a}_2 f_{21}^{1+\prime} + a_2 \tilde{f}_{21}^{1-\prime}) \right] \tag{A.1b}$$

$$(v + 2i\omega_2) f_{22}^2 = \frac{v}{3} \frac{\partial}{\partial v} \left[\frac{1}{v} (a_2 f_{21}^1 + \tilde{a}_2 f_{23}^1 + \tilde{a}_1 f_{12}^{1+\prime} + a_1 \tilde{f}_{12}^{1-\prime}) \right] \tag{A.1c}$$

$$[v + i(\omega_1 + \omega_2)] f_{12}^{2+} = \frac{v}{3} \frac{\partial}{\partial v} \left[\frac{1}{v} (a_2 f_{11}^1 + a_1 f_{21}^1 + \tilde{a}_1 f_{21}^{1+\prime} + \tilde{a}_2 f_{12}^{1+\prime}) \right] \tag{A.1d}$$

and

$$[v + i(\omega_1 - \omega_2)] f_{12}^{2-} = \frac{v}{3} \frac{\partial}{\partial v} \left[\frac{1}{v} (\tilde{a}_2 f_{11}^1 + a_1 \tilde{f}_{21}^1 + a_2 f_{12}^{1-\prime} + \tilde{a}_1 \tilde{f}_{21}^{1-\prime}) \right] \tag{A.1e}$$

where the symbol $\sim$ over any letter denotes its complex conjugate. The second of Eqs. (2.13) and (3.3b) similarly give

$$(v + i\omega_1) f_{11}^1 = a_1 \frac{\partial}{\partial v} \left(f_0^0 + \frac{2}{5} f_0^2 \right) + \frac{6a_1}{5v} f_0^2$$

$$+ \frac{\tilde{a}_1}{2} \frac{\partial}{\partial v}\left(f_{12}^0 + \frac{2}{5} f_{12}^2\right) + \frac{3\tilde{a}_1}{5v} f_{12}^2$$

$$+ \frac{\tilde{a}_2}{2} \frac{\partial}{\partial v}\left(f_{12}^{0+} + \frac{2}{5} f_{12}^{2+}\right) + \frac{3\tilde{a}_2}{5v} f_{12}^{2+}$$

$$+ \frac{a_2}{2} \frac{\partial}{\partial v}\left(f_{12}^{0-} + \frac{2}{5} f_{12}^{2-}\right) + \frac{3a_2}{5v} f_{12}^{2-} \qquad \text{(A.2a)}$$

$$(v + i\omega_2)f_{21}^1 = a_2 \frac{\partial}{\partial v}\left(f_0^0 + \frac{2}{5} f_0^0\right) + \frac{6a_2}{5v} f_0^0$$

$$+ \frac{\tilde{a}_2}{2} \frac{\partial}{\partial v}\left(f_{22}^0 + \frac{2}{5} f_{22}^2\right) + \frac{3\tilde{a}_2}{5v} f_{22}^2$$

$$+ \frac{\tilde{a}_1}{2} \frac{\partial}{\partial v}\left(f_{12}^{0+} + \frac{2}{5} f_{12}^{2+}\right) + \frac{3\tilde{a}_1}{5v} f_{12}^{2+}$$

$$+ \frac{a_1}{2} \frac{\partial}{\partial v}\left(f_{12}^{0-} + \frac{2}{5} \tilde{f}_{12}^{2-}\right) + \frac{3a_1}{5v} \tilde{f}_{12}^{2-} \qquad \text{(A.2b)}$$

$$(v + 3i\omega_1)f_{13}^1 = \frac{a_1}{2} \frac{\partial}{\partial v}\left(f_{12}^0 + \frac{2}{5} f_{12}^2\right) + \frac{3a_1}{5v} f_{12}^2 \qquad \text{(A.2c)}$$

$$(v + 3i\omega_2)f_{23}^1 = \frac{a_2}{2} \frac{\partial}{\partial v}\left(f_{22}^0 + \frac{2}{5} f_{22}^2\right) + \frac{3a_2}{5v} f_{22}^2 \qquad \text{(A.2d)}$$

$$[v + i(\omega_1 + 2\omega_2)]f_{12}^{1+\prime} = \frac{a_1}{2} \frac{\partial}{\partial v}\left(f_{22}^0 + \frac{2}{5} f_{22}^2\right) + \frac{3a_1}{5v} f_{22}^2$$

$$+ \frac{a_2}{2} \frac{\partial}{\partial v}\left(f_{12}^{0+} + \frac{2}{5} f_{12}^{2+}\right) + \frac{3a_2}{5v} f_{12}^{2+} \qquad \text{(A.2e)}$$

$$[v + i(\omega_1 - 2\omega_2)]f_{12}^{1-\prime} = \frac{a_1}{2} \frac{\partial}{\partial v}\left(\tilde{f}_{22}^0 + \frac{2}{5} \tilde{f}_{22}^2\right) + \frac{3a_1}{5v} \tilde{f}_{22}^2$$

$$+ \frac{\tilde{a}_2}{2} \frac{\partial}{\partial v}\left(f_{12}^{0-} + \frac{2}{5} f_{12}^{2-}\right) + \frac{3\tilde{a}_2}{5v} f_{12}^{2-} \qquad \text{(A.2f)}$$

$$[v + i(\omega_2 + 2\omega_1)]f_{21}^{1+\prime} = \frac{a_2}{2} \frac{\partial}{\partial v}\left(f_{12}^0 + \frac{2}{5} f_{12}^2\right) + \frac{3a_2}{5v} f_{12}^2$$

$$+ \frac{a_1}{2} \frac{\partial}{\partial v}\left(f_{12}^{0+} + \frac{2}{5} f_{12}^{2+}\right) + \frac{3a_1}{5v} f_{12}^{2+} \qquad \text{(A.2g)}$$

and

$$[v + i(\omega_2 - 2\omega_1)]f_{21}^{1-\prime} = \frac{a_2}{2}\frac{\partial}{\partial v}\left(\tilde{f}_{12}^0 + \frac{2}{5}\tilde{f}_{12}^2\right) + \frac{3a_2}{5v}\tilde{f}_{12}^2$$

$$+ \frac{\tilde{a}_1}{2}\frac{\partial}{\partial v}\left(\tilde{f}_{12}^{0-} + \frac{2}{5}\tilde{f}_{12}^{2-}\right) + \frac{3\tilde{a}_1}{5v}\tilde{f}_{12}^{2-} \quad \text{(A.2h)}$$

and the first of Eqs. (2.13) and (3.3a) give

$$-\frac{1}{6v^2}\frac{\partial}{\partial v}[v^2(\tilde{a}_1 f_{11}^1 + \tilde{a}_2 f_{21}^1)] = \frac{m}{Mv^2}\frac{\partial}{\partial v}(vv^3 f_0^0) + \frac{kT}{Mv^2}\frac{\partial}{\partial v}\left(vv^2\frac{\partial f_0^0}{\partial v}\right)$$

$$\text{(A.3a)}$$

$$2i\omega_1 f_{12}^0 = \frac{1}{6v^2}\frac{\partial}{\partial v}[v^2(a_1 f_{11}^1 + \tilde{a}_1 f_{13}^1 + \tilde{a}_2 f_{21}^{1+\prime} + a_2 \tilde{f}_{21}^{1-\prime})] \quad \text{(A.3b)}$$

$$2i\omega_2 f_{22}^0 = \frac{1}{6v^2}\frac{\partial}{\partial v}[v^2(a_2 f_{21}^1 + \tilde{a}_2 f_{23}^1 + \tilde{a}_1 f_{12}^{1+\prime} + a_1 \tilde{f}_{12}^{1-\prime})] \quad \text{(A.3c)}$$

$$i(\omega_1 + \omega_2)f_{12}^{0+} = \frac{1}{6v^2}\frac{\partial}{\partial v}[v^2(a_1 f_{21}^1 + \tilde{a}_1 f_{21}^{1+\prime} + a_2 f_{11}^1 + \tilde{a}_2 f_{21}^{1+\prime})] \quad \text{(A.3d)}$$

$$i(\omega_1 - \omega_2)f_{12}^{0-} = \frac{1}{6v^2}\frac{\partial}{\partial v}[v^2(a_1 \tilde{f}_{21}^1 + \tilde{a}_1 \tilde{f}_{21}^{1-\prime} + \tilde{a}_2 f_{11}^1 + a_2 f_{12}^{1-\prime})] \quad \text{(A.3e)}$$

where it has been assumed that

$$\frac{m}{Mv^2}\frac{\partial}{\partial v}(vv^3 F) + \frac{kT}{Mv^2}\frac{\partial}{\partial v}\left(vv^2\frac{\partial F}{\partial v}\right) \qquad F = f_{12}^0, f_{22}^0, f_{12}^{0+}, \text{and } f_{12}^{0-} \quad \text{(A.3f)}$$

appearing on the right-hand sides of Eq. (A.3a–e) are negligibly small. Using the approximations given in the text, one can now evaluate the components of f^0 and f^2 from Eqs. (A.1) and (A.3); these, on substitution in Eq. (A.2) lead to Eqs. (3.4) of the text.

Let us now try to examine the justification for neglecting expression (A.3f); consider the case when $F = f_{12}^0$. From Eqs. (A.3b) and (3.4a),

$$f_{12}^0 = \frac{a_1^{\,2}}{12i\omega_1 v^2}\frac{\partial}{\partial v}\left(\frac{v^2}{v + i\omega_1}\frac{\partial f_0^0}{\partial v}\right)$$

which, for a Maxwellian f_0^0 and negligible v, may be written as

$$|f_{12}^0| = (m/2M)\alpha(3 - 2u^2)f_0^0 \simeq (m/2M)\alpha f_0^0$$

where $\alpha = e^2 E_{00}^2 M/6m^2\omega_1^{\,2}kT$, and $u = (m/2kT)^{1/2}v$ is the dimensionless electron velocity; the last approximate equality follows from the fact that the most probable value of u is 1. Now, the neglect of expression (A.3f) in (A.3b) is equivalent to the inequality

$$\frac{m}{Mv^2}\frac{\partial}{\partial v}(vv^3 f^0_{12}) + \frac{kT}{Mv^2}\frac{\partial}{\partial v}\left(vv^2\frac{\partial f^0_{12}}{\partial v}\right) \ll \frac{a_1^2}{6v^2}\frac{\partial}{\partial v}\left[\frac{v^2}{v+i\omega_1}\frac{\partial f_0^0}{\partial v}\right]$$

which, for $\omega_1 \gg v$, and on integration, reduces to

$$\frac{\partial f^0_{12}}{\partial v} + \frac{mv}{kT} f^0_{12} \ll \frac{\omega_1}{v}\alpha\frac{\partial f_0^0}{\partial v}$$

On further integration, this gives

$$f^0_{12} \ll \frac{\omega_1}{v}\alpha u^2 f_0^0 \simeq \frac{\omega_1}{v}\alpha f_0^0$$

which is equivalent to stating that $m/2M \ll \omega_1/v$. This inequality is obviously justified, in view of the extremely small magnitude of m/M (10^{-4}) and the large values of ω_1/v $(10, 10^2,$ etc.$)$.

For a general dependence of the electron collision frequency on electron velocity (i.e., for a general value of n) and for Case I mentioned in the text, the expressions for the A defined by Eqs. (3.7a,b) are given by

$$\frac{\pi^{3/2} A^{+\prime}_{12}}{\omega_p^2 \omega_0^2}$$

$$= \frac{v_0^3}{(\omega_1 + 2\omega_2)^2}\left[\left\{\frac{n(2n+3)}{15}\frac{\omega_1^2 + \omega_2^2}{\omega_1^2 \omega_2^2(\omega_1 + \omega_2)^2} + \frac{n(2n+3)}{60\omega_2^4}\right\}\right.$$

$$\times \Pi\left(\frac{3n+1}{2}\right) + \left\{\frac{1}{15\omega_2^4} + \frac{4}{15}\frac{\omega_1^2 + \omega_2^2}{\omega_1^2 \omega_2^2(\omega_1 + \omega_2)^2}\right\}\Pi\left(\frac{3n+5}{2}\right)$$

$$\left. - \left\{\frac{6n+10}{15}\frac{\omega_1^2 + \omega_2^2}{\omega_1^2 \omega_2^2(\omega_1 + \omega_2)^2} + \frac{3n+5}{30\omega_2^4}\right\}\Pi\left(\frac{3n+3}{2}\right)\right]$$

$$+ \frac{iv_0^2}{(\omega_1 + 2\omega_2)^2}\left[\left\{\frac{2}{15}(n+5)\frac{1}{\omega_1\omega_2(\omega_1 + \omega_2)} + \frac{3}{10}(2n+5)\right.\right.$$

$$\times \frac{\omega_1^2 + \omega_2^2}{\omega_1^2 \omega_2^2(\omega_1 + \omega_2)} + \frac{4n+11}{12\omega_2^3}\right\}\Pi\left(\frac{2n+3}{2}\right)$$

$$- \left\{\frac{3n(n+3)(\omega_1^2 + \omega_2^2)}{20\omega_1^2 \omega_2^2(\omega_1 + \omega_2)} + \frac{3n(n+3)}{40\omega_2^3}\right\}\Pi\left(\frac{2n+1}{2}\right)$$

$$\left. - \left\{\frac{4}{15\omega_1\omega_2(\omega_1 + \omega_2)} + \frac{3(\omega_1^2 + \omega_2^2)}{5\omega_1^2 \omega_2^2(\omega_1 + \omega_2)} + \frac{11}{30\omega_2^3}\right\}\Pi\left(\frac{2n+5}{2}\right)\right]$$

$$-\frac{iv_c{}^2}{\omega_1 + 2\omega_2}\left[\left\{\frac{n(2n+3)(\omega_1{}^2 + \omega_2{}^2)}{15\omega_1{}^2\omega_2{}^2(\omega_1 + \omega_2)^2} + \frac{n(2n+3)}{60\omega_2{}^4}\right\}\Pi\left(\frac{2n+1}{2}\right)\right.$$

$$+\left\{\frac{1}{15\omega_2{}^4} + \frac{4(\omega_1{}^2 + \omega_2{}^2)}{15\omega_1{}^2\omega_2{}^2(\omega_1 + \omega_2)^2}\right\}\Pi\left(\frac{2n+5}{2}\right)$$

$$\left.-\left\{\frac{(6n+10)(\omega_1{}^2 + \omega_2{}^2)}{15\omega_1{}^2\omega_2{}^2(\omega_1 + \omega_2)^2} + \frac{3n+5}{30\omega_2{}^4}\right\}\Pi\left(\frac{2n+3}{2}\right)\right] + \frac{v_0}{(\omega_1 + 2\omega_2)^2}$$

$$\times\left[\left(\frac{3}{2\omega_1\omega_2} + \frac{3}{4\omega_2{}^2}\right)\Pi\left(\frac{n+3}{2}\right) - \left(\frac{3}{5\omega_1\omega_2} + \frac{3}{10\omega_2{}^2}\right)\Pi\left(\frac{n+5}{2}\right)\right]$$

$$+\frac{v_0}{\omega_1 + 2\omega_2}\left[\left\{\frac{2(n+5)}{15\omega_1\omega_2(\omega_1 + \omega_2)} + \frac{3(2n+5)(\omega_1{}^2 + \omega_2{}^2)}{10\omega_1{}^2\omega_2{}^2(\omega_1 + \omega_2)} + \frac{4n+11}{12\omega_2{}^3}\right\}\right.$$

$$\times \Pi\left(\frac{n+3}{2}\right) - \left\{\frac{3n(n+3)(\omega_1{}^2 + \omega_2{}^2)}{20\omega_1{}^2\omega_2{}^2(\omega_1 + \omega_2)} + \frac{3n(n+3)}{40\omega_2{}^3}\right\}\Pi\left(\frac{n+1}{2}\right)$$

$$\left.-\left\{\frac{4}{15\omega_1\omega_2(\omega_1 + \omega_2)} + \frac{3}{5}\frac{\omega_1{}^2 + \omega_2{}^2}{\omega_1{}^2\omega_2{}^2(\omega_1 + \omega_2)} + \frac{11}{30\omega_2{}^3}\right\}\Pi\left(\frac{n+5}{2}\right)\right]$$

$$\tag{A.4a}$$

$$A_{13} = \frac{\omega_p{}^2\omega_0{}^2}{\omega_1{}^3\pi^{3/2}}\left[\frac{v_0{}^3}{180\omega_1{}^3}\frac{n}{3}(2n+3)\Pi\left(\frac{3n+1}{2}\right) + \frac{iv_0{}^2}{90\omega_1{}^2}\right.$$

$$\times\left\{\frac{38n+85}{6}\Pi\left(\frac{2n+3}{2}\right) - \frac{17}{3}\Pi\left(\frac{2n+5}{2}\right) - \frac{n}{4}(7n+15)\Pi\left(\frac{2n+1}{2}\right)\right\}$$

$$\left.+\frac{v_0}{3\omega_1}\left\{\frac{2n+7}{6}\Pi\left(\frac{n+3}{2}\right) - \frac{7}{16}\Pi\left(\frac{n+5}{2}\right) - \frac{3n}{40}(n+3)\Pi\left(\frac{n+1}{2}\right)\right\}\right]$$

$$\tag{A.4b}$$

where

$$\Pi(\zeta) = \int_0^\infty \eta^\zeta \exp(-\eta)\,d\eta = \Gamma(\zeta + 1)$$

APPENDIX B. LIMITATION OF THE ANALYSIS IN THE PRESENCE OF A DC FIELD

Our analysis is valid only when the dc electric field is limited by the inequality

$$(m/M)\alpha\mathscr{E}_d{}^2 \ll v/\omega_1 \qquad\text{or}\qquad v/\omega_2$$

or

$$E_d{}^2 \ll (6m\omega_0{}^2 kT/e^2)\left(\frac{v}{\omega_1}\right)$$

Taking the left-hand side of the inequality to be about one-tenth of the right-hand side, and choosing the parameters $v/\omega_1 = 0.1$, $\omega_0 = \omega_1 = 10^7/\mathrm{sec}$, $T = 300°\,\mathrm{K}$, we have $E_d \cong 3 \times 10^{-5}\ \mathrm{esu/cm}(= 10^{-2}\ \mathrm{volts/cm})$. Thus we see that our analysis is valid only for very low dc fields. However, this limitation need not be there in any experiments that may be conducted in this direction.

Appendix C. Components of the Distribution Function in an Homogeneous Magnetoplasma

Substituting Eqs. (3.17a–e) in Eq. (2.4d,e), remembering that $\partial f/\partial x = \partial f/\partial y = 0$ for a homogeneous plasma, and equating the time-independent terms and the coefficients of various frequency terms on both sides of the final equation, one obtains

$$vf^2_{0x,y} = \frac{v}{3}\frac{\partial}{\partial v}\left[\frac{1}{v}(\tilde{a}_{1x,y}f^1_{11x,y} + \tilde{a}_{2x,y}f^1_{21x,y})\right] \tag{C.1a}$$

$$(v + 2i\omega_1)f^2_{12x,y}$$
$$= \frac{v}{3}\frac{\partial}{\partial v}\left[\frac{1}{v}(a_{1x,y}f^1_{11x,y} + \tilde{a}_{1x,y}f^1_{13x,y} + \tilde{a}_{2x}f^{1+\prime}_{21x,y} + a_{2x,y}\tilde{f}^{1-\prime}_{21x,y})\right] \tag{C.1b}$$

$$(v + 2i\omega_2)f^2_{22x,y}$$
$$= \frac{v}{3}\frac{\partial}{\partial v}\left[\frac{1}{v}(a_{2x,y}f^1_{21x,y} + \tilde{a}_{2x,y}f^1_{23x,y} + \tilde{a}_{1x,y}f^{1+\prime}_{12x,y} + a_{1x,y}f^{1-\prime}_{12x,y})\right] \tag{C.1c}$$

$$[v + i(\omega_1 + \omega_2)]f^{2+}_{12x,y}$$
$$= \frac{v}{3}\frac{\partial}{\partial v}\left[\frac{1}{v}(a_{1x,y}f^1_{21x,y} + a_{2x,y}f^1_{11x,y} + \tilde{a}_{1x,y}f^{1+\prime}_{21x,y} + \tilde{a}_{2x,y}f^{1+\prime}_{12x,y})\right] \tag{C.1d}$$

$$[v + i(\omega_1 - \omega_2)]f^{2-}_{12x,y}$$
$$= \frac{v}{3}\frac{\partial}{\partial v}\left[\frac{1}{v}(a_{1x,y}\tilde{f}^1_{21x,y} + \tilde{a}_{2x,y}f^1_{11x,y} + a_{1x,y}\tilde{f}^{1-\prime}_{21x,y} + a_{2x,y}f^{1-\prime}_{12x,y})\right] \tag{C.1e}$$

Similarly, from Eqs. (2.4b,c) and (3.17a–c), one obtains

$$(v + i\omega_1)f^1_{11x} + \omega_B f^1_{11y} = a_{1x}\left[\frac{\partial f_0^0}{\partial v} + \frac{2}{5}\frac{\partial f^2_{0x}}{\partial v} + \frac{6}{5v}f^2_{0x}\right]$$
$$+ \frac{\tilde{a}_{1x}}{2}\left[\frac{\partial f^0_{12}}{\partial v} + \frac{2}{5}\frac{\partial f^2_{12x}}{\partial v} + \frac{6}{5v}f^2_{12x}\right]$$
$$+ \frac{a_{2x}}{2}\left[\frac{\partial f^{0-}_{12}}{\partial v} + \frac{2}{5}\frac{\partial f^{2-}_{12x}}{\partial v} + \frac{6}{5v}f^{2-}_{12x}\right]$$
$$+ \frac{\tilde{a}_{2x}}{2}\left[\frac{\partial f^{0+}_{12}}{\partial v} + \frac{2}{5}\frac{\partial f^{2+}_{12x}}{\partial v} + \frac{6}{5v}f^{2+}_{12x}\right] \tag{C.2a}$$

$$(v + 3i\omega_1)f_{13x}^1 + \omega_B f_{13y}^1 = \frac{a_{1x}}{2}\frac{\partial f_{12}^0}{\partial v} + \frac{3a_{1x}}{5v}f_{12x}^2 + \frac{a_{1x}}{5}\frac{\partial f_{12x}^2}{\partial v} \quad \text{(C.2b)}$$

$$[v + i(\omega_1 + 2\omega_2)]f_{12x}^{1+\prime} + \omega_B f_{12y}^{1+\prime} = \frac{a_{1x}}{2}\frac{\partial f_{22}^0}{\partial v} + \frac{a_{1x}}{5}\frac{\partial f_{22x}^2}{\partial v} + \frac{3a_{1x}}{5v}f_{22x}^2$$

$$+ \frac{a_{2x}}{2}\left[\frac{\partial f_{12}^{0+}}{\partial v} + \frac{6}{5v}f_{12x}^{2+} + \frac{2}{5}\frac{\partial f_{12x}^{2+}}{\partial v}\right]$$

$$\text{(C.2c)}$$

$$[v + i(\omega_1 - 2\omega_2)]f_{12x}^{1-\prime} + \omega_B f_{12y}^{1-\prime} = \frac{a_{1x}}{2}\frac{\partial \tilde{f}_{22}^0}{\partial v} + \frac{a_{1x}}{5}\frac{\partial \tilde{f}_{22x}^2}{\partial v} + \frac{3a_{1x}}{5v}\tilde{f}_{22x}^2$$

$$+ \frac{\tilde{a}_{2x}}{2}\left[\frac{\partial f_{12}^{0-}}{\partial v} + \frac{6}{5v}f_{12x}^{2-} + \frac{2}{5}\frac{\partial f_{12x}^{2-}}{\partial v}\right]$$

$$\text{(C.2d)}$$

and corresponding equations with x and y interchanged and ω_B replaced by $-(\omega_B)$.

Similar equations may be written for the x and y components of f_{21}^1, f_{23}^1, $f_{12}^{1+\prime}$ and $f_{12}^{1-\prime}$. Finally, from Eqs. (2.4a) and (3.17a–c), one obtains

$$-\frac{1}{6v^2}\frac{\partial}{\partial v}[v^2(\tilde{a}_{1x}f_{11x}^1 + \tilde{a}_{1y}f_{11y}^1 + \tilde{a}_{2x}f_{21x}^1 + \tilde{a}_{2y}f_{21y}^1)]$$

$$= \frac{m}{Mv^2}\frac{\partial}{\partial v}(vv^3f_0^0) + \frac{kT}{Mv^2}\frac{\partial}{\partial v}\left(vv^2\frac{\partial f_0^0}{\partial v}\right) \quad \text{(C.3a)}$$

$$2i\omega_1 f_{12}^0 = \frac{1}{6v^2}\frac{\partial}{\partial v}[v^2(a_{1x}f_{11x}^1 + \tilde{a}_{1x}f_{13x}^1 + a_{1y}f_{11y}^1 + \tilde{a}_{1y}f_{13y}^1$$

$$+ \tilde{a}_{2y}f_{21y}^{1+\prime} + \tilde{a}_{2x}f_{21x}^{1+\prime} + a_{2x}\tilde{f}_{21x}^{1-\prime} + a_{2y}\tilde{f}_{21y}^{1-\prime})] \quad \text{(C.3b)}$$

$$2i\omega_2 f_{22}^0 = \frac{1}{6v^2}\frac{\partial}{\partial v}[v^2(a_{2x}f_{21x}^1 + \tilde{a}_{2x}f_{23x}^1 + a_{2y}f_{21y}^1 + \tilde{a}_{2y}f_{23y}^1$$

$$+ \tilde{a}_{1x}f_{12x}^{1+\prime} + a_{1y}\tilde{f}_{12y}^{1-\prime} + \tilde{a}_{1y}f_{12y}^{1+\prime} + a_{1x}\tilde{f}_{12x}^{1-\prime})] \quad \text{(C.3c)}$$

$$i(\omega_1 + \omega_2)f_{12}^{0+} = \frac{1}{6v^2}\frac{\partial}{\partial v}[v^2(a_{1x}f_{21x}^1 + a_{1y}f_{21y}^1 + \tilde{a}_{1x}f_{21x}^{1+\prime} + \tilde{a}_{1y}f_{21y}^{1+\prime}$$

$$+ a_{2x}f_{11x}^1 + a_{2y}f_{11y}^1 + \tilde{a}_{2x}f_{12x}^{1+\prime} + \tilde{a}_{2y}f_{12y}^{1+\prime})] \quad \text{(C.3d)}$$

$$i(\omega_1 - \omega_2)f_{12}^{0-} = \frac{1}{6v^2}\frac{\partial}{\partial v}[v^2(a_{1x}\tilde{f}_{21x}^1 + a_{1y}\tilde{f}_{21y}^1 + \tilde{a}_{2x}f_{11x}^1 + \tilde{a}_{2y}f_{11y}^1$$

$$+ a_{2x}f_{12x}^{1-\prime} + a_{2y}f_{12y}^{1-\prime} + \tilde{a}_{1y}\tilde{f}_{21y}^{1-\prime} + \tilde{a}_{1x}\tilde{f}_{21x}^{1-\prime})] \quad \text{(C.3e)}$$

where the expansion for $(\partial f^0/\partial t)_c$ for a Lorentzian plasma [Desloge and Matthysse (63)] has been used.

Using the approximations

$$f^1_{13x,y}, \; f^1_{23x,y}, \; f^{1\pm\prime}_{12x,y}, \; f^{1\pm\prime}_{21x,y} \ll f^1_{11x,y}, \; f^1_{21x,y}, \; \text{etc.}$$

etc. which correspond to the usual assumption that the magnitudes of the harmonic components are considerably smaller than that of the fundamental, one gets the equations given in the text for the components of f^1.

APPENDIX D. COMPONENTS OF THE DISTRIBUTION FUNCTION FOR AN INHOMOGENEOUS PLASMA

Substituting Eqs. (3.38a–c) into Eq. (3.37c) and equating the time-independent terms and the coefficients of other frequency terms, one obtains:

$$v f_0^2 = -\frac{2v}{3}\left(\gamma N \frac{\partial f_0{}^1}{\partial N} + CT \frac{\partial f_0{}^1}{\partial T}\right) + \frac{v}{3}\frac{\partial}{\partial v}\left[\frac{1}{v}(\tilde{a}_1 f^1_{11} + \tilde{a}_2 f^1_{21})\right] \quad \text{(D.1a)}$$

$$(v + i\omega_1)f^2_{11} = -\frac{2v}{3}\left(\gamma N \frac{\partial f^1_{11}}{\partial N} + CT \frac{\partial f^1_{11}}{\partial T}\right)$$

$$+ \frac{v}{3}\frac{\partial}{\partial v}\left[\frac{1}{v}(2a_1 f_0{}^1 + \tilde{a}_1 f^1_{12} + \tilde{a}_2 f^{1+}_{12} + a_2 f^{1-}_{12})\right] \quad \text{(D.1b)}$$

$$(v + 2i\omega_1)f^2_{12} = -\frac{2v}{3}\left(\gamma N \frac{\partial f^1_{12}}{\partial N} + CT \frac{\partial f^1_{12}}{\partial T}\right)$$

$$+ \frac{v}{3}\frac{\partial}{\partial v}\left[\frac{1}{v}(a_1 f^1_{11} + \tilde{a}_1 f^1_{13} + \tilde{a}_2 f^{1+\prime}_{21} + a_2 \tilde{f}^{1-\prime}_{21})\right] \quad \text{(D.1c)}$$

$$[v + i(\omega_1 + \omega_2)]f^{2+}_{12} = -\frac{2v}{3}\left(vN \frac{\partial f^{1+}_{12}}{\partial N} + CT \frac{\partial f^{1+}_{12}}{\partial T}\right)$$

$$+ \frac{v}{3}\frac{\partial}{\partial v}\left[\frac{1}{v}(a_1 f^1_{21} + a_2 f^1_{11} + \tilde{a}_1 \tilde{f}^{1+\prime}_{21} + \tilde{a}_2 f^{1+\prime}_{12})\right]$$

$$\text{(D.1d)}$$

$$[v + i(\omega_1 - \omega_2)]f^{2-}_{12} = -\frac{2v}{3}\left(\gamma N \frac{\partial f^{1-}_{12}}{\partial N} + CT \frac{\partial f^{1-}_{12}}{\partial T}\right)$$

$$+ \frac{v}{3}\frac{\partial}{\partial v}\left[\frac{1}{v}(a_1 \tilde{f}^1_{21} + \tilde{a}_2 f^1_{11} + \tilde{a}_1 \tilde{f}^{1-\prime}_{21} + a_2 f^{1-\prime}_{21})\right]$$

$$\text{(D.1e)}$$

Expressions for f_{21}^2, and f_{22}^2 are analogous to those for f_{11}^2 and f_{12}^2. Equation (3.37b) yields, after a similar treatment,

$$vf_0{}^1 = -v\left[\gamma N \frac{\partial}{\partial N}\left(f_0{}^0 + \frac{2}{5}f_0{}^2\right) + CT\frac{\partial}{\partial T}\left(f_0{}^0 + \frac{2}{5}f_0{}^2\right)\right]$$

$$+ \frac{\tilde{a}_1}{2}\frac{\partial}{\partial v}\left(f_{11}^0 + \frac{2}{5}f_{11}^2\right) + \frac{3\tilde{a}_1}{5v}f_{11}^2 + \frac{\tilde{a}_2}{2}\frac{\partial}{\partial v}\left(f_{21}^0 + \frac{2}{5}f_{21}^2\right) + \frac{3\tilde{a}_2}{5v}f_{21}^2$$

$$\text{(D.2a)}$$

$$(v + i\omega_1)f_{11}^1 = -v\left[\gamma N \frac{\partial}{\partial N}\left(f_{11}^0 + \frac{2}{5}f_{11}^2\right) + CT\frac{\partial}{\partial T}\left(f_{11}^0 + \frac{2}{5}f_{11}^2\right)\right]$$

$$+ a_1\left(\frac{\partial f_0{}^0}{\partial v} + \frac{2}{5}\frac{\partial f_0{}^2}{\partial v} + \frac{6}{5v}f_0{}^2\right)$$

$$+ \frac{\tilde{a}_1}{2}\left(\frac{\partial f_{12}^0}{\partial v} + \frac{2}{5}\frac{\partial f_{12}^2}{\partial v} + \frac{6}{5v}f_{12}^2\right)$$

$$+ \frac{\tilde{a}_2}{2}\left(\frac{\partial f_{12}^{0+}}{\partial v} + \frac{2}{5}\frac{\partial f_{12}^{2+}}{\partial v} + \frac{6}{5v}f_{12}^{2+}\right)$$

$$+ \frac{a_2}{2}\left(\frac{\partial f_{12}^{0-}}{\partial v} + \frac{2}{5}\frac{\partial f_{12}^{2-}}{\partial v} + \frac{6}{5v}f_{12}^{2-}\right) \qquad \text{(D.2b)}$$

$$(v + 2i\omega_1)f_{12}^1 = -v\left[\gamma N \frac{\partial}{\partial N}\left(f_{12}^0 + \frac{2}{5}f_{12}^2\right) + CT\frac{\partial}{\partial T}\left(f_{12}^0 + \frac{2}{5}f_{12}^2\right)\right]$$

$$+ \frac{a_1}{2}\left(\frac{\partial f_{11}^0}{\partial v} + \frac{2}{5}\frac{\partial f_{11}^2}{\partial v} + \frac{6}{5v}f_{11}^2\right) \qquad \text{(D.2c)}$$

$$(v + 3i\omega_1)f_{13}^1 = \frac{a_1}{2}\left(\frac{\partial f_{12}^0}{\partial v} + \frac{2}{5}\frac{\partial f_{12}^2}{\partial v} + \frac{6}{5v}f_{12}^2\right) \qquad \text{(D.2d)}$$

$$[v + i(\omega_1 + \omega_2)]f_{12}^{1+} = -v\left[\gamma N \frac{\partial}{\partial N}\left(f_{12}^{0+} + \frac{2}{5}f_{12}^{2+}\right)\right.$$

$$\left. + CT\frac{\partial}{\partial T}\left(f_{12}^{0+} + \frac{2}{5}f_{12}^{2+}\right)\right]$$

$$+ \frac{a_1}{2}\left(\frac{\partial f_{21}^0}{\partial v} + \frac{2}{5}\frac{\partial f_{21}^2}{\partial v} + \frac{6}{5v}f_{21}^2\right) \qquad \text{(D.2e)}$$

$$[v + i(\omega_1 - \omega_2)]f_{12}^{1-} = -v\left[\gamma N \frac{\partial}{\partial N}\left(f_{12}^{0-} + \frac{2}{5}f_{12}^{2-}\right)\right.$$

$$\left. + CT\frac{\partial}{\partial T}\left(f_{12}^{0-} + \frac{2}{5}f_{12}^{2-}\right)\right]$$

$$+ \frac{a_1}{2}\left(\frac{\partial \tilde{f}_{21}^{0}}{\partial v} + \frac{2}{5}\frac{\partial \tilde{f}_{21}^{2}}{\partial v} + \frac{6}{5v}\tilde{f}_{21}^{2}\right) \qquad \text{(D.2f)}$$

$$[v + i(\omega_1 + 2\omega_2)]f_{21}^{1+\prime} = \frac{a_1}{2}\left(\frac{\partial f_{21}^{0}}{\partial v} + \frac{2}{5}\frac{\partial f_{21}^{2}}{\partial v} + \frac{6}{5v}f_{21}^{2}\right)$$

$$+ \frac{a_2}{2}\left(\frac{\partial f_{12}^{0+}}{\partial v} + \frac{2}{5}\frac{\partial f_{12}^{2+}}{\partial v} + \frac{6}{5v}f_{12}^{2+}\right) \qquad \text{(D.2g)}$$

$$[v + i(\omega_1 - 2\omega_2)]f_{12}^{1-\prime} = \frac{a_1}{2}\left(\frac{\partial \tilde{f}_{21}^{0}}{\partial v} + \frac{2}{5}\frac{\partial \tilde{f}_{21}^{2}}{\partial v} + \frac{6}{5v}\tilde{f}_{21}^{2}\right)$$

$$+ \frac{\tilde{a}_2}{2}\left(\frac{\partial f_{12}^{0-}}{\partial v} + \frac{2}{5}\frac{\partial f_{12}^{2-}}{\partial v} + \frac{6}{5v}f_{12}^{2-}\right) \qquad \text{(D.2h)}$$

Expressions for f_{12}^1, f_{22}^1, f_{23}^1, $f_{21}^{1+\prime}$, and $f_{21}^{1-\prime}$ are analogous to those for $f_{11}^1, f_{12}^1, f_{13}^1, f_{12}^{1+\prime}$, and $f_{12}^{1-\prime}$, respectively. Finally Eqs. (3.37a) and (3.38a–c) lead to the following equations:

$$\frac{v}{3}\left[\gamma N \frac{\partial f_0^{1}}{\partial N} + CT\frac{\partial f_0^{1}}{\partial T}\right] - \frac{1}{6v^2}\frac{\partial}{\partial v}[v^2(\tilde{a}_1 f_{11}^1 + \tilde{a}_2 f_{21}^1)]$$

$$= \frac{m}{Mv^2}\frac{\partial}{\partial v}(vv^3 f_0^{0}) + \frac{kT}{Mv^2}\frac{\partial}{\partial v}\left(vv^2 \frac{\partial f_0^{0}}{\partial v}\right) \qquad \text{(D.3a)}$$

$$i\omega_1 f_{11}^0 = -\frac{v}{3}\left[\gamma N \frac{\partial f_{11}^1}{\partial N} + CT\frac{\partial f_{11}^1}{\partial T}\right]$$

$$+ \frac{1}{6v^2}\frac{\partial}{\partial v}[v^2(2a_1 f_0^1 + \tilde{a}_1 f_{12}^1 + a_2 f_{12}^{1-} + \tilde{a}_2 f_{12}^{1+})] \qquad \text{(D.3b)}$$

$$2i\omega_1 f_{12}^0 = -\frac{v}{3}\left[\gamma N \frac{\partial f_{12}^1}{\partial N} + CT\frac{\partial f_{12}^1}{\partial T}\right]$$

$$+ \frac{1}{6v^2}\frac{\partial}{\partial v}[v^2(a_1 f_{11}^1 + \tilde{a}_1 f_{13}^1 + \tilde{a}_2 f_{21}^{1+\prime} + a_2 \tilde{f}_{21}^{1-\prime})] \qquad \text{(D.3c)}$$

$$i(\omega_1 + \omega_2)f_{12}^{0+} = -\frac{v}{3}\left[\gamma N \frac{\partial f_{12}^{1+}}{\partial N} + CT\frac{\partial f_{12}^{1+}}{\partial T}\right]$$

$$+ \frac{1}{6v^2}\frac{\partial}{\partial v}[v^2(a_1 f_{21}^1 + a_2 f_{11}^1 + \tilde{a}_1 \tilde{f}_{21}^{1+\prime} + \tilde{a}_2 f_{12}^{1+\prime})]$$

$$\text{(D.3d)}$$

$$i(\omega_1 - \omega_2)f_{12}^{0-} = -\frac{v}{3}\left[\gamma N \frac{\partial f_{12}^{1-}}{\partial N} + CT \frac{\partial f_{12}^{1-}}{\partial T}\right]$$

$$+ \frac{1}{6v^2}\frac{\partial}{\partial v}\left[v^2(a_1\tilde{f}_{21}^1 + \tilde{a}_2 f_{11}^1 + \tilde{a}_1 \tilde{f}_{21}^{1-\prime} + a_2 f_{12}^{1-\prime})\right]$$

$$\text{(D.3e)}$$

Expressions for f_{21}^0, f_{22}^0, are again analogous to those for f_{11}^0 and f_{12}^0. Using the two approximations mentioned in the text, the above set of coupled equations can be solved and lead to Eq. (3.39) of the text.

The expressions for the A (which are integrals involving the collision frequency) for Case I and when $n = 1$, i.e., when the electron collision frequency is proportional to the first power of electron velocity, are as follows:

$$A_{12d} = \frac{kT}{e}\frac{\omega_p^2}{24i\pi^{3/2}\omega_2}\left[\frac{45}{16}\pi^{1/2}\frac{v_0^2}{\omega_2^2} - 7i\frac{v_0}{\omega_2} - \frac{3}{2}\pi^{1/2}\right]$$

$$+ \frac{kT}{e}\frac{\omega_p^2}{15\pi^{3/2}\omega_2}\left[-\frac{45}{8}i\pi^{1/2}\frac{v_0^2}{\omega_2^2} - 11\frac{v_0}{\omega_2} - \frac{15i}{8}\pi^{1/2}\right]$$

$$+ \frac{kT}{e}\frac{\omega_p^2}{12i\pi^{3/2}\omega_2}\left[\frac{15}{4}\pi^{1/2}\frac{v_0^2}{\omega_2^2} - 10i\frac{v_0}{\omega_2} - \frac{9}{4}\pi^{1/2}\right]$$

$$+ \frac{kT}{e}\frac{\omega_p^2}{24i\pi\omega_2}$$

$$+ \frac{kT}{e}\frac{\omega_p^2}{15\pi^{3/2}\omega_2}\left[-\frac{15}{8}i\pi^{1/2}\frac{v_0^2}{\omega_2^2} - \frac{v_0}{\omega_2}\right]$$

$$+ \frac{kT}{e}\frac{\omega_p^2}{15\pi^{3/2}\omega_2}\left[\frac{5}{4}\frac{v_0}{\omega_2} - \frac{i\pi^{1/2}}{2}\right]$$

$$\text{(D.4a)}$$

$$A_{12t} = \frac{kT}{e}\frac{\omega_p^2}{48i\pi^{3/2}\omega_2}\left[\frac{225}{16}\pi^{1/2}\frac{v_0^2}{\omega_2^2} - 28i\frac{v_0}{\omega_2} - \frac{9}{2}\pi^{1/2}\right]$$

$$+ \frac{kT}{e}\frac{\omega_p^2}{30\pi^{3/2}\omega_2}\left[-\frac{225}{8}i\pi^{1/2}\frac{v_0^2}{\omega_2^2} - 44\frac{v_0}{\omega_2} - \frac{45}{8}i\pi^{1/2}\right]$$

$$+ \frac{kT}{e}\frac{\omega_p^2}{24i\pi^{3/2}\omega_2}\left[\frac{75}{4}\pi^{1/2}\frac{v_0^2}{\omega_2^2} - 40i\frac{v_0}{\omega_2} - \frac{27}{4}\pi^{1/2}\right] + \frac{kT}{e}\frac{\omega_p^2}{16i\pi\omega_2}$$

$$+ \frac{kT}{e}\frac{\omega_p^2}{30\pi^{3/2}\omega_2}\left[-\frac{4v_0}{\omega_2} - \frac{75}{8}i\pi^{1/2}\frac{v_0^2}{\omega_2^2}\right]$$

$$+ \frac{kT}{e}\frac{\omega_p^2}{30\pi^{3/2}\omega_2}\left[\frac{5v_0}{\omega_2} - \frac{3}{2}i\pi^{1/2}\right]$$

$$\text{(D.4b)}$$

$$A_{12d}^{+} = \frac{kT}{e} \frac{\omega_p{}^2\omega_0{}^2}{6i\pi^{3/2}(\omega_1+\omega_2)^2}$$

$$\times \left[\frac{45}{4}\pi^{1/2}\frac{v_0{}^2}{\omega_1{}^2(\omega_1+\omega_2)} - \left\{\frac{12iv_0}{\omega_1(\omega_1+\omega_2)} + \frac{8iv_0{}^2}{\omega_1{}^2}\right\} - \frac{6\pi^{1/2}}{2\omega_1}\right]$$

$$+ \frac{kT}{e}\frac{\omega_p{}^2\omega_0{}^2}{6i\pi^{3/2}(\omega_1+\omega_2)^2}$$

$$\times \left[\frac{45}{4}\pi^{1/2}\frac{v_0{}^2}{\omega_2{}^2(\omega_1+\omega_2)} - \left\{\frac{12iv_0}{\omega_2(\omega_1+\omega_2)} + \frac{8iv_0}{\omega_2{}^2}\right\} - \frac{3\pi^{1/2}}{2\omega_2}\right]$$

$$+ \frac{2}{15}\frac{kT}{e}\frac{\omega_p{}^2\omega_0{}^2}{\pi^{3/2}(\omega_1+\omega_2)^2}$$

$$\times \left[-45i\pi^{1/2}\frac{v_0{}^2}{\omega_1{}^2(\omega_1+\omega_2)} - \left\{\frac{20v_0}{\omega_1{}^2} + \frac{48v_0}{\omega_1(\omega_1+\omega_2)}\right\} + \frac{15i\pi^{1/2}}{2\omega_1}\right]$$

$$+ \frac{2}{15}\frac{kT}{e}\frac{\omega_p{}^2\omega_0{}^2}{\pi^{3/2}(\omega_1+\omega_2)^2}$$

$$\times \left[-45i\pi^{1/2}\frac{v_0{}^2}{\omega_2{}^2(\omega_1+\omega_2)} - \left\{\frac{20v_0}{\omega_2{}^2} + \frac{48v_0}{\omega_2(\omega_1+\omega_2)}\right\} + \frac{15i\pi^{1/2}}{2\omega_2}\right]$$

$$+ \frac{2}{15}\frac{kT}{e}\frac{\omega_p{}^2\omega_0{}^2}{\pi^{3/2}(\omega_1+\omega_2)^2}\left[-\frac{15}{2}i\pi^{1/2}\frac{v_0{}^2}{\omega_2{}^3} - \frac{4v_0}{\omega_2{}^2}\right]$$

$$+ \frac{2}{15}\frac{kT}{e}\frac{\omega_p{}^2\omega_0{}^2}{\pi^{3/2}(\omega_1+\omega_2)^2}\left[-\frac{15}{2}i\pi^{1/2}\frac{v_0{}^2}{\omega_1{}^3} - \frac{4v_0}{\omega_1{}^2}\right]$$

$$+ \frac{2}{15}\frac{kT}{e}\frac{\omega_p{}^2\omega_0{}^2}{\pi^{3/2}(\omega_1+\omega_2)^2}\left[\frac{5v_0}{\omega_2{}^2} - \frac{2i\pi^{1/2}}{\omega_2}\right]$$

$$+ \frac{2}{15}\frac{kT}{e}\frac{\omega_p{}^2\omega_0{}^2}{\pi^{3/2}(\omega_1+\omega_2)^2}\left[\frac{5v_0}{\omega_1{}^2} - \frac{2i\pi^{1/2}}{\omega_1}\right]$$

$$+ \frac{kT}{e}\frac{\omega_p{}^2\omega_0{}^2}{6i\pi^{3/2}\omega_2(\omega_1+\omega_2)}$$

$$\times \left[15\pi^{1/2}\frac{v_0{}^2}{\omega_2{}^2(\omega_1+\omega_2)} - \left\{\frac{16iv_0}{\omega_2(\omega_1+\omega_2)} + \frac{12iv_0}{\omega_2{}^2}\right\} - \frac{9}{2}\frac{\pi^{1/2}}{\omega_2}\right]$$

$$+ \frac{kT}{e}\frac{\omega_p{}^2\omega_0{}^2}{6i\pi^{3/2}\omega_1(\omega_1+\omega_2)}$$

$$\times \left[15\pi^{1/2}\frac{v_0{}^2}{\omega_1{}^2(\omega_1+\omega_2)} - \left\{\frac{16iv_0}{\omega_1(\omega_1+\omega_2)} + \frac{12iv_0}{\omega_1{}^2}\right\} - \frac{9}{2}\frac{\pi^{1/2}}{\omega_1}\right]$$

$$+ \frac{kT}{e}\frac{\omega_p{}^2\omega_0{}^2}{3i\pi\omega_2(\omega_1+\omega_2)^2} + \frac{kT}{e}\frac{\omega_p{}^2\omega_0{}^2}{3i\pi\omega_1(\omega_1+\omega_2)^2} \tag{D.4c}$$

$$A^+_{12t} = \frac{kT}{e} \frac{\omega_p{}^2\omega_0{}^2}{12i\pi^{3/2}(\omega_1 + \omega_2)^2}$$

$$\times \left[\frac{225}{4} \pi^{1/2} \frac{v_0{}^2}{\omega_1{}^2(\omega_1 + \omega_2)} - \frac{9\pi^{1/2}}{\omega_1} - \left\{ \frac{48iv_0}{\omega_1(\omega_1 + \omega_2)} + \frac{32iv_0}{\omega_1{}^2} \right\} \right]$$

$$+ \frac{kT}{e} \frac{\omega_p{}^2\omega_0{}^2}{12i\pi^{3/2}(\omega_1 + \omega_2)^2}$$

$$\times \left[\frac{225}{4} \pi^{1/2} \frac{v_0{}^2}{\omega_2{}^2(\omega_1 + \omega_2)} - \frac{9\pi^{1/2}}{\omega_2} - \left\{ \frac{48iv_0}{\omega_2(\omega_1 + \omega_2)} + \frac{32iv_0}{\omega_2{}^2} \right\} \right]$$

$$+ \frac{kT}{e} \frac{\omega_p{}^2\omega_0{}^2}{15\pi^{3/2}(\omega_1 + \omega_2)^2} \left[-225i\pi^{1/2} \frac{v_0{}^2}{\omega_1{}^2(\omega_1 + \omega_2)} \right.$$

$$\left. + \frac{45}{2} \frac{i\pi^{1/2}}{\omega_1} - \left\{ \frac{80v_0}{\omega_1{}^2(\omega_1 + \omega_2)} + \frac{192v_0}{\omega_1(\omega_1 + \omega_2)} \right\} \right]$$

$$+ \frac{kT}{e} \frac{\omega_p{}^2\omega_0{}^2}{15\pi^{3/2}(\omega_1 + \omega_2)^2}$$

$$\times \left[-225i\pi^{1/2} \frac{v_0{}^2}{\omega_2{}^2(\omega_1 + \omega_2)} + \frac{45}{2} \frac{i\pi^{1/2}}{\omega_2} - \left\{ \frac{80v_0}{\omega_2{}^2} + \frac{192v_0}{\omega_2(\omega_1 + \omega_2)} \right\} \right]$$

$$+ \frac{kT}{e} \frac{\omega_p{}^2\omega_0{}^2}{15\pi^{3/2}(\omega_1 + \omega_2)^2}$$

$$\times \left[\frac{75}{2} i\pi^{1/2}v_0{}^2 \left(\frac{1}{\omega_2{}^3} + \frac{1}{\omega_1{}^3} \right) - 16v_0 \left(\frac{1}{\omega_2{}^2} + \frac{1}{\omega_1{}^2} \right) \right]$$

$$+ \frac{kT}{e} \frac{\omega_p{}^2\omega_0{}^2}{15\pi^{3/2}(\omega_1 + \omega_2)^2} \left[20v_0 \left(\frac{1}{\omega_1{}^2} + \frac{1}{\omega_2{}^2} \right) - 6i\pi^{1/2} \left(\frac{1}{\omega_2} + \frac{1}{\omega_1} \right) \right]$$

$$+ \frac{kT}{2e} \frac{\omega_p{}^2\omega_0{}^2}{i(\omega_1 + \omega_2)^2} \left(\frac{1}{\omega_2} + \frac{1}{\omega_1} \right)$$

$$+ \frac{kT}{e} \frac{\omega_p{}^2\omega_0{}^2}{12i\pi^{3/2}\omega_2(\omega_1 + \omega_2)}$$

$$\times \left[75\pi^{1/2} \frac{v_0{}^2}{\omega_2{}^2(\omega_1 + \omega_2)} - \frac{27}{2} \frac{\pi^{1/2}}{\omega_2} - \left\{ \frac{64iv_0}{\omega_2(\omega_1 + \omega_2)} + \frac{48iv_0}{\omega_2{}^2} \right\} \right]$$

$$+ \frac{kT}{e} \frac{\omega_p{}^2\omega_0{}^2}{12i\pi^{3/2}\omega_1(\omega_1 + \omega_2)}$$

$$\times \left[75\pi^{1/2} \frac{v_0{}^2}{\omega_1{}^2(\omega_1 + \omega_2)} - \frac{27}{2} \frac{\pi^{1/2}}{\omega_1} - \left\{ \frac{64iv_0}{\omega_1(\omega_1 + \omega_2)} + 48i \frac{v_0}{\omega_1{}^2} \right\} \right]$$

$$\tag{D.4d}$$

The expressions for the remaining A can be obtained in a manner outlined in the text.

APPENDIX E. COMPONENTS OF THE COLLISION INTEGRALS FOR AN INHOMOGENEOUS MAGNETOPLASMA

We present here the expressions for the A used in Eqs. (3.46a,b) for the case $n = 0$, i.e., for an electron collision frequency independent of electron velocity. The expressions are

$$A_{1d} = \frac{kT}{e}\left(\frac{\omega_p{}^2\omega}{\pi i}\right)\left[\frac{(v + i\omega)(v + 2i\omega)}{\{(v + i\omega)^2 + \omega_B{}^2\}\{(v + 2i\omega)^2 + \omega_B{}^2\}}\right]$$
$$+ \frac{kT}{e}\left(\frac{\omega_p{}^2\omega^2}{\pi}\right)\left[\frac{(v + i\omega)}{\{(v + i\omega)^2 + \omega_B{}^2\}\{(v + 2i\omega)^2 + \omega_B{}^2\}}\right] \tag{E.1a}$$

$$A_{2d} = -\frac{kT}{4e}\left(\frac{\omega_p{}^2\omega}{\pi i}\right)\frac{(v + i\omega)\omega_B}{[(v + i\omega)^2 + \omega_B{}^2][(v + 2i\omega)^2 + \omega_B{}^2]} \tag{E.1b}$$

$$A_{2d}' = \frac{3}{4}\frac{kT}{e}\left(\frac{\omega_p{}^2\omega}{\pi i}\right)\frac{(v + 2i\omega)\omega_B}{[(v + i\omega)^2 + \omega_B{}^2][(v + 2i\omega)^2 + \omega_B{}^2]} \tag{E.1c}$$

$$A_{3d} = \frac{kT}{4e}\left(\frac{\omega_p{}^2\omega}{\pi i}\right)\frac{(v + i\omega)(v + 2i\omega)}{[(v + i\omega)^2 + \omega_B{}^2][(v + 2i\omega)^2 + \omega_B{}^2]}$$
$$+ \frac{3}{4}\frac{kT}{e}\left(\frac{\omega_p{}^2\omega}{\pi i}\right)\frac{\omega_B{}^2}{[(v + i\omega)^2 + \omega_B{}^2][(v + 2i\omega)^2 + \omega_B{}^2]} \tag{E.1d}$$

$$A_{4d} = -\frac{kT}{4e}\left(\frac{\omega_p{}^2\omega}{\pi i}\right)\frac{(v + i\omega)\omega_B}{[(v + i\omega)^2 + \omega_B{}^2][(v + 2i\omega)^2 + \omega_B{}^2]}$$
$$- \frac{kT}{e}\left(\frac{\omega_p{}^2\omega^2}{\pi}\right)\frac{(v + i\omega)\omega_B}{[(v + i\omega)^2 + \omega_B{}^2][(v + 2i\omega)^2 + \omega_B{}^2](v + 2i\omega)} \tag{E.1e}$$

$$A_{4d}' = -\frac{3}{4}\frac{kT}{e}\left(\frac{\omega_p{}^2\omega}{\pi i}\right)\frac{(v + i\omega)\omega_B}{[(v + i\omega)^2 + \omega_B{}^2][(v + 2i\omega)^2 + \omega_B{}^2]} \tag{E.1f}$$

$$A_{5d} = -\frac{3}{4}\frac{kT}{e}\left(\frac{\omega_p{}^2\omega}{\pi i}\right)\left[\frac{(v + i\omega)\omega_B + \omega_B(v + 2i\omega)}{[(v + i\omega)^2 + \omega_B{}^2][(v + 2i\omega)^2 + \omega_B{}^2]}\right]$$
$$- \frac{kT}{e}\left(\frac{\omega_p{}^2\omega^2}{\pi}\right)\frac{\omega_B}{[(v + i\omega)^2 + \omega_B{}^2][(v + 2i\omega)^2 + \omega_B{}^2]} \tag{E.1g}$$

$$A_{6d} = -\frac{kT}{e}\left(\frac{\omega_p{}^2\omega^2}{\pi}\right)\frac{\omega_B{}^2}{[(v + i\omega)^2 + \omega_B{}^2](v + 2i\omega)[(v + 2i\omega)^2 + \omega_B{}^2]} \tag{E.1h}$$

$$A_{6d}' = \frac{3}{4}\frac{kT}{e}\left(\frac{\omega_p{}^2\omega}{\pi i}\right)\frac{(v + i\omega)(v + 2i\omega)}{[(v + i\omega)^2 + \omega_B{}^2][(v + 2i\omega)^2 + \omega_B{}^2]}$$
$$- \frac{3}{4}\frac{kT}{e}\left(\frac{\omega_p{}^2\omega}{\pi i}\right)\frac{\omega_B{}^2}{[(v + i\omega)^2 + \omega_B{}^2][(v + 2i\omega)^2 + \omega_B{}^2]} \tag{E.1i}$$

Appendix F. Justification of Retaining Terms up to f^2 Only

It was mentioned in Section II-A that retention of terms up to f^2 in the Legendre polynomial expansion of the distribution function f is a necessary and sufficient condition for evaluating correctly the magnitudes of the higher order components in the current density, up to the third harmonic and the second-order combination frequencies. In this Appendix, we give a proof of this statement for third-harmonic generation in a homogeneous isotropic plasma.

It is readily understandable that because of the orthogonality properties of the Legendre polynomials, the components $f^s(s > 1)$ will not contribute directly to the current density. The effect of a higher order term f^s is considered only because its inclusion in analysis affects the value of f^{s-1}, which in turn affects the value of f^{s-2}, and so on, until f^1 (and through it the current density) is affected. We proceed to show that the inclusion of f^3 in the analysis (after neglecting f^4 and higher terms) makes a negligible contribution to f^1 and the current density; the inclusion of f^4 and higher order terms (whose presence will be felt through f^3) will obviously make a still lesser contribution to f^1.

Because we are discussing the case of a homogeneous isotropic plasma, we can use Eqs. (3.3a–c) suitably modified to include contributions due to f^3. It can be shown that Eqs. (3.3a–b) remain unaltered, and that Eq. (3.3c) takes the form

$$\frac{\partial f^2}{\partial t} + vf^2 = a\left[\frac{2}{3} v \frac{\partial}{\partial v}\left(\frac{f^1}{v}\right) + \frac{3}{7v^4}\frac{\partial}{\partial v}(v^4 f^3)\right] \tag{F.1}$$

and that the equation for f^3

$$\frac{\partial f^3}{\partial t} + vf^3 = a\left[\frac{3}{5} v^2 \frac{\partial}{\partial v}\left(\frac{f^2}{v^2}\right)\right] \tag{F.2}$$

Assuming $\mathbf{a} = \mathbf{a}_0\, \exp(i\omega t)$ and the time-dependences for the components of f discussed in section 2.2 one obtains

$$vf_0^2 = \left[\frac{\tilde{a}_0}{3} v \frac{\partial}{\partial v}\left(\frac{f_1^1}{v}\right) + \frac{3}{14}\frac{\tilde{a}_0}{v^4}\frac{\partial}{\partial v}(v^4 f_1^3)\right] \tag{F.3}$$

$$(v + 2i\omega)f_2^2 = \left[\frac{a_0}{3} v \frac{\partial}{\partial v}\left(\frac{f_1^1}{v}\right) + \frac{3}{14}\frac{a_0}{v^4}\frac{\partial}{\partial v}(v^4 f_1^3) + \frac{3}{14}\frac{\tilde{a}_0}{v^4}\frac{\partial}{\partial v}(v^4 f_3^3)\right] \tag{F.4}$$

$$(v + i\omega)f_1^3 = \frac{3}{5} v^2 \frac{\partial}{\partial v}\left[\frac{1}{v^2}\left(a_0 f_0^2 + \frac{\tilde{a}_0}{2} f_2^2\right)\right] \tag{F.5}$$

$$(v + 3i\omega)f_3^3 = \frac{3}{10} v^2 \frac{\partial}{\partial v}\left[\frac{a_0}{v^2} f_2^2\right] \tag{F.6}$$

Equations (F.3) and (F.4) show that if $f_1{}^3, f_3{}^3 \ll f_1{}^1$, then the contribution of f^3 to f^2 (and hence f^1) will be negligible. A comparison of Eqs. (F.5), (F.6), and (3.4a) shows that this is indeed so. Thus we are justified in retaining terms up to f^2 only in an expansion of f. It may be pointed out here that the contributions of $f_3{}^3$ and $f_1{}^3$ to $f_4{}^2$ (and hence the higher harmonics) in the current density is by no means negligible, and so retention of terms up to f^2 only is not justified if one is interested in evaluating the magnitudes of harmonics higher than the third.

For inhomogeneous plasmas also, a similar analysis can be made. It can be shown (76) that an additional condition for the contribution of f^3 to be negligible is that the density and temperature gradients must be small.

Note Added in Proof

Several papers (77–89) on the generation of harmonics and combination frequencies in plasmas have appeared in the literature since the time this article was originally written. We shall briefly discuss some of the more relevant ones here.

The generation of second harmonic in the presence of a dc electric field of arbitrary magnitude was investigated by Kaw (77). Qualitatively, the difference between this and the earlier work lies in the fact that the distortion of $f_0{}^0$ due to the strong dc electric field has been taken into account. An interesting conclusion is that there is an optimum value of the dc electric field for which the second harmonic generation is most efficient; for stronger fields, the generated second harmonic decreases with the increase of the field [this result is in contrast to the one obtained using weak dc field theory in Sec. 3.1(b)]. The details of this paper will not be given here because it is primarily devoted to harmonic generation in semiconductors.

A kinetic theory analysis of the second harmonic generation due to a non-uniformity in the applied electric field has been given by Shkarofsky (78), and Kaw and Mittal (79). Shkarofsky's analysis (78) is also applicable to partially and fully ionized plasmas because he takes account of the Coulomb collisions between the electrons and ions: he also discusses the mixing of two waves because of this effect. For simplicity, in the following discussion, however, we shall limit ourselves to the problem of second harmonic generation in slightly ionized plasmas. (79) We write the basic system of Eqs. (2.4a–e) in its full tensorial form as

$$\frac{\partial f^0}{\partial t} + \frac{v}{3} \nabla \cdot \mathbf{f}^1 = \left(\frac{\partial f^0}{\partial t}\right)_c + \frac{1}{3v^2} \frac{\partial}{\partial v}(v^2 \mathbf{a} \cdot \mathbf{f}^1) \tag{1a}$$

$$\frac{\partial \overline{\mathbf{f}}^1}{\partial t} + v\mathbf{f}^1 = v\,\nabla f^0 + \mathbf{a}\,\frac{\partial f^0}{\partial v} + \boldsymbol{\omega}_b \times \mathbf{f}^1 - \frac{2v}{5}\nabla \cdot \mathbf{f}^2 + \frac{2}{5v^3}\frac{\partial}{\partial v}(v^3 \mathbf{a} \cdot \mathbf{f}^2) \tag{1b}$$

and

$$\left\{\frac{\partial \mathbf{f}^2}{\partial t} + v\mathbf{f}^2\right\}_2$$

$$= \left\{-v[\nabla \mathbf{f}^1 - \tfrac{1}{3}(\nabla \cdot \mathbf{f}^1)I_2] + 2\boldsymbol{\omega}_b \times \mathbf{f}^2 + v\frac{\partial}{\partial v}\left[\frac{\mathbf{a}\mathbf{f}^1}{v} - \frac{1}{3v}(\mathbf{a}\cdot\mathbf{f}^1)I_2\right]\right\}_2 \quad (1c)$$

where I_2 is the unit dyadic, the symbol $\{\ \}_2$ introduced by Shkarofsky *et al.* (*62a*) means that we take both permutations of the second-order tensors in the equation, add them, and divide by two; and $\boldsymbol{\omega}_b \equiv \boldsymbol{\omega}_b(\mathbf{r}, t) = \boldsymbol{\omega}_{bo}(\mathbf{r}) \exp(i\omega t)$ denotes the electron cyclotron frequency in the magnetic field of the electromagnetic wave itself. Writing down the explicit time dependences of f^0, $\mathbf{f}^1$, and $\mathbf{f}^2$, and proceeding as usual to make assumptions regarding the smallness of the fields and their gradients (which permit us to neglect terms containing their third and higher powers in comparison to those containing only their single power), one obtains for $\mathbf{f}_2^1$, the second harmonic component of $\mathbf{f}^1$, the expression

$$(v + 2i\omega)\mathbf{f}_2^1$$

$$= \nabla a_1^2\left(\frac{i}{12\omega v}\right)\frac{\partial}{\partial v}\left\{\frac{v^2}{v + i\omega}\frac{\partial f_0^0}{\partial v}\right\} + \mathbf{a}_1(\nabla \cdot \mathbf{a}_1)\left(\frac{i}{6\omega}\right)\frac{\partial}{\partial v}\left\{\frac{v}{v + i\omega}\frac{\partial f_0^0}{\partial v}\right\}$$

$$- \mathbf{a}_1 \times (\nabla \times \mathbf{a}_1)\left\{\frac{i}{2\omega(v + i\omega)}\frac{\partial f_0^0}{\partial v}\right\}$$

$$- \frac{1}{5}\left[(\mathbf{a}_1 \cdot \nabla)\mathbf{a}_1 + \mathbf{a}_1(\nabla \cdot \mathbf{a}_1) - \frac{1}{3}\nabla a_1^2\right]\left(\frac{v^2}{v + 2i\omega}\right)\frac{\partial}{\partial v}\left\{\frac{1}{v + i\omega}\frac{1}{v}\frac{\partial f_0^0}{\partial v}\right\}$$

$$- \frac{1}{5}\left[\frac{1}{2}(\mathbf{a}_1 \cdot \nabla)\mathbf{a}_1 + \frac{1}{4}\nabla a_1^2 - \frac{1}{3}\mathbf{a}_1(\nabla \cdot \mathbf{a}_1)\right]\frac{1}{v^3}\frac{\partial}{\partial v}\left\{\frac{v^4}{(v + i\omega)^2}\frac{\partial f_0^0}{\partial v}\right\} \quad (2)$$

The expression for $\mathbf{f}_3^1$ does not differ from the uniform electric field case [Sec. 3.1(a)] because of our assumption of the low gradients of the applied electric field. Using (2) one readily obtains the following expression for the second harmonic component in the current density

$$\mathbf{J}_2 = \frac{m}{M}\alpha[A_1(\mathscr{E} \cdot \nabla)\mathscr{E} + A_2\mathscr{E}(\nabla \cdot \mathscr{E}) + A_3\mathscr{E} \times (\nabla \times \mathscr{E})]\exp(2i\omega t) \quad (3)$$

where $\mathscr{E} = \mathbf{E}_1/E_{00}$ and the A's are integrals involving the collision frequency. The collision integrals are evaluated as usual for the special cases $\omega \gg v$ and $v \gg \omega$, taking f_0^0 to be Maxwellian and assuming $v = v_0 u^n$; the relevant expressions are

Case 1: $\omega \gg v$

$$A_1 = \frac{kT}{e}\frac{\omega_p^2}{\omega\pi^{3/2}}\left\{\frac{i}{2}\Pi\left(\frac{3}{2}\right) - \frac{v_0}{\omega}\left[\frac{(13n + 50)}{60}\Pi\left(\frac{n + 3}{2}\right)\right]\right\} \quad (4a)$$

$$A_2 = \frac{kT}{e}\frac{\omega_p^{\,2}}{\omega\pi^{3/2}}\left\{i\Pi\!\left(\frac{3}{2}\right) - \frac{v_0}{\omega}\left[\frac{3n+35}{20}\,\Pi\!\left(\frac{n+3}{2}\right)\right]\right\} \tag{4b}$$

$$A_3 = \frac{kT}{e}\frac{\omega_p^{\,2}}{\omega\pi^{3/2}}\left\{\frac{i}{2}\,\Pi\!\left(\frac{3}{2}\right) - \frac{v_0}{\omega}\left[\frac{4n+35}{60}\,\Pi\!\left(\frac{n+3}{2}\right)\right]\right\} \tag{4c}$$

Case II: $(m/M)v \ll \omega \ll v$

$$A_1 = \frac{kT}{e}\frac{\omega_p^{\,2}\omega}{v_0^{\,2}\pi^{3/2}}\left\{i\,\frac{(n-2)}{3}\,\Pi\!\left(\frac{3-2n}{2}\right) + \frac{\omega}{v_0}\frac{(3n-4)}{3}\,\Pi\!\left(\frac{3-3n}{2}\right)\right\} \tag{4d}$$

$$A_2 = \frac{kT}{e}\frac{\omega_p^{\,2}\omega}{v_0^{\,2}\pi^{3/2}}\left\{i\,\frac{(n-3)}{3}\,\Pi\!\left(\frac{3-2n}{2}\right) + \frac{\omega}{v_0}(n-1)\Pi\!\left(\frac{3-3n}{2}\right)\right\} \tag{4e}$$

$$A_3 = \frac{kT}{e}\frac{\omega_p^{\,2}\omega}{v_0^{\,2}\pi^{3/2}}\left\{i\,\frac{n-5}{3}\,\Pi\!\left(\frac{3-2n}{2}\right) + \frac{\omega}{v_0}\frac{6n-19}{3}\,\Pi\!\left(\frac{3-3n}{2}\right)\right\} \tag{4f}$$

Equation (3) is the general expression for the second harmonic component generated in the current density because of this mechanism. For $v = 0$, it reduces to

$$\mathbf{J}_2 = (iNe^3/2m^2\omega^3)(\mathbf{E}_1\,\mathrm{div}\,\mathbf{E}_1 + \tfrac{1}{4}\nabla E_1^{\,2})\exp(2i\omega t) \tag{5}$$

This is the Krenz and Kino (*37*) expression for the second harmonic generation in a collisionless plasma with a uniform electron density. [This expression differs from what we derived in Sec. 3.1(e) by a term proportional to div $\mathbf{E}_1$. We lost the div $\mathbf{E}_1$ term in our earlier analysis because we had unjustifiably neglected the electron density fluctuations arising through the self-consistent fields in the plasma; this was pointed out recently by Shkarofsky (*78*). The analysis of Sec. 3.1(e) can be properly modified by writing an electron continuity equation as

$$\partial n/\partial t + \nabla \cdot (n\mathbf{v}) = 0$$

which shows that even for a homogeneous plasma, one has

$$n = n_0 + n_1 \exp(i\omega t)$$

where

$$n_1 = (in_0/\omega)\,\nabla \cdot \mathbf{v}_1$$

is finite because $\mathbf{v}_1$ is nonuniform. The total second harmonic current density is then given by

$$\mathbf{J} = n_0\,e\mathbf{v}_2 + n_1\,e\mathbf{v}_1$$

thus we immediately see what the origin of the $\mathbf{E}_1$ div $\mathbf{E}_1$ term is.]

The above kinetic theory analysis further reveals that the effect of collisions on harmonic generation process is significant primarily in Case II

$(v \gg \omega)$. To illustrate the variation of the magnitude of second harmonic generated with the collision frequency and its dependence on electron velocity, Kaw and Mittal (79) have carried out some numerical calculations for $n = 0$ and $n = 1$, i.e., respectively for an electron collision frequency independent of and proportional to the electron velocity. It was found that the magnitude of the second harmonic decreases as the collision frequency increases. Further, the results for $n = 0$ and $n = 1$ were found to be very different, thereby showing that in order to illustrate the effect of collisions on the harmonic generation process, it is necessary to take account of the dependence of electron collision frequency on electron velocity (and hence the distribution of electron velocities).

The effect of a static magnetic field on the generation of combination frequencies, due to non uniformities in the applied electric fields, has recently been investigated by Mittal and Gupta (80). Proceeding in a manner similar to that discussed above, they found that $\mathbf{f}_{12}^{1+}$ (the $\omega_1 + \omega_2$ frequency component of $\mathbf{f}^1$) is given by

$$\{[v + i(\omega_1 + \omega_2)] - \boldsymbol{\omega}_B \times \}\mathbf{f}_{12}^{1+}$$

$$= -\frac{2v^2}{5\{v + i(\omega_1 + \omega_2)\}} \frac{\partial}{\partial v} \left[\frac{(\nabla \cdot \mathbf{a}_1)\mathbf{f}_{21}^1 + (\mathbf{a}_1 \cdot \nabla)\mathbf{f}_{21}^1}{4v} \right.$$

$$+ \frac{(\nabla \cdot \mathbf{f}_{21}^1)\mathbf{a}_1 + (\mathbf{f}_{21}^1 \cdot \nabla)\mathbf{a}_1}{4v}$$

$$+ \frac{(\nabla \cdot \mathbf{a}_2)\mathbf{f}_{11}^1 + (\mathbf{a}_2 \cdot \nabla)\mathbf{f}_{11}^1 + (\nabla \cdot \mathbf{f}_{11}^1)\mathbf{a}_2 + (\mathbf{f}_{11}^1 \cdot \nabla)\mathbf{a}_2}{4v}$$

$$\left. - \frac{\nabla(\mathbf{a}_1 \cdot \mathbf{f}_{21}^1) + \nabla(\mathbf{a}_2 \cdot \mathbf{f}_{11}^1)}{6v} \right]$$

$$- \frac{1}{2}\frac{\hat{c}}{\partial v}\left[\frac{v}{3i\omega_2}\mathbf{a}_1(\nabla \cdot \mathbf{f}_{21}^1) + \frac{v}{3i\omega_1}\mathbf{a}_2(\nabla \cdot \mathbf{f}_{11}^1) \right] + \frac{i}{2\omega_1}(\nabla \times \mathbf{a}_1) \times \mathbf{f}_{21}^1$$

$$+ \frac{i}{2\omega_2}(\nabla \times \mathbf{a}_2) \times \mathbf{f}_{11}^1$$

$$- \frac{1}{5v^3}\frac{\partial}{\partial v}\left[\frac{v^4}{v + i\omega_2}\left\{ \frac{1}{2}\mathbf{a}_1 \cdot (\nabla\mathbf{f}_{21}^1) + \frac{1}{2}\mathbf{a}_1 \cdot (\widetilde{\nabla\mathbf{f}_{21}^1}) - \frac{1}{3}\mathbf{a}_1(\nabla \cdot \mathbf{f}_{21}^1) \right\} \right.$$

$$\left. + \frac{v^4}{v + i\omega_1}\left\{ \frac{1}{2}\mathbf{a}_2 \cdot (\nabla\mathbf{f}_{11}^1) + \frac{1}{2}\mathbf{a}_2 \cdot (\nabla\mathbf{f}_{11}^1) - \frac{1}{3}\mathbf{a}_2(\nabla \cdot \mathbf{f}_{11}^1) \right\} \right]$$

$$- \frac{1}{6i(\omega_1 + \omega_2)v}\frac{\partial}{\partial v}[v^2\{\nabla(\mathbf{a}_1 \cdot \mathbf{f}_{21}^1 + \mathbf{a}_2 \cdot \mathbf{f}_{11}^1)\}] \tag{6}$$

where the symbol $\sim$ over any term represents its transpose. The sum frequency component in the current density is then given by

$$
\begin{aligned}
J_{12}^{+} = \frac{m}{M}\,\alpha\{ &A_1\mathscr{E}_2(\nabla \cdot \mathscr{E}_1) + A_2\mathscr{E}_1(\nabla \cdot \mathscr{E}_2) + A_3(\mathscr{E}_1 \cdot \nabla)\mathscr{E}_2 + A_4(\mathscr{E}_2 \cdot \nabla)\mathscr{E}_1 \\
&+ A_5(\nabla \times \mathscr{E}_1) \times \mathscr{E}_2 + A_6(\nabla \times \mathscr{E}_2) \times \mathscr{E}_1 + A_7(\nabla \cdot \mathscr{E}_1)(\omega_B \times \mathscr{E}_2) \\
&+ A_8(\nabla \cdot \mathscr{E}_2)(\omega_B \times \mathscr{E}_1) + A_9(\mathscr{E}_1 \cdot \nabla)(\omega_B \times \mathscr{E}_2) \\
&+ A_{10}(\mathscr{E}_2 \cdot \nabla)(\omega_B \times \mathscr{E}_1) + A_{11}[\nabla \cdot (\omega_B \times \mathscr{E}_1)]\mathscr{E}_2 \\
&+ A_{12}[\nabla \cdot (\omega_B \times \mathscr{E}_2)]\mathscr{E}_1 + A_{13}[(\omega_B \times \mathscr{E}_2) \cdot \nabla]\mathscr{E}_1 \\
&+ A_{14}[(\omega_B \times \mathscr{E}_1) \cdot \nabla]\mathscr{E}_2 + A_{15}\nabla[\mathscr{E}_1 \cdot (\omega_B \times \mathscr{E}_2)] \\
&+ A_{16}\nabla[\mathscr{E}_2 \cdot (\omega_B \times \mathscr{E}_1)] + A_{17}(\nabla \times \mathscr{E}_1) \times (\omega_B \times \mathscr{E}_2) \\
&+ A_{18}(\nabla \times \mathscr{E}_2) \times (\omega_B \times \mathscr{E}_1) + A_{19}\mathscr{E}_1[\widetilde{\nabla(\omega_B \times \mathscr{E}_2)}] \\
&+ A_{20}\mathscr{E}_2 \cdot [\widetilde{\nabla(\omega_B \times \mathscr{E}_1)}]\}\{\exp(i\overline{\omega_1 + \omega_2}\,t)
\end{aligned}
\tag{7}
$$

where the A's as usual are integrals involving the velocity-dependent collision frequency. Mittal and Gupta *(80)* evaluated these integrals for the special cases with ω_1, ω_2, $|\omega_1 - \omega_2| \gg \nu$ and $\omega_B = 0$, ω_1, ω_2, $\omega_1 + \omega_2$, etc. The results have been used for some numerical work which illustrates the variations of the generated sum frequency components with (v_0/ω) and (ω_1/ω_2) around the various resonant and nonresonant cases. The conclusions are: (a) The resonances around $\omega_B = \omega_1$ and $\omega_B = \omega_2$ are comparatively weak, whereas the one around $\omega_B = (\omega_1 + \omega_2)$ is significant. (b) Around $\omega_B = (\omega_1 + \omega_2)$, the sum frequency component decreases with the increase of collision frequency and the ratio (ω_1/ω_2). (c) The results for $n = 0$ and $n = 1$ are quite different, illustrating as before the need for taking the electron velocity distribution into account.

Gupta *(81)* has recently investigated the harmonic components reflected from a plasma-free space interface around plasma resonance. He used the expressions for the current density components derived by Moriyama and Sumi for such a situation [Sec. 3.2(c)]. The main motivation behind the analysis was to obtain realistic estimates of the second and third harmonics generated by the Moriyama–Sumi mechanism without having to make any arbitrary choice for the values of radiation resistance, etc. Following the usual method discussed in Sec. 4, Gupta *(81)* obtained the following expressions for the second and third harmonic components in the reflected wave:

$$
B_{r2}/B_{i1}^2 = 4\alpha R_2/(\beta_2 + 2A)(1 - A)^2(2 - \beta_2)
$$

$$
B_{r3}/B_{i1}^3 = 4\alpha R_3/(\beta_3 + 2A)(1 - A)^3(3 - \beta_3)
$$

where

$$A = (1 + 4\pi i \sigma_1/\varepsilon\omega + 4\pi i K_1/\varepsilon\omega)^{1/2}$$

$$R_2 = 8\pi i K_2/\varepsilon\omega E_{00}, \qquad R_3 = 12\pi i K_3/\varepsilon\omega E_{00}$$

and K_1, K_2, and K_3 are expressions involving the collision frequency [see Gupta (*81*)]. Using these equations Gupta (*81*) found that the magnitudes of the generated second and third harmonics decrease with the increase of collision frequency. As an example of the typical order of magnitude of the harmonics generated, he considered the cases where $\omega = 10^9$ sec^{-1}, $(v_0/\omega) = 0.01$, and $E_i = 3$ V/cm, and found that the second and third harmonics are, respectively, 11.7 and 11% of the fundmental.

Chiyoda (*82*) has considered the generation of harmonics in an inhomogeneous plasma in the presence of a magnetic field; this problem is similar to that discussed in Sec. 3.2(b). Chiyoda, however, neglected the important f^2 terms in the expansion of f and also dropped the odd frequency components of f^0 so that "the theory will agree qualitatively with experimental results"; it is not clear as to which experimental results are qualitatively disagreed with if $f_1{}^0$ is included.

Ellis and Porkolab (*83*) considered the situation when the external microwave signals excite Bernstein-like electrostatic modes in a plasma in the presence of a magnetic field, which may then interact and lead to some mixing. They have used a quasilinear-type analysis to evaluate the magnitude of the mixed output; the analysis is not, however, completely presented, and only qualitative conclusions (which agree with experiments) regarding resonances in which the magnitude of the external magnetic field is varied are presented.

Some work on the harmonic generation in fully ionized plasmas by strong laser beams has been done by Berk and Birmingham (*84*) and Kaw and Salat (*85*). It is found that the harmonic generated is quite small, in general.

REFERENCES

1. L. Wetzel and T. Tang, *J. Res. Natl. Bur. Std. Radio Propagation* **69D**, 599 (1965).

2. H. Margenau and L. M. Hartman, *Phys. Rev.* **73**, 309 (1948).

3. P. Rosen, *Phys. Fluids* **4**, 341 (1961).

4. M. S. Sodha and C. J. Palumbo, *Can. J. Phys.* **42**, 1236 (1964).

5. I. M. Vilenskii, *Zh. Eksperim. i Teor. Fiz.* **22**, 544 (1952).

6. I. M. Vilenskii, *Zh. Eksperim i Teor. Fiz.* **26**, 42 (1954).

7. V. M. Fain, *Zh. Eksperim i Teor. Fiz.* **28**, 422 (1955).

8. A. V. Gurevich, Dissertation, Phys. Inst. Acad. M. (1957). (Quoted in Ref. 9.)

9. V. L. Ginzburg and A. V. Gurevich, *Soviet Phys.–Usp.* (*English Transl.*) **3**, 115, 175 (1960).

10. L. Wetzel, *Phys. Rev.* **123**, 722 (1961).

11. V. P. Silin, *Soviet Phys. JETP* (*English Transl.*) **20**, 135 (1965).

12. K. Chiyoda, *J. Phys. Soc. Japan* **20**, 290 (1965).

13. J. H. Krenz, *Phys. Fluids* **8**, 1871 (1965).

14. G. P. Gupta, *Proc. Phys. Soc.* **92**, 1122 (1967).

15. G. P. Gupta, Private communication, 1967.

16. R. S. Mittal and P. K. Kaw, *Can. J. Phys.* **45**, 1587 (1967).

17. M. S. Sodha and P. K. Kaw, *Phys. Fluids* **8**, 1402 (1965).

18. M. S. Sodha and P. K. Kaw, *Prog. Theoret. Phys.* (*Kyoto*) **34**, 557 (1965).

19. M. S. Sodha and P. K. Kaw, *Proc. Phys. Soc.* **88**, 373 (1966).

20. M. S. Sodha and P. K. Kaw, *J. Phys. Soc. Japan* **21**, 134 (1966).

21. M. S. Sodha and P. K. Kaw, *Phys. Fluids* **9**, 603 (1966).

22. M. S. Sodha and P. K. Kaw, *J. Phys. Soc. Japan* **21**, 2674 (1966).

23. M. S. Sodha and P. K. Kaw, *J. Phys. Soc. Japan* **21**, 2684 (1966).

24. M. S. Sodha and P. K. Kaw, *J. Appl. Phys.* **37**, 3632 (1966).

25. V. V. Paranjape, Private communication (1966).

26. B. Murphy, *Phys. Fluids* **8**, 1534 (1965).

27. T. Tang, *Proc. IEEE Trans.* **AP-14**, 54 (1966).

28. E. A. Desloge and P. D. Coleman, *Phys. Fluids* **9**, 1389 (1966).

29. L. Wetzel, *J. Appl. Phys.* **32**, 327 (1961).

30. L. Wetzel, *Symposium Electromagnetic Theory and Antennas* (E. C. Jordan, ed.), p. 365. Pergamon Press, New York, 1963.

31. B. Blachier, N. Guyen Van Tran, and E. Leiba, *CRVI Conf. Intern. Phenomenon Ionisation dan les Gaz.* S.E.R.M.A., Paris, 1963.

32. R. F. Whitmer and E. B. Barrett, *Phys. Rev.* **121**, 661 (1961).

33. R. F. Whitmer and E. B. Barrett, *Phys. Rev.* **125**, 1478 (1962).

34. E. B. Barrett, R. F. Whitmer, and S. J. Tetenbaum, *Phys. Rev.* **135**, 369 (1964).

35. J. H. Krenz, MW Lab. Rep. No. 1055, July. Stanford Univ. (1963).

36. J. H. Krenz, and G. S. Kino, MW Lab Rep. No. 948, Sept. Stanford Univ. (1963).

37. J. H. Krenz and G. S. Kino, *J. Appl. Phys.* **36**, 2387 (1965).

38. S. F. Smerd, *Nature* **175**, 297 (1955).

39. L. S. Taylor, *J. Appl. Phys.* **33**, 2913 (1962).

40. S. Visvanathan, *J. Appl. Phys.* **33**, 2481 (1962).

41. O. E. H. Rydbeck, *J. Res. Natl. Bur. Std. Radio Propagation* (*English Transl.*) **69D**, 111 (1965).

42. V. L. Ginzburg, *Soviet Phys. JETP* (*English Transl.*) **35**, 1100 (1959).

43. J. R. Baird and P. D. Coleman, *Proc. IRE* **49**, 1890 (1961).

44. M. Moriyama and M. Sumi, *J. Phys. Soc. Japan* **17**, 397 (1962).

45. K. Chiyoda and T. Tamaru, *Proc. IEEE* **53**, 1257 (1965).

46. K. Forsterling and H. O. Wüster, *Compt. Rend.* **231**, 8312 (1950).

47. K. Forsterling and H. O. Wüster, *J. of Atmospher. and Terrest. Phys.* **2**, 22 (1952).

48. M. Moriyama and M. Sumi, *J. Phys. Soc. Japan* **18**, 1656 (1963).

49. M. Moriyama and M. Sumi, *J. Phys. Soc. Japan* **20**, 138 (1965).

50. Mz. Uenohara, My. Uenohara, T. Masutani, and K. Inada, *Proc. IRE* **45**, 1419 (1957).

51. K. D. Froome, *Nature* **184**, 808 (1959).

52. K. D. Froome, *Nature* **186**, 959 (1960).

53. K. D. Froome, *Nature* **188**, 43 (1960).

54. K. D. Froome, *Nature* **193**, 1169 (1960).

55. N. R. Bierrum and D. Walsch, *J. Electron. Control* **8**, 81 (1960).

56. N. R. Bierrum and D. Walsch, *Nature* **186**, 626 (1960).

57. C. B. Swan, *Proc. IRE* **49**, 1941 (1961).

58. R. M. Hill and S. J. Tetenbaum, *J. Appl. Phys.* **30**, 1610 (1959).

59. H. Dreicer, *Bull. Am. Phys. Soc.* **7**, 151 (1962).

60. S. J. Tetenbaum, R. F. Whitmer, and E. B. Barrett, *Phys. Rev.* **135**, 374 (1964).

61. R. A. Stern, *Phys. Rev. Letters* **14**, 538 (1965).

62. J. A. Green, *J. Geophys. Res.* **70**, 3244 (1965).

62a. I. P. Shkarosfky, T. W. Johnston, and M. P. Bachynski, "Particle Kinetics of a Plasma," Addison-Wesley, Cambridge, Massachusetts, 1966.

63. E. A. Desloge and S. W. Matthysse, *Am. J. Phys.* **28**, 1 (1960).

64. M. S. Sodha and C. J. Palumbo, *Can. J. Phys.* **41**, 1702 (1963).

65. C. G. Darwin, *Proc. Roy. Soc.* **182**, 152 (1943); **146**, 17 (1934).

66. V. L. Ginzburg, *Izv. Akad. Nauk. SSSR, Ser. Fiz.* **8**, 76 (1944).

67. V. L. Ginzburg, "Theory of Radio Wave Propagation in Ionosphere," Gostekhizdat, 1949.

68. B. B. Kadomtsev, *Zh. Eksperim. i Teor. Fiz.* **33**, 151 (1957).

69. V. L. Ginzburg, "Propagation of Electromagnetic Waves in a Plasma," Gordon and Breach, New York, 1961.

70. M. S. Sodha and B. K. Sawhney, Communication 1967.

71. M. Moriyama, Private communication to P. K. Kaw (1966).

72. M. S. Sodha and C. J. Palumbo, *Proc. Phys. Soc.* **80**, 1155 (1962).

73. M. Epstein, *Phys. Fluids* **5**, 492 (1962).

74. M. S. Sodha and C. J. Palumbo, *Can. J. Phys.* **41**, 2155 (1963).

75. M. S. Sodha and C. J. Palumbo, *Can. J. Phys.* **42**, 349 (1964).

76. G. P. Gupta, Private Communication.

77. P. K. Kaw, *J. Appl. Phys.* **40**, 793 (1969).

78. I. P. Shkarosfky, *Plasma Phys.* **10**, 169 (1968).

79. P. K. Kaw and R. S. Mittal, *J. Appl. Phys.* **39**, 1975 (1968).

80. R. S. Mittal and G. P. Gupta. To be published.

81. G. P. Gupta, *J. Phys. Soc. Japan*, **24**, 201 (1968).

82. K. Chiyoda, *J. Phys. Soc. Japan* **22**, 910 (1967).

83. R. A. Ellis and M. Porkolab, *Phys. Rev. Letters* **21**, 529 (1968).

84. H. L. Berk and T. J. Birmingham, *Bull. Am. Phys. Soc.* **9**, 482 (1964).

85. P. K. Kaw and A. R. Salat, *Phys. Fluids* **10**, 2223 (1968).

86. S. J. Tetenbaum, *Phys. Fluids* **10**, 1855 (1967).

87. R. E. McIntosh, National Aeronautics and Space Administration Tech. Note D-4309 (1968).

88. J. L. Delcroix and A. M. Pointu, *Compt. Rend. Ser B.* **262**, 351, 472 (1966).

89. T. Tang, *Phys. Fluids*, **9**, 415 (1966); *IEEE Trans. Antennas Propagation* **AP-14**, 54 (1966).

A Review of Reflex and Penning Discharges*

E. B. HOOPER, JR.

*Department of Engineering and Applied Science
Yale University, New Haven, Connecticut*

Reflex and Penning discharges are reviewed, with attention to the various modes of operation. The review considers wide ranges of gas pressures, magnetic fields, and discharge voltages, and both cold and thermionic cathodes. The important physical mechanisms operating in the different modes are discussed and compared with experiments. Practical applications of the discharge are described.

I. INTRODUCTION

The reflex discharge, also called the Penning discharge or Philips ionization gage (PIG), has received a great deal of attention in the scientific and engineering literature because of the many interesting physical phenomena

* This work was primarily supported by NASA grant SAR/NGR 07-004-028.

295

that occur in it and because of its many applications. The device operates on the principle, first observed in the afterglow of a discharge by Phillips (*118, 119*), that charged particles in crossed electric and magnetic fields move in the $E \times B$ direction. The effect was further studied by Strutt (*136*), who apparently operated the first active discharge taking advantage of the phenomena, and by Wehrli (*146*), who studied particle trajectories.

The discharge was popularized by Penning's experiments (*116, 202*). He showed that for gas pressures below about 10^{-3} torr, the current through the discharge is proportional to the pressure so that it can be used as a manometer. Such a gage was sold by the Philips Company of The Netherlands, whence the name PIG; it is perhaps of historical interest that this name is unrelated to C. E. S. Phillips whose work first inspired the discharge.

A geometry that takes advantage of the magnetic field confinement properties is shown in Fig. 1. There are many variations of these geometries; the

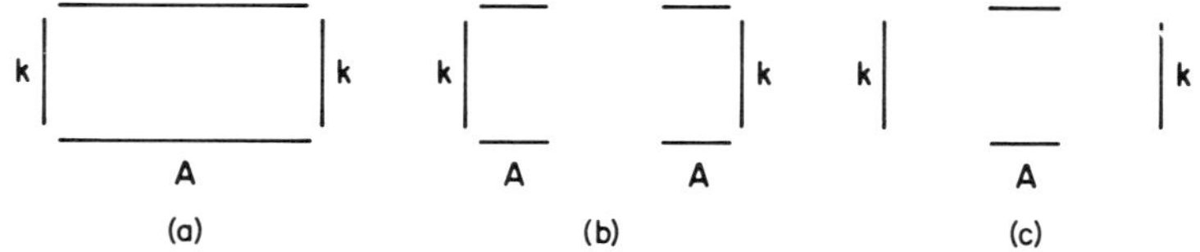

FIG. 1. Typical Penning discharge geometries. (a) Cylindrical anode. (b) Two ring anodes. (c) Ring anode. The magnetic field is applied parallel to the tube axis.

cathodes may be thermionic or cold and may have diameters smaller than or approximately equal to the anode diameter; the anode length may be intermediate between the cases (a) and (c), etc., in the figure. Such modifications are included in the present review. The magnetron geometry, which consists of coaxial cathode and anode, is not considered here; although it shows many similarities to the reflex geometry, its inclusion would have considerably lengthened the review.

A number of authors have studied breakdown of the cold cathode discharge (*134, 123, 130, 131*). These results are essentially an extension of glow discharge breakdown analysis, taking into account the magnetic field and the more complicated electric field. Breakdown will not be discussed here; instead, the reader is referred to the cited papers.

Many complicating effects occur in a reflex discharge, among which are secondary emission at the cathodes, ionization phenomena, highly non-Maxwellian particle distributions, complicated electric fields, and plasma instabilities. No adequate theories of the discharge exist and, as a result, one is forced into a semiempirical study of discharge behavior. It is found that the discharge behavior takes on qualitatively different characteristics for different values of the voltage, current, pressure, magnetic field, etc.; this permits one to identify modes of the discharge and discuss the behavior with appropriate approximations for each mode. To identify these modes, we generally must draw from information about many discharges; consequently there is corresponding uncertainty in the general description. The modes of the cold cathode discharge with cylindrical anode (Fig. 2) have been described by

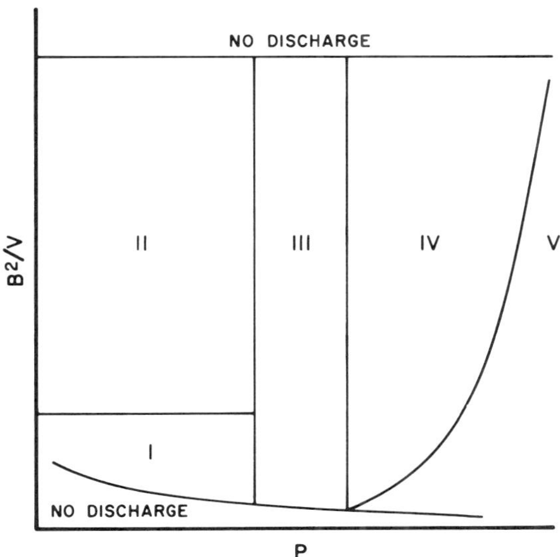

FIG. 2. Modes of the cold cathode discharge geometry (a). The mode boundaries are schematic and not precise. B is the magnetic field, V the discharge voltage, and p the gas pressure.

Schuurman, whose procedure is followed here with some modifications; additional identification is provided by the reviewer. The reader is warned that because of the complexity of the Penning discharge, any given discharge will differ more or less from the mode description given here, which is only a guideline.

The following discussion is divided into sections on cold cathode and thermionic cathode discharges. These sections are followed by short discussions about discharges and experiments that do not fit naturally into the main sections, practical uses of the discharge, the important instabilities present in the discharge, and electric fields parallel to magnetic fields. Several articles of interest but not referred to in the text are included in the reference list under Additional References.

II. Cold Cathode Discharge

A. General Description

The simplest form of the Penning discharge uses a cold cathode with ion bombardment supplying the electron emission. Although the three geometries have common features, the potential distribution depends on the electrode configuration, leading to differences in behavior. Consequently the three geometries will be discussed separately.

B. Geometry (a)

The modes of operation are illustrated in Fig. 2, which is a modification of a chart due to Schuurman (*130, 131*). The shapes of the boundaries are only approximate. Additional modes are occasionally claimed to exist (*47*), but these are the only ones that are well documented.

The qualitative features of the modes are as follows: In Modes I and II the ion density is much lower than the electron density throughout much of the discharge volume because the time for an average ion to reach a cathode is much less than the electron confinement time. In Mode I the electrons fill the discharge with an approximately uniform density. This mode, interestingly enough, is often observed to be stable, and theoretical predictions are moderately successful. In Mode II the electron density is sufficiently large that most of the discharge voltage drop occurs in an anode sheath; the center of the discharge is filled with a quasineutral plasma with a potential close to that of the cathode and an electron density much below that in the sheath. These modes apparently have no low pressure limit; Klopfer (*156*) has operated discharges at pressures as low as 10^{-12} torr.

As the pressure is raised, the electron loss rate increases and the plasma switches to a quasineutral regime (Mode III) in which the electron and ion densities are approximately equal throughout the discharge volume. As the particle mean-free paths are still much longer than the discharge dimension, the velocity distribution functions are far from Maxwellian, and large electric fields may exist parallel to the magnetic field; the discharge potential drop is partly along the axis and partly across the magnetic field. This mode is observed to be unstable, but the instability has not been identified.

At still higher pressures, the ion mean-free path becomes comparable to the discharge length. This leads to a thermalization of the ion distribution, with the consequence that the large axial fields may no longer be maintained. The discharge switches to a Mode IV in which the voltage drop occurs in cathode sheaths. Electrons are emitted from the cathode by ion bombardment and accelerated across the sheaths. The resulting interpenetrating beams are highly unstable, and strong oscillations often occur, resulting in a beam-plasma type of discharge. It is observed that for sufficiently low magnetic fields, the discharge is stable, leading to the curved boundary between Modes IV and V. When the pressure rises sufficiently either to destroy the beams or to damp the oscillations, or when the magnetic field is low enough for radial losses to become large, the discharge enters a high pressure regime (Mode V) which is essentially a glow (or arc) discharge of unusual geometry.

Because the discharge character changes considerably from one mode to another, there is considerable hysteresis associated with the mode changes. In addition, because strong electric fields exist inside the discharge at low pressures, the actual field configuration is quite sensitive to the electrode geometry. These facts plus the strong instability effects make it difficult to compare different experiments. The reviewer believes, nevertheless, that most of the observed phenomena can be fitted into the qualitative scheme described above. He wishes to point out to future experimenters, however, that current-voltage characteristics as a function of magnetic field and gas pressure are insufficient to provide a description of a reflex discharge; also needed are measurements of the particle and field volume distributions, the particle velocity distributions, and quantitative measures of the fields, densities, and their correlations associated with instabilities. This is asking a lot of the experimenters, but without all this information, and more, it is almost impossible to obtain a unique description of the discharge characteristics.

A detailed description of the modes follows below.

1. Mode I

In Mode I the ions leave the discharge so readily that the discharge contains essentially an electron gas. For the long anode case the axial variation of density, etc., apparently is small, and in a first approximation only the radial variation need be considered; the potential variation with radius illustrated in Fig. 3a has been observed experimentally by Helmer and Jepsen (63) and by Dow (47). In addition, it is observed that this mode is often, although not always, stable.

A detailed theory of the discharge requires consideration of ionization, which is complicated by the coupling with the potential and density distributions through secondary emission at the cathode. Analysis including these effects as well as the complicated particle orbits has been attempted by several authors. Theories including space charge have been made by Jepsen (76), Schuurman (130, 131), and Reikhrudel et al. (127), and it is found (130, 131) that the assumption of an ionization frequency independent of position yields

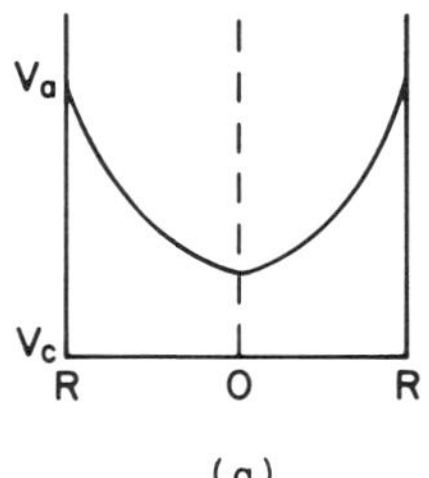
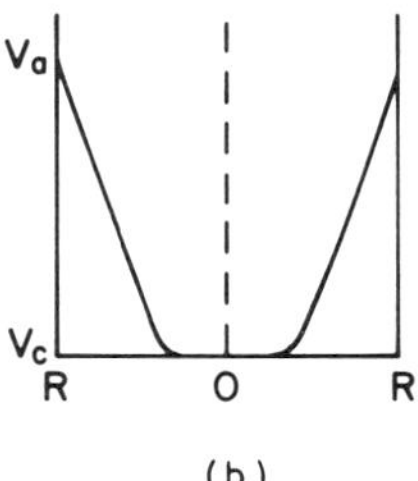

Fig. 3. The radial potential distribution. (a) Mode I; (b) Mode II. V_a and V_c are the cathode and anode potentials.

results that compare as well with theory as the more complicated theories, although such a result is difficult to justify.

We therefore start with Poisson's equation and the equation of continuity. Consider the case in which the anode cylinder has a length comparable to the cathode-cathode spacing, and assume the axial electric field to be small except near the discharge ends; a change in geometry will invalidate this approximation. Some axial effects will be discussed later. For the present case we write Poisson's equation and the electron continuity equation:

$$\frac{1}{r}\frac{d}{dr} rE_r = -\frac{en}{\varepsilon_0} \tag{1}$$

and

$$\frac{1}{r}\frac{d}{dr}(rnv_r) = v_i(1+\gamma)n \tag{2}$$

where E_r is the radial electric field, v_r is the radial electron velocity averaged over the velocity distribution, n is the electron density, v_i is the ionization frequency due to electron-neutral collisions, and γ is the cathode secondary emission coefficient. The factor proportional to γ (also assumed independent of radius) arose from the z component of the electron flux; ions are assumed to fall freely to the cathode where they liberate electrons which flow freely into the plasma. Equations (1) and (2) combine to yield

$$\frac{1}{r}\frac{d}{dr}\left[r\left(nv_r - \frac{v_i(1+\gamma)\varepsilon_0 E_r}{e}\right)\right] = 0 \tag{3}$$

There are three mechanisms causing radial transport in a stable discharge. They are the radial electric field, the centrifugal force due to the rotation in crossed fields, and diffusion. If for the moment we assume a parabolic potential of depth V, a magnetic field B, and a discharge radius R, the ratio of centrifugal and electric forces is approximately

$$\frac{F_c}{F_e} = \frac{2mV}{eB^2R^2}$$

Typical discharge parameters place this as 10^{-1}, so it may be neglected in an approximate theory. Comparison with Eq. (1) indicates that we are requiring $\omega_c^2 \gg \omega_p^2$, with ω_c and ω_p the electron cyclotron and plasma frequencies, respectively.

Consideration of diffusion effects is difficult because an appropriate diffusion coefficient is not known. This is particularly true as finite Larmor radius effects become important for high random electron energy perpendicular to the magnetic field. It is the reviewer's feeling that consideration of these effects requires the proper application of kinetic theory and that electric field transport will give as good a semiquantitative theory as is possible. In support of this, note that Popov (121) has included approximations to diffusion in an analysis of Mode II and has obtained results that are no closer to experiment than those of the simple model neglecting it.

Therefore, as the collision frequency is much smaller than the cyclotron frequency, the electron equation of motion is

$$v_r = \frac{-mv_c}{eB^2} E_r \tag{4}$$

Here v_c is the electron-neutral collision frequency. From Eqs. (3) and (4) we find a constant electron density

$$n = \frac{\varepsilon_0 B^2}{m} \left[\frac{v_i(1 + \gamma)}{v_c} \right] \tag{5}$$

independent of the discharge voltage and current. This result must fail within an electron Larmor radius of the anode, at which point the density will drop rapidly.

Poisson's equation immediately integrates to yield

$$V = \frac{eB^2r^2}{4m} \frac{v_i(1 + \gamma)}{v_c} \tag{6}$$

Finally, the discharge current is found from the ion and electron currents at the cathode:

$$J = \pi R^2 Lenv_i(1 + \gamma)$$
$$= \frac{\pi R^2 \varepsilon_0 LeB^2}{m} \frac{v_i^2(1 + \gamma)^2}{v_c} \tag{7}$$

with L the discharge length.

These results are in general agreement with the many experiments. If the shape of the electron and ion distribution functions are independent of pressure, $v_i^2(1 + \gamma)^2/v_c$ (and hence the current) is proportional to pressure, in excellent agreement with experiment. The vacuum gage, which generally operates in this mode, utilizes this dependence; the reader may find numerous papers in the appropriate reference section. The parabolic potential variation is in good agreement with the qualitative measurements of Helmer and Jepsen (*63*) and of Dow (*47*).

Perhaps the most systematic experimental examination of the operating characteristics was made by Schuurman (*130, 131*) in a discharge of 2 cm diameter and 4 cm length; several of his figures are reproduced in Figs. 3 through 5. For Mode I we note the following: On the J/B diagram (Fig. 4),

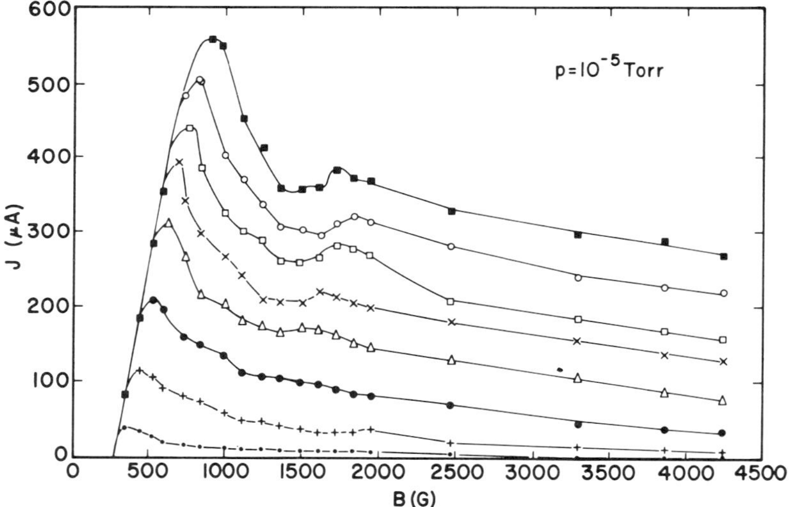

FIG 4. Current versus magnetic field in a cold cathode discharge. Mode I is to the left of the peak, Mode II to the right; J is the current, B the magnetic field, and p the pressure. The discharge voltage varies in steps of 500 volts from 500 volts (lowest current) to 4000 volts (highest current). (From Schuurman, *131*.)

Mode I occurs below the peak in the current. The current is proportional to $B - B_0$ (with B_0 a fixed magnetic field) rather than B^2; however, if we pick $v_i(1 + \gamma)v_c \sim 0.1$, we obtain good numerical agreement. The discrepancy presumably arises from the neglect of finite orbit effects, which become most important at low magnetic fields. In Fig. 5 the dependence is slightly stronger than p; for other operating parameters the dependence is closer to linear. Finally, in V_a versus J (Fig. 6) the vertical part of the curves (i.e., high V_a) is

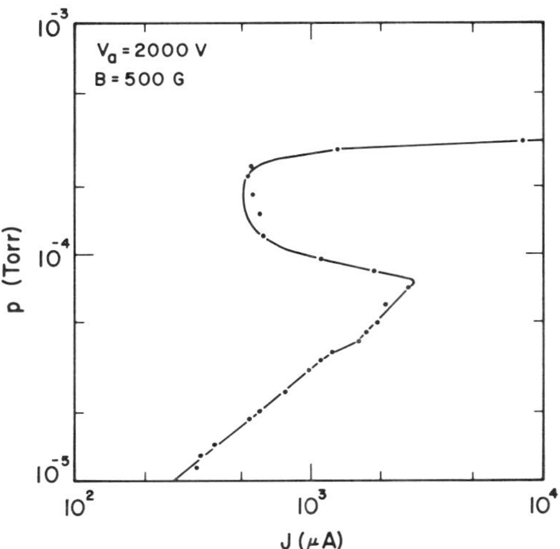

FIG. 5. Pressure versus current for a cold cathode discharge. The straight line is Mode I; the discharge then shifts through Mode III ($p \sim 10^{-4}$ torr) into Mode IV. (From Schuurman, *131*.)

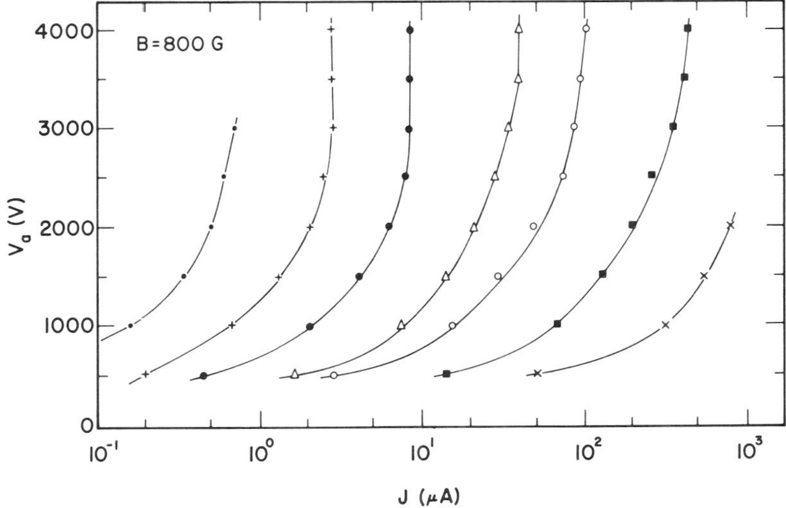

FIG. 6. Current versus voltage for a cold cathode discharge. Mode I is at high voltage ($\gtrsim 2000$ volts), mode II at low voltage. The different symbols represent pressure of 3×10^{-8} torr (lowest current), 10^{-7} torr, 3×10^{-7} torr, 10^{-6} torr, 3×10^{-6} torr, 10^{-5} torr, and 3×10^{-5} torr (highest current). (From Schuurman, *130*.)

Mode I; as predicted by Eq . 7, at fixed B and p the current is independent of applied voltage. The other features of these curves are discussed later.

When the short anode of Schuurman's discharge is considered, the results of the experiment are in surprisingly good agreement with such a simplified theory. Thus, although several points can be cleared up (notably the effects of axial dependencies and a definitive study of ionization effects), we may conclude that the physics of this mode is reasonably well understood.

Several further points are of interest. First, simple analysis of electron trajectories in a magnetic field and a cylindrical electron cloud of constant density indicates that cycloid motion is stable only for $\omega_c^2 > 2\omega_p^2$; for densities that violate this condition, the electron is pushed out to the anode. This condition may be written as

$$n < \frac{\varepsilon_0 B^2}{2m} \tag{8}$$

However, this is greater than the density in the discharge (Eq. 5); Schuurman's experimental results indicate a factor of 4 to 5, justifying the assumption used in Eq. (4). Such an instability may have been observed by Poliak (120) in a discharge with a small heated cathode (on axis) surrounded by a cold plate. The assumption that the electron cathode current is proportional to γv_i is not valid in this discharge. Poliak observes a mode transition when Eq. (8) is violated. Similar results were observed by Hirsch (64), who injected electrons from an auxiliary filament into a Penning geometry at a pressure of 10^{-9} torr. He found electrons reaching his cathodes when Eq. (8) was not satisfied, which he attributed to acceleration by an unobserved instability. These results have doubtful applicability to an active discharge.

A second point of interest concerns the axial electron motion. Kucherenko and co-workers (96–100) point out that although the orbits of electrons emitted from the cathodes generally become effectively randomized by $E \times B$ drifts after a few transits of the discharge, if a discrete number of rotational orbits occur in a transit the motion in succeeding transits is effectively coherent. When this is coupled with a parabolic radial potential, both radial and axial motions of the electrons are harmonic, and the motions of particles emitted at different radii are correlated. One would expect this to be important if the orbit radius were comparable to the discharge radius or, roughly speaking, for

$$\frac{1}{RB}\left(\frac{2mV}{e}\right)^{1/2} \sim 1 \tag{9}$$

In Schuurman's discharge ($130, 131$), the left-hand side is of the order of 0.1 and one would expect a small effect, but in Kucherenko and Saenko's

(*98–101*) discharge it is 0.4 and a nonnegligible effect might be expected. They find the effect to be more pronounced, the larger the left-hand side of Eq. (9), which is consistent with the presentation given above.

Because the satisfaction of Eq. (9) implies a low electron density (electrons escape in one collision time), the discharge potential is close to the vacuum potential. For a potential parabolic along the axis as well as radially, which is found if the anode length is fairly short (Ianova *et al.*, *75*), the electron motion is harmonic both radially and axially. One easily solves the equation of motion to predict the effect to occur at

$$B_n = 4\left[\frac{2m}{e}\left(\frac{V_0}{L^2}\,n^2 + \frac{V - V_0}{4R^2}\right)\right]^{1/2} \qquad n = 1, 2, \ldots \qquad (10)$$

where V_0 is the voltage in the discharge center. Kucherenko *et al.* (*98–101*) observed the discharge current to oscillate with the magnetic field in excellent agreement with Eq. (10), as shown in Fig. 7. The dependence on L is also

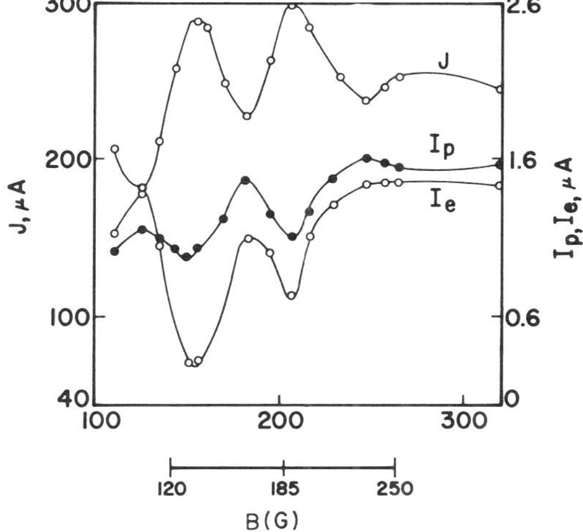

FIG. 7. Current versus magnetic field for a cold cathode discharge, showing the orbit resonances. Resonances are predicted at 120, 185, and 350 gauss. I_e and I_p are the electron and ion currents passing through a hole in the cathode. (From Kucherenko, *100*.)

verified. Similar effects were observed in both cold and hot cathode discharges. The resonance conditions are the conditions under which electrons emitted from one cathode can reach the other, and Knauer (*85*) has pointed out that if this occurs, it will have a strong effect on the discharge; such an effect is observed in the figure.

Dow (*47*) observed a transition from Mode I to Mode II which showed considerable pressure dependence. As his cathodes were constructed of grids to permit an electron beam to pass through the discharge, it is difficult to know to what extent comparison with other work is valid. In particular, secondary emission phenomena may have been quite different from those with solid cathodes, resulting in density and potential distribution differing from the usual.

Phillips (*117*) observed an instability in a discharge operating in Mode I. He did not report the frequency of oscillation, but it was probably low as it was present in the current. There is insufficient evidence to attempt to identify the nature of the instability.

2. *Mode I–Mode II Transition*

The transition from Mode I to Mode II occurs when the potential in the discharge center reaches the cathode voltage. From Eq. (6) we see that this occurs for

$$V_T = \frac{eB^2R^2}{4m} \frac{v_i(1 + \gamma)}{v_c} \tag{11}$$

As long as the electron distribution is independent of pressure, we therefore expect V_T to be independent of p and proportional to B^2. Schuurman (*130, 131*) takes the peak currents to be the transition point (Fig. 4) and verifies both predictions; for the magnetic field result, see Fig. 8. The R^2 dependence has

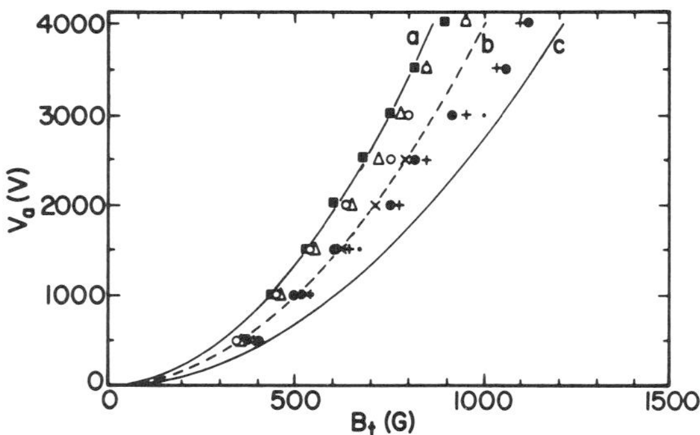

FIG. 8. The transition magnetic field. The curves are Eq. (11) with $v_i(1 + \gamma)/v_c$ equal to (a), 0.12, (b), 0.09, and (c), 0.06. The symbols represent pressures ranging from 3×10^{-8} torr to 10^{-5} torr. (From Schuurman, *131*).

not been checked. It may be noted that if the value for $v_i(1 + \gamma)/v_c$ found in this manner is used in the formulas listed for Mode I, excellent agreement with experiment is obtained; the values are generally about 0.1.

3. *Mode II*

In Mode II the discharge voltage appears across an anode sheath of thickness less than the discharge radius, so the potential profile has the shape illustrated in Fig. 3b. This mode has been studied extensively by Knauer and co-workers (*85–89*), who measured the potential profile by means of the Stark shift (*87*). Dow's (*47*) measurements using an electron beam provided qualitative information yielding the same result. One attains an approximate solution to the profile in this mode by using Eq. (3) outside the radius r_0, at which the sheath drop equals the discharge voltage; note, however, that centrifugal force, diffusion, and Larmor radius effects will be more important than in Mode I because of the smaller sheath thickness. Popov (*121*) has considered diffusion effects, but his assumptions are somewhat arbitrary and will not be reproduced here.

Mode II is often unstable to the diocotron instability, but for the present we consider a stable model. This can be justified in part by the evidence that the cross field transport is of the magnitude of that predicted classically (*85, 87, 89*) even in the presence of the instability. Instability effects will be discussed later.

Hirsch (*65*) inserted a plane sheet through a slit in the anode, parallel to the discharge axis. Because this intercepts the rotating anode sheath, he claims that the resulting discharge behavior indicates the presence of anomalous transport. However, the sheet changes the discharge character so completely that the conclusions are suspect. Hirsch also observes electrons reaching the cathode centers; these are attributed to acceleration by oscillations, but as the discharge center is close to cathode potential, little acceleration is necessary.

Assuming classical transport and noting that the axial electric field inside a long conducting cylinder is zero, we find a constant electron density outside r_0:

$$n = \frac{\varepsilon_0}{m} \frac{v_i(1 + \gamma)}{v_c} \tag{12}$$

The potential is given by

$$V = 0, \qquad r < r_0 \tag{13a}$$

$$V = \frac{en}{2\varepsilon_0} \left[\frac{r^2 - r_0^2}{2} - r_0^2 \ln \frac{r}{r_0} \right], \qquad r > r_0 \tag{13b}$$

and the value of r_0 may be obtained from

$$V_a = \frac{en}{2\varepsilon_0} \left[\frac{R^2 - r_0^2}{2} - r_0^2 \ln \frac{R}{r_0} \right] \tag{14}$$

The current depend on the value of r_0; for large values of r_0 one finds

$$J = \frac{4\pi Le}{\varepsilon_0} v_i(1 + \gamma)V_a \tag{15}$$

For small values of r_0, Eq. (14) is transcendental and will not be solved here; but it is easy to see that J is predicted to increase monotonically with V_a.

These results are surprisingly well satisfied by experiment, especially when one considers not just the main assumptions but also the existence of instabilities. The current is approximately independent of B and proportional to p and V_a as predicted (Figs. 4, 6, 9). The peak and ripples in the J/B diagram

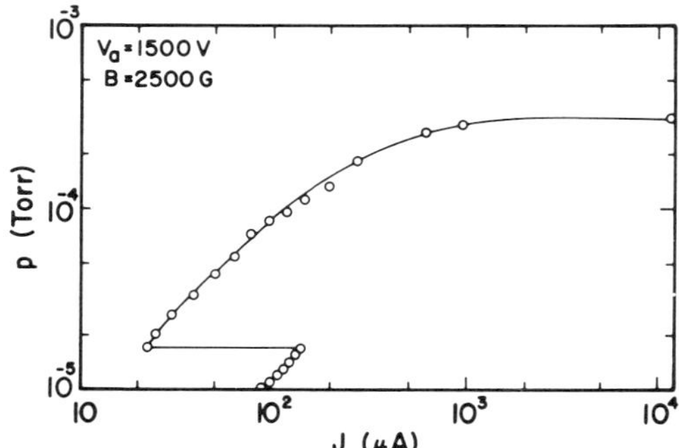

FIG. 9. Current versus pressure for a cold cathode discharge. Mode I occurs at low pressures; the mode switches to Mode II and then Mode IV as the pressure rises. (From Schuurman, *130*.)

are thought to be associated with different modes of instability.

Knauer (*85*) has examined the cathode sputtering pattern that is observed in a discharge operating in this region. It is observed that the sputtering occurs on a ring with a slightly sputtered center section and a diffuse outer boundary. He ascribes this pattern to the fact that ionization occurs only in the anode sheath, so that the ion trajectories all originate near the anode. The typical trajectories shown in Fig. 10 clearly will miss the discharge axis. Knauer ascribes the outer boundary as representing the radius at which ions have the minimum energy for sputtering. Measurement of the boundaries is in quantitative agreement with this model, as is the energy and angular distribution of ions collected through a hole in the cathode. It is interesting to note that some of Knauer's results suggest that ions are often reflected from the cathode rather than absorbed by it, and on the average may traverse the discharge several times before being collected.

Kurbatov (*102*) studied the ions passing through a slot in the cathode and found that on the discharge axis they are highly collimated and that they

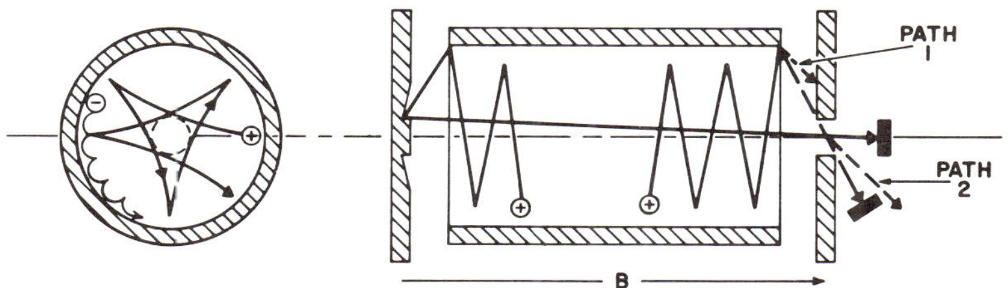

FIG. 10. Particle orbits in Mode II, demonstrating the origin of the sputtering pattern. (From Knauer, 85.)

diverge from the cathode normal if off axis. This is in agreement with the conclusions of Knauer and has obvious consequences for ion guns. See also Smirnitskaya and Babertsan (*133*).

Helmer and Jepsen (*63*) obtained complicated sputtering patterns in noncircular discharges. They suspect that this arises from instabilities with spatial modes of corresponding geometry, but it is possible that the static fields could explain the effect.

The discharge is unstable to the diocotron instability if the anode sheath becomes thin enough (see the Appendix A). Oscillations have been observed by Knauer *et al.* (*89*) by electromagnetic coupling to the plasma through a small hole in the anode, and by Conn and Daglish (*38*) with a small antenna placed close to the discharge. Kühn (*103*) apparently also observed the oscillations. In discharges with split anodes the instability is observed as an oscillating voltage between the plates (*63, 84, 85*); if sufficiently low power is coupled out, the split anode apparently does not affect the oscillations (*89*). That the instability is diocotron in nature has been shown by Knauer and Poeschel (*88*), who compare the conditions for onset with theory.

4. Transition to Mode III

The transition to Mode III is generally considered to occur when the electron loss time becomes comparable to the ion loss time and the discharge becomes quasineutral. There is no quantitative theory predicting this; the best one can do is to note that the electron loss time is

$$t_e = [v_i(1 + \gamma)]^{-1}$$

and is thus proportional to p^{-1}. As the ions stream freely to the cathodes, their loss time has no direct pressure dependence. Experimentally, the transition depends only on the pressure (*130, 131*), which is in agreement with the above picture; the transition generally occurs in the 10^{-5} to 10^{-4} torr range. Mode III is unstable, and the transition may well indicate the onset of the instability rather than the establishment of charge neutrality.

5. Mode III

This mode has received very little attention, although it has often been noted in ionization gages where it manifests itself as the drop in discharge current shown in Figs. 5 and 9. The mode is unstable to low frequency oscillations.

The discharge potential drop occurs partially along the axis and partially radially (*47*); apparently the drop in radial field causes the current drop (*130, 131*). The nonthermal character of the ion and electron distributions make this possible, as discussed in Appendix B.

Dow's electron beam studies indicate that the instability is rotational in nature, suggesting that it may be a variation of the neutral drag instability often observed in thermionic cathode discharges (*cf.* Appendix A). The oscillating fields are very large: Knauer *et al.* (*87*) observe a Stark broadening in the discharge corresponding to 4000 volts/cm. These fields can accelerate electrons sufficiently for them to reach the outer rims of the cathodes (*64*).

Kaganskii *et al.* (*78*) apparently operated in this mode. They found that the oscillation frequency was proportional to V/B, suggesting a rotational mode. However, when they split the anode, the signal on the halves was in phase, indicating that the mode was azimuthally symmetric or had an even wave number.

6. Transition to Mode IV

For a pressure in the vicinity of 10^{-3} torr, the discharge changes to a mode in which most of the discharge potential appears as a cathode sheath (*47*). There is no adequate theory to describe the change, but one may hypothesize that it occurs when the ion mean-free path becomes comparable to the discharge length. Under such a condition, large electric fields along the magnetic field may no longer be maintained. The nature of the discharge instability apparently changes, however, and it may play a role in the transition.

7. Mode IV

In this mode. electrons are emitted from the cathodes and accelerated across the sheaths. As the sheath drops are in the 100- to 1000-volt range, the discharge contains two interpenetrating beams. This is a very unstable situation, and although the instabilities have not been studied in detail, they are probably identical with those observed in geometry (b) where both ion and electron oscillations have been observed. Batten *et al.* (*12*), who studied oscillations in a discharge with cathode diameters small compared with anode diameters, observed frequencies in the range 1 to 10^3 mHz whose characteristics are consistent with such beam-plasma instabilities.

Backus (7–9) and Backus and Huston (10) have studied the average characteristics of the discharge. A typical description of their discharge is given in Table I.

The cathode material is observed to affect the discharge voltage (7, 8, 59). Aluminum and beryllium, which have stable oxide coatings, operate at relatively low voltages. They also resist sputtering, but the coating will eventually be removed and the discharge voltage rise.

Laffineur and Pecker (104) and Bayet and Dumas (13, 51) made brief studies of the emission of radiation at the cyclotron frequency from this mode.

8. Transition to Mode V

This mode is not of much interest, since it is actually just a glow discharge. Roughly speaking, a discharge enters this regime when the magnetic field loses its effectiveness; which is to say, when the electron collision and cyclotron frequencies are comparable.

C. Geometry (b)

The dependences on discharge parameters of the modes in this geometry have not been studied in sufficient detail to permit the drawing of a mode diagram. Generally speaking, however, the discharge behavior is very similar to that of geometry (a). Brachet and Vasseur (25) observe a transition at a pressure of about 6×10^{-5} torr, which appears quite similar to the transition to mode III in geometry (a), although the nature of the discharge is not established in detail.

Strong interest has arisen concerning the several instabilities that occur. These instabilities have three primary forms: beam-plasma oscillations, cyclotron harmonic radiation, and a rotational mode. We discuss these effects below. As with geometry (a), the discharge is quiescent at high pressures and low magnetic fields; Klan (83) reports that there are no observable oscillations at $p > 4 \times 10^{-2}$ torr in helium and $B < 500$ gauss. These values are surprisingly low, since the collision frequency is much less than the cyclotron frequency.

1. Beam-Plasma Instabilities

These occur in a mode corresponding to IV of Fig. 2. Most of the discharge voltage appears across the cathode sheath; electrons accelerated across these sheaths form interpenetrating beams. The resulting oscillations fall into

TABLE I. Typical Parameters of Some Cold Cathode Arcs

Author	Pressure (torr)	Voltage (volts)	Current (amperes)	Magnetic Field (gauss)	Dimensions (cm)	Electron Density (cm^{-3})	Temperatures (eV)	Oscillation frequency
Briffod *et al.* (*27, 29*) Bonnal *et al.* (*22*)	Ar, H_2, N_2 7×10^{-4}–10^{-2}	400–800	0.2–3	2000	90 long, 3–5 diam	5×10^{10}– 5×10^{11}		5–20 Hz 100 MHz — f_c
Pavlichenko *et al.* (*113, 115*)	5×10^{-4}–10^{-2}	<2000	0.5–3	<3400	70 long, 3.6 diam	10^{11}– 3×10^{12}	$T_e = 15$–60	5–1000 Hz 80–200 kHz 0.3–4 MHz
Backus (*7–9*) Backus and Huston (*10*)	A, 6×10^{-4}–3×10^{-3} N_2, 9×10^{-4}–4×10^{-3} He, 11×10^{-3}–55×10^{-3} H_2, 3×10^{-3}–30×10^{-3}	1000	0.5	<3500	15 long, 2.5 diam	2×10^{13}	$T_e = 2$–10 $T_i = 0.5$	
Dubovoi and Popov (*49*)	10^{-3}	3000	100 (pulsed)	600–1000	40 long, 4 diam	1–6×10^{13}	$T_e = 50$–70 (unstable) $= 3$–5 (stable)	10–100 kHz and 1–30 MHz
Agdur and Ternström (*1*) and private communication	Air, 1–10×10^{-3}	400–10,000	10^{-3}–10^{-2}	2500	3–20 long 1.2–2.1 diam	10^{12}		0.1–3 GHz
Bliman (*19*)	He, 10^{-3}	700–1800	$<2 \times 10^{-2}$	200–700	25 long, 1 diam		$T_e = 1 - 6$	0.5–3 GHz
Thomassen (*142*)	5×10^{-4}	400–900	5×10^{-2}– 4×10^{-1}	<1000	126 long, 3 diam	10^{10}–10^{11}	$T_e = 2$–4	0–10 Hz 0.01–1 GHz
Klan (*83*)	He, 10^{-3}–10^{-1}	500	0.15	500	58 long, 5 diam	10^{12}	$T_e \gtrsim 1$	1 kHz

rf oscillations (0.5–20 MHz) and oscillations at higher frequencies (100 to 3000 MHz). Both oscillations have strong effects on the discharge, which becomes essentially a beam-plasma discharge in which the transfer of energy from the power supply is via the electron beam; the beam excites strong oscillations and in turn heats a background plasma to cause ionization. The oscillations come in pulses with lengths and repetition times of the order of 1 to 10 μsec. During the bursts, the discharge potential drops rapidly and the beams from the cathodes are destroyed. The large increase in electron density during the pulse (27) contributes to the increase in discharge conductivity. Although the instabilities excited in this discharge do not directly lead to enhanced particle radial transport, as do those in the hot cathode discharge (see below), the common beam and oscillation axial velocities should lead to strong resonant particle effects (48, 135).

The discharges may be divided into those for which the rf oscillations are important, and those for which they are not. Vhf oscillations are usually found in the first class along with rf oscillations; the relative importances of the two are not clear and probably vary from discharge to discharge. We consider discharges falling into the two classes that follow:

a. *Radio frequency oscillations dominate.* A number of rf oscillation discharges are described in Table I. The discharge noted by Briffod and co-workers (27, 28) and by Bonnal *et al.* (22) is the one most thoroughly studied; some of their results are illustrated in Figs. 11 and 12. The spectrum of emit-

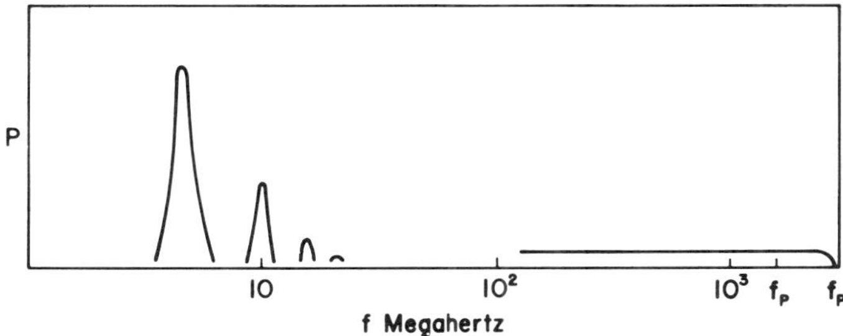

FIG. 11. Power spectrum generated by discharge. (From Briffod *et al.*, 1964.)

ted power is shown in Fig. 11. Note that both rf and vhf oscillations were observed. The rf oscillations are standing waves along the discharge column, with frequencies $f = nv_b/2L$, where $v_b = (2\,eV_{\mathrm{arc}}/m_e)^{1/2}$, L is the discharge length, and $n = 1, 2, \ldots$. The dispersion relation for ion acoustic waves is examined and it is concluded that such oscillations are excited by the beam. The instability is excited at a critical field (Fig. 12), at which point the radial voltage drop (which is always $\gtrsim 0.1\ V_{\mathrm{arc}}$) begins to decrease. The ratio of density at a point just exterior to the anodes to that on the axis also increases,

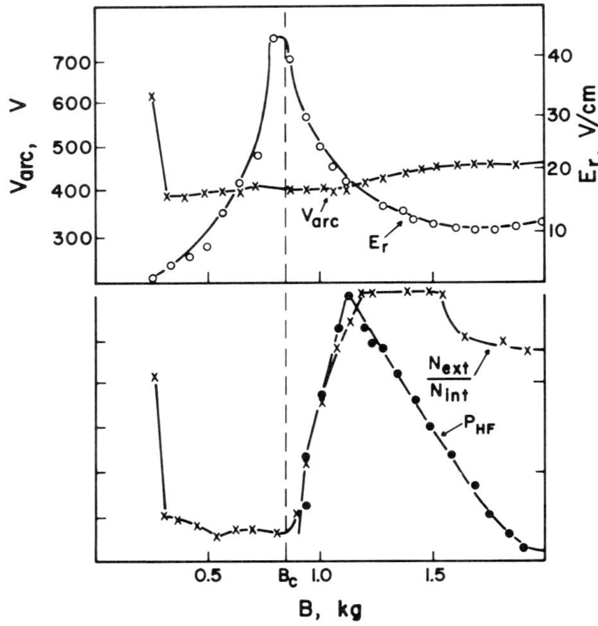

FIG. 12. Top: Variation of discharge voltage V_{arc} and radial electric field E_r with magnetic field. Bottom: Figure indicating the onset of noise with magnetic field and its relationship with the change in the density ratio at the critical magnetic field. N_{ext} and N_{int} are the plasma densities at radii greater than and less than the cathode; P_{HF} is the high frequency power. (From Briffod *et al.*, *29*.)

as illustrated in the figure. A grid placed across the arc does not affect the oscillations. The pulse rate of the oscillations, which is approximately 10 μsec, is observed to be a function of magnetic field and gas pressure; for details the reader is referred to Briffod *et al.* (*27, 28*).

Radio frequency was apparently present in the discharges investigated by Pavlichenko *et al.* (*113–115*), who conclude that the excited frequencies are proportional to the square root of the electron temperature as expected for ion acoustic waves. It is unclear why the length did not determine the frequencies in this experiment. They observed a very small radial electric field (0.01–0.1 volts/cm) with an applied arc voltage of 1000 volts! The plasma was unstable to a rotational instability, discussed later, and the authors claim that this rotating mode is responsible for the enhanced radial transport.

b. Vhf and other oscillations dominate. If the discharge is short enough, ion acoustic waves cannot resonate in the column, but the vhf oscillations can. In such cases, strong standing wave patterns may be excited if the discharge parameters are properly adjusted, although at arbitrary parameters the pattern is generally not so sharp. The waves are excited only at frequencies below the electron plasma or cyclotron frequency, whichever is lower, suggesting that they are on the whistler branch of plasma waves.

The standing-wave pattern observed by Agdur and Ternström (1) is shown in Fig. 13. The oscillations occur at discrete frequencies, corresponding

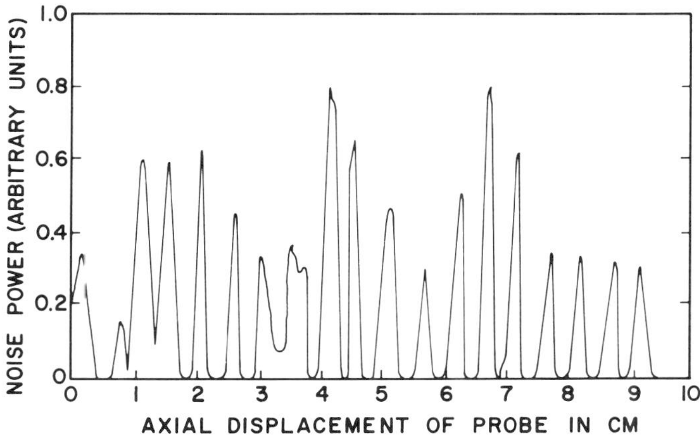

FIG. 13. Standing-wave pattern measured in tube. (From Agdur and Ternström, 1.)

to $f = nv_b/2L$, as in the acoustic wave case. The discharge was operated at 400 to 2000 volts, 1 to 100 mA, 2500 gauss, and 10^{-3} to 10^{-2} torr of air. Discharges were typically 12 cm long and 1 cm in diameter. Low frequency waves (>20 kHz) were present, but apparently not important. Agdur and Ternström (private communication) have also operated very high voltage discharges (10,000 volts at 10 mA) and by surrounding the arc with a helix have coupled out over 40% of the input power as vhf power.

Bliman et al. (19–21) studied a discharge in which the anode is a microwave cavity, resonant at 3020 mHz in the absence of the plasma. This discharge operated in He at 10^{-3} torr, voltage between 700 and 1800 volts, and less than 20 mA current. The discharge was 25 cm long and 1 cm in diameter. They measured an electron temperature of 1 to 6 eV, and observed that the density, temperature, and floating potential varied spatially with the vhf power, as shown in Fig. 14.

Thomassen (142) studied these oscillations, noting that at high power levels the condition $f = nv_b/2L$ is violated. He claims that a low frequency instability also present in his discharge causes the enhanced radial transport, and that there appears to be no direct association with the high frequency oscillations. The pulsed arc observed by Dubovi and Popov (49; Table II) had characteristics of both hot and cold cathode discharges. They observed that all oscillations, including the ion acoustic waves, were greatly reduced in

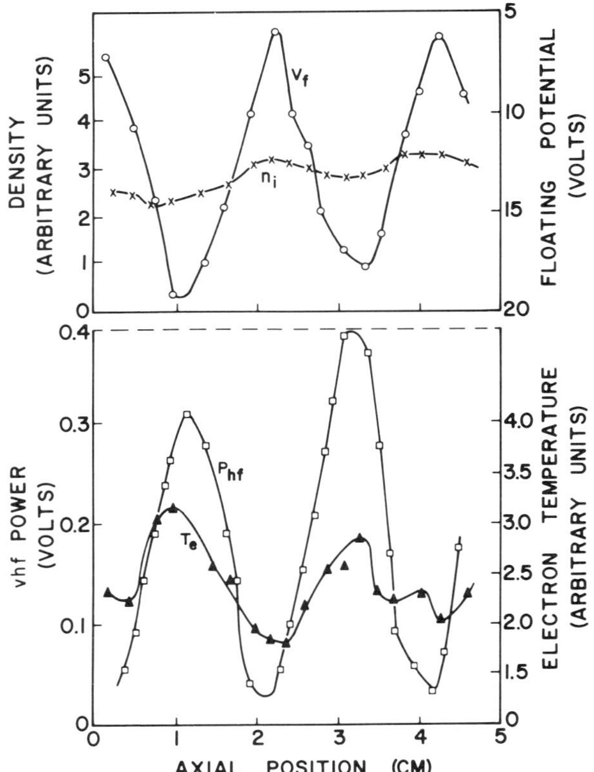

Fig. 14. Axial modulation of the plasma characteristics: □, high frequency oscillations; ○, floating potential; ×, electron density; ▲, electron temperature.

amplitude in a minimum magnetic field configuration. As it is difficult to understand why acoustic waves should be so reduced, it would be of interest to apply a minimum field configuration to a steady state arc to see if the effect occurs there also.

2. Cyclotron Radiation

Strong oscillations at the harmonics of the cyclotron frequency are observed in cold cathode discharges (*105, 106, 114*). The fast electrons excite the Bernstein cyclotron harmonic modes and as many as 45 harmonics have been observed. Radiation from these modes has been observed in many types of discharges; *cf.* Crawford (*40*) for a review. The radiation does not appear to be strong enough to have a large effect on the discharge, although experiments have not been directed to this point. Waniek *et al.* (*145*) observed strong cyclotron radiation from a high power (20 kV, 90 kA, pulsed 10 μsec) discharge in a magnetic mirror, apparently due to extremely hot electrons (~ 100 keV) formed in the arc.

3. Rotational Mode

Several authors have observed a rotational oscillation ($m = 1$, $f \approx$ 1–100 kHz), often coexisting with other oscillations (*113, 114, 26, 142*). The mode may exhibit a coherent oscillation or a broad "hashy" spectrum. Anomalous diffusion in the plasma is usually attributed to this mode rather than to other instabilities. The rotating "rod" has no axial dependence and contains a strong temperature maximum. Klan (*83*), working at 2×10^{-2} torr, measures a temperature variation from 0.2 eV outside the "rod" to 1.2 eV within. He also claims that cyclotron harmonic emission is strongly correlated with the rotation. The nature of the mode is unclear: it may be a form of the neutral drag instability (see Appendix A) or it may be some other form of the drift instability (*26*).

D. Geometry (c)

These discharges have not been studied so extensively as those with cylindrical anodes, although the discharges studied by Penning and co-workers were of this type.

Photographs by Kreindel and Ionov (*95*) offer the best evidence of the mode character. As may be seen in Fig. 15, one can distinguish two modes; one in which the glowing plasma does not touch the cathode and one in which

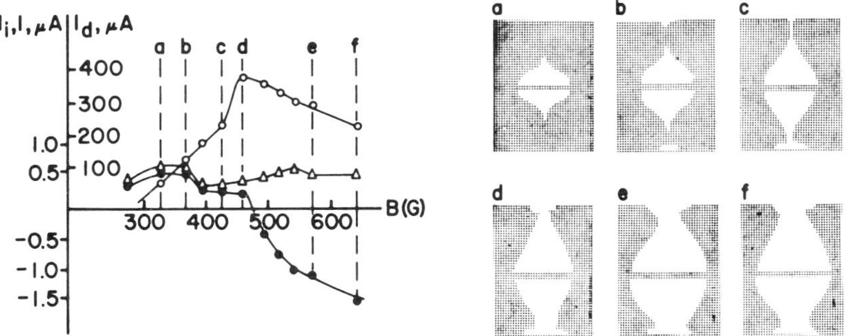

FIG. 15. Photographs and *I–B* plots indicating a mode change in a ring anode discharge. I_d (O) is the total current, I_i ($\triangle$) and I (●) the ion and total currents through an anode hole. The pictures show the visual appearance of the discharge for the magnetic fields indicated by the corresponding letters (a–f) on the graph. (From Kreindel and Ionov, *95*.)

it does. The first mode is associated with low magnetic field or high pressures (but not too high, of course), and the other with high field and low pressures; the shape of the latter is not stable, but pulsates in time. The discharge current may be observed to increase with the magnetic field for low fields and then decrease again after the transition field. This behavior is very reminiscent of the behavior of the discharges with cylindrical anodes, although the strong axial dependence and the photographic evidence suggest that the details of the transition mechanisms must differ.

Reikhrudel *et al.* (*122, 125, 123, 126*) have studied the ions passing through a hole in the cathode and related this to the potential distribution in the discharge. They find modes that appear to correlate with those of Kreindel and Ionov (*95*). They find the discharge is highly negative at low pressure, but becomes space-charge limited at pressures near 10^{-4} torr. This change apparently agrees with the observations of Kreindel.

Kreindel (*92*) and Kreindel and Ionov (*95*) have studied current emitted through a hole in the cathode. In the high magnetic field mode they find that the current collected by a Faraday cup is electronic. However, as the net current approaches zero as the collector approaches the hole, it appears that a quasineutral plasma is flowing through the rather large hole (3.25 mm diameter). Because the ions have large Larmor radii, they will be lost from the beam before the electrons so that the net current will become electronic away from the hole. See also Vasileva and Reikhrudel (*143*).

Very high frequency oscillations have been observed in ring anode discharges (*69, 124*); these oscillations appear to be beam-plasma instabilities with properties similar to those observed in geometry (b).

III. Thermionic Cathode Discharge

If the cathode is a copious source of electrons, ion bombardment is not required to supply the discharge current, and the discharge voltage may have a considerably different volume distribution.

Low pressure discharges ($p < 10^{-4}$ torr) have not been studied extensively, presumably because of the possibility of cathode damage due to ion bombardment. At low pressures the discharge contains an electron cloud, just as in the cold cathode discharges (*73, 74*). In addition, electron densities higher than in the cold cathode case are possible; the resulting electric field has a radial dependence that must be taken into account in analyzing the particle orbits (*120, 64, 65*).

In discharges with ring anodes (geometry c) considerable axial electric field can exist because of nonthermal particle distributions. The effect is described in Appendix B. Meyerand *et al.* (*110*) and Salz *et al.* (*129*) studied the field and developed a computer technique to predict it. The problem is not a simple one because the ionization is coupled strongly to the field distribution and quantitative results apparently require numerical work. For an example of their results, see Fig. 16.

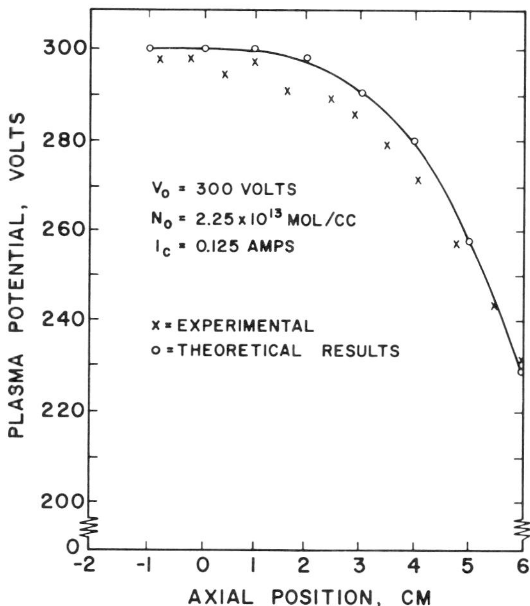

FIG. 16. Comparison of experimental and theoretical plasma potentials in the Penning discharge. (From Salz, *et al.*, *129*.)

The major interest in the hot cathode reflex discharge has arisen from the unstable character of discharges with geometry (b) and pressures greater than about 10^{-3} torr. Because oscillations in these discharges cause considerable enhanced transport, they have been studied by several people. The oscillations may be coherent, turbulent, or in between, and thus the discharge offers a number of possible transport mechanisms. The different regimes are determined by the magnetic field, pressure, current, etc.

In discharges with space-charge limited cathodes, the plasma potential is tied strongly to the cathode potential. Chen (*30*) has studied the time independent characteristics of arcs in which the cathode diameter equals the discharge cross section and obtained the potential, density, and temperature profiles shown in Fig. 17. Note that the axial plasma potential is only a few volts above the cathode; the sheath potential then increases radially. Axial variations (outside the sheath) are small. There is a minimum in density on the

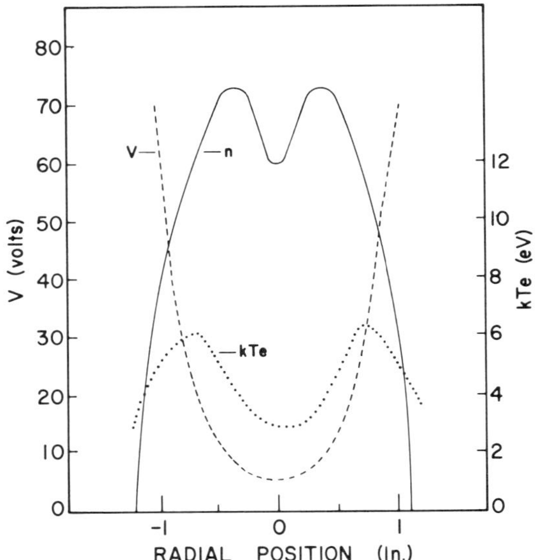

FIG. 17. Typical radial profiles of potential, density, and electron temperature in a reflex arc with thermionic cathode. The density scale is uncalibrated. (From Chen, *30*.)

axis because some electrons have sufficient energy to reach the cathode from the discharge. The effect is strongest on the axis where the sheath potential is low.

The combination of the radial density variation and the radial electric field drive oscillations that are believed to arise from a drift instability which is sometimes called the "neutral drag" instability (*132, 66, 67, 16, 17, 31, 32*). The experimental evidence that this is the correct instability is strong but not completely conclusive. The nature of the instability is discussed in Appendix A. The reflex arc usually oscillates with an azimuthal variation of $m = 1$ or $m = 3$. With one exception (*79*; see below), the oscillations show no phase change along the axis, in agreement with the theory that predicts that the mode with $k_{\parallel} = 0$ should have the fastest growth rate.

Several pieces of evidence indicate that the instability is the neutral drag one. First, the instability has critical magnetic fields (Fig. 18) of the same order as those predicted by the model (*139, 52, 43*). Second, Chen and Cooper (*35*) have shown by axial correlation methods that the instabilities are not ion acoustic waves; note that there is no reason to extend this conclusion to cold cathode discharges where acoustic waves appear to be important. Finally, the radial electric field and density gradient clearly drive the instability; either

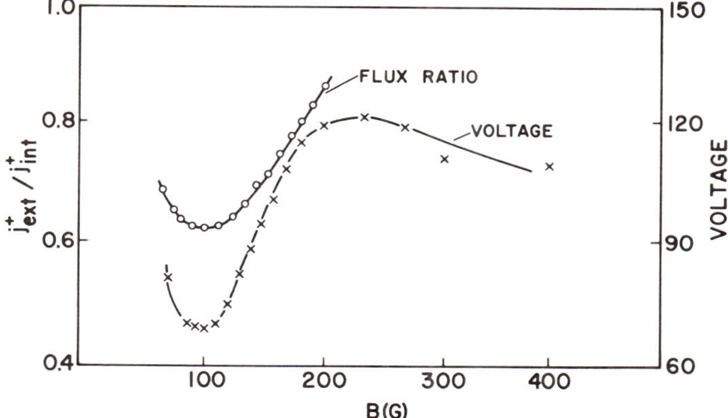

FIG. 18. Ion current collected at the wall normalized to its value at the center, and discharge voltage at constant current, indicating a change in the particle loss rate at 100 gauss. (From Thomassen, *140*.)

grids placed across the arc (Allen *et al.*, *2*) or a minimum magnetic field configuration (*49, 141*) supresses the oscillations. The instability acts so as to reduce the instability mechanism; the reduction in density gradient after onset is clearly indicated by the flux ratio illustrated in Fig. 18.

Although the instability is clearly driven by the radial field and density gradient, its detailed nature is uncertain. A radial electron temperature gradient exists in the plasma (Fig. 17) and presumably should not be neglected in proper theory. In addition, as Chen has pointed out (private communication) the nature of the instability depends strongly on the behavior of the oscillation in the sheath because the electron mean-free path is long and provides a large electric conductivity along the axis. A study of the characteristics of the oscillation in the sheath is thus needed for a complete description, although volume measurements of the correlation between the oscillating potential and density are of considerable assistance (*33*). Such measurements in the reviewer's laboratory indicate that the flow of electrons across the magnetic field in the sheath is small; therefore the lack of axial variation is true there as well as in the discharge volume.

The electric current drawn across the magnetic field in an arc is often greater than can be explained by classical processes. The mechanism for this is not clear, since the simple $E \times B$ instability transports electrons and ions at an equal rate. A difference in the fluxes may arise if the ions are retarded, e.g., by collisions with neutrals or by inertial effects. In addition. electrons whose axial velocities are close to the axial phase velocity of the oscillation effectively see a low frequency oscillation which may drive them from the discharge (*48, 135*), but this effect is apparently precluded by the instability nature. Order of magnitude estimates by the reviewer suggest that the ion-neutral collisions usually dominate, in which case the average radial current density in the discharge is (neglecting all axial effects)

$$\langle j_r \rangle = \frac{e}{1 + \mu_+^2 B^2} \left[\frac{\langle nE_\theta \rangle}{B} - \mu_+ \left(\langle nE_r \rangle + \frac{\partial}{\partial r} \left\langle \frac{nkT_i}{e} \right\rangle \right) \right]$$

In the same approximation the continuity equation becomes

$$\frac{1}{r} \frac{\partial}{\partial r} \left[\frac{r \langle nE_\theta \rangle}{B} \right] = \langle v_i n \rangle$$

Note that we have used the quasineutral approximation, which may not be proper for the calculation of current because it equals the difference in fluxes of the two particles; some experimental evidence suggests that this may be the case (70). Experimental examination of the current flow mechanism is being made in the reviewer's laboratory.

A number of plasmas operating in this mode are described in Table II; corresponding comments are given below.

The discharge investigated by Chen *et al.* (34, 35) had one hot tungsten cathode and a cold reflector (the Princeton L-2 machine). The particle transport across the magnetic field was sufficiently high that the plasma was not symmetric about the midplane of the discharge. The discharge current was quite large, and the resulting oscillation spectrum was very broad, indicating a turbulent nature.

Thomassen's discharge had one heated cathode, consisting of an oxide-coated mesh formed in a cylinder, of diameter slightly less than the anode, and a cold reflecting plate. The noise power of the discharge was estimated to be large enough to cause particle fluxes across the field of the same order as those found in the discharge, verifying the theoretical conclusion that the instability should enhance the plasma transport.

Datlov *et al.* (43) have studied a discharge with one heated cathode (a hot tungsten wire) and one cold one. Their results are in general agreement with those of other authors with regard to discharge parameters and the flute instability. In addition, they observe high frequency (2–30 MHz) oscillations which appear in bursts. They identify these tentatively as products of diocotron instability, but since their characteristics are very similar to those of the oscillations observed in cold cathode discharges, the reviewer suspects that they are the ion waves identified there.

Using a discharge smaller than, but otherwise similar to the one mentioned above, Koons and Fiocco (90) scattered laser light from the electrons. They measured peak electron densities of 4.8×10^{13} cm^{-3}, but more significantly found the electron temperature to be anisotropic. For motion parallel to the magnetic field they obtained a temperature of 30,800°K, and for motion perpendicular, 13,570°K. In both cases the distribution functions were

TABLE II. Typical Parameters of Some Thermionic Cathode Discharges

Author	Pressure (torr)	Voltage (volts)	Current (amperes)	Magnetic Field (gauss)	Dimensions (cm)	Electron density (cm^{-3})	Temperatures (eV)	Oscillation frequencies
Bingham and Chen (17) Chen (30) Chen and Cooper (35) Chen (32)	He, 1–5×10^{-3}	40–500	1–45	1000–4000	300 long, 2.5 diam	2×10^{12}	$T_e = 5$	0–1 kHz
Thomassen (139–141)	H$_2$, 1–3×10^{-3}	50–150	0.1–40.4	50–500	66 long, 4.1 diam	10^{10}–10^{11}	$T_e = 5$ $T_i = 0.5$	20–100 kHz
Swartz and Napoli (137)	Cs, 10^{-4}–10^{-2}	3–15	0.1–2	5000	3.5 long 0.5 diam	10^{14}–10^{15}	$T_e = 1$–3 $T_i = 0.17$	50–250 kHz
Kerr (79)	Ar, N$_2$, H$_2$, 10^{-4}–10^{-2}		<1	50–500	16 long, 1.6 diam	10^{-9}–10^{11}	$T_e = 1.5$–2.5	2–50 kHz (spiral)
Fishkova et al. (52)	Ar, 5×10^{-4}–1×10^{-3}	<500	<10	<2100	105 long, 5 diam	10^{12}–10^{13}		
Allen et al. (2) (Ring anode)	Ar, Kr, Xe, 10^{-3}–10^{-1}		0.5	0–800	5 long, 4.8 diam	10^{11}		
Datlov et al. (43)	H$_2$, 10^{-4}–2×10^{-3}	50–500	1–5	10–1500	100 long 10 diam	10^{11}	$T_e = 2.5$ $T_i = 0.2$–0.5	10–100 kHz 2–30 MHz

approximately Maxwellian. The discharge operated in helium at a pressure of 10^{-2} torr, and a magnetic field of 1000 gauss. Oscillations were not studied.

The anode used in the discharge by Swartz and Napoli (*137*) was heated to 1900°K, sufficiently hot to cause considerable surface ionization of the cesium vapor. Additional ionization was caused by the discharge current, and at peak density (10^{15} cm^{-3}) the cesium was 50% ionized. Oscillations in the 1 to 10 MHz range were observed in addition to those identified as a neutral drag instability, but were not studied in detail.

The discharge examined by Fishkova *et al.* (*52*) had a hot plane cathode and a reflector. They did not look for oscillations, but observed changes in discharge voltage, density profiles, etc., which were similar to those observed in other arcs.

The arc of Allen *et al.* (*2*) was used to demonstrate the stabilizing characteristics of a grid placed in the plasma. Oscillations were not looked for, but the gross characteristics of the discharge were similar to those observed in other arcs.

The discharge studied by Dubovoĭ and Popov (*49*) had cold cathodes but showed characteristics of both hot and cold cathode discharges. Perhaps the cathode sheaths differed from those of steady state, cold cathode discharges because of the high pulsed currents so that sufficient radial electric field existed to excite the neutral drag instability.

Kerr's (*79*) discharge was operated with a reflector and a hot cathode emitting in the temperature limited regime. He observed spiral oscillations rather than the axially independent oscillations observed by others. The reasons for this are unclear, but there may have been considerable axial electric field because of the cathode operation. It is well known that the axial field in the positive column excites a spiral instability [see Hoh and Lehnert (*68*) for a physical explanation], and Guest and Simon (*61*) predicted such an instability for the reflex arc arising from the axial particle drifts. It seems likely that the instability noted by Kerr is basically neutral drag in nature and that the screw arises because of a modification of the instability by an axial field.

Kistemaker and Snieder (*82*) and Gabovich *et al.* (*53*) have also studied discharges with thermionic cathodes. Both references report that the potential difference between the anode and the discharge axis increases with the discharge current, presumably because the voltage is driving the current. In addition, Kistemaker and Snieder point out that negative ions, if formed, could have an effect on the potential distribution; this point should be considered by anyone operating a discharge in air.

Tanaka and Yamamoto (*138*), Nagao *et al.* (*112*), and Berezin *et al.* (*15*) have studied the oscillations. Their observations are consistent with the neutral drag instability.

IV. Additional Discharges and Experiments of Interest

A. Pulsed Discharges

In order to obtain high densities at low pressures, pulsed high current discharges may be used. The discharge of Dubovoĭ and Popov (*49*), described earlier, was a cold cathode arc; for its description see Table I.

Geller and Pigache (*57, 58*), Consoli *et al.* (*39*), and Geller and Hosea (*56*) have constructed pulsed hot cathode discharges. They operated with pressures around 10^{-3} torr and currents of 10 to 100 amperes, and obtained electron densities in the 10^{13} cm^{-3} range. At low fields (up to a few hundred gauss) the discharge is quiescent and can be explained on a classical basis. At higher fields the discharge goes unstable and the electron density drops. The nature of the instabilities is not described, but they are probably of the neutral drag type.

Ayphassorho and Loup (*6*) observed X-ray emission from a high voltage, pulsed discharge. The emission apparently arose from the anode and was correlated to instabilities in the discharge. The X-rays had energies up to about the discharge voltage (100 kV).

Barbian, *et al.* (*11*) studied the growth of a pulsed, high voltage (5 kV) discharge. They considered ionization effects, and by taking time-resolved pictures, observed the buildup of the plasma column.

B. Discharges in Mirror Magnetic Fields

Reflex discharges have been operated in mirror magnetic fields. The arc used by Angerth *et al.* (*4*) had a mirror ratio of 30–50:1 with maximum fields of 5 to 10 kG, and was operated in hydrogen and deuterium at pressures in the range 10^{-4} to 10^{-3} torr. High voltages were used (up to 17 kV); a high and low pressure mode were observed ($p < 10^{-4}$ torr, current of milliamperes; $p > 10^{-4}$, current of amperes). A hot ion plasma (10 keV) was formed, with a density in the cathode region (magnetic mirror) of 10^{10} to 10^{11} cm^{-3}.

Waniek *et al.* (*145*) formed a pulsed discharge at higher pressures (2×10^{-2} torr) in air. They operated in a 2:1 mirror ratio, with an average field up to 10 kG. The discharge voltage was less than 20 kV, and currents were as high as 90 kA. A hot plasma was formed ($T_e \sim 10$ eV) with a large tail out to 100 keV, with densities of about 10^{12} cm^{-3}. They studied the strong microwave power emitted from the discharge, particularly near the cyclotron frequency.

Bottiglioni and co-workers (*23, 24*) operated a discharge in which the magnetic field had several maxima between the cathodes, with a mirror ratio of

2:1. They injected gas near the cathodes and differentially pumped at the minimum of the field, obtaining a gas pressure of about 10^{-4} torr, electron density 10^{11} to 10^{12} cm^{-3}, and electron temperature of 10 eV. The discharge had a rotational instability.

Doucet (*46*) operated a discharge in a mirrorlike field of rather unusual geometry, and observed good plasma confinement and a high electron temperature (30 eV).

C. Afterglow Studies

The reflex arc generates a dense plasma that is very useful for afterglow studies. The toroidal discharge of Geller (*54, 55*) was used for this purpose, as was the arc of Geller and Pigache (*58*). An experiment of Hooper and Bekefi (*71*) used a discharge of unusual geometry, chosen to permit the discharge to be inserted into a laser interferometer. The oxide cathodes were outside the main discharge region, but end caps acted as virtual cathodes during the current pulse (150 amperes). Electron densities in excess of 10^{14} cm^{-3} were generated in argon at 10^{-2} to 2×10^{-1} torr. The studies of the afterglow concluded that the decay of density could be explained as purely classical if the diffusion decay of the electron temperature were taken into account; temperature measurements were not made, however, so the conclusion was not definite.

Grigorenko *et al.* (*60*) studied the afterglow of a pulsed, cold cathode discharge. For sufficiently high magnetic field the discharge was unstable during the pulse; they observed that the oscillations persisted into the afterglow, where they affected the plasma decay.

D. Discharge Response to Applied ac Signals

Crownfield (*41*) and Crownfield *et al.* (*42*) have applied a vhf signal to the anode of a cold cathode discharge. They observe that the discharge current changes by very small amounts when the applied frequency equals an oscillation frequency of the discharge. Similar results were obtained by Pavlichenko *et al.* (*113*), who coupled to a very low frequency instability (*5–50* Hz).

E. Toroidal Discharges

Geller (*54, 55*) and Berezin *et al.* (*14*) have generated pulsed toroidal plasmas with a Penning geometry. Geller's discharge (Bagel) had a "bumpy" magnetic field to help minimize the drift losses in the toroidal geometry and consequently increase the plasma density. The discharge operated in the presure range of 10^{-4} to 10^{-3}, with magnetic fields of several hundred gauss.

The cathode was oxide coated, and emitted current pulsed up to 100 amperes. Ionization as high as 87% is claimed. The loss mechanisms in the afterglow of the discharge were studied, and it was tentatively concluded that anomalous effects were absent.

The discharge of Berezin *et al.* (*14*) was similar to Geller's. Magnetic fields up to 7000 gauss were used, and pressures were in the range of 2×10^{-4} to 2×10^{-2} torr. Two modes of operation were observed during the current pulse: low current and high current. The electron density was higher in the low current part of a pulse (1–10 amperes, $n > 1.7 \times 10^{13}$) than in the high current part (200 amperes, $n < 1.7 \times 10^{13}$). The mode change is tentatively associated with an increase in plasma turbulence during the high current pulse.

V. Practical Applications of the Penning Discharge

The Penning discharge has found several practical uses. It is not intended to discuss these fully here, but references have been collected for the use of the interested reader. In addition to the specific references, we note that Lundgren *et al.* (*108*) have reviewed applications of the discharge, with particular attention to otherwise unpublished work at the University of California (Berkeley).

1. Vacuum gage (*165–176*). The discharge is useful as a vacuum gage if operating in geometry (a), Mode I, for which the current drawn is proportional to the pressure. It has also been used at higher pressures where precise calibration is necessary. If the discharge switches modes during experiments, one must be careful because hysteresis effects can occur. The discharge also acts as a pump so that the pressure measured is below that of the system to which the gage is attached. Literature on the subject is listed under a separate heading in the references; the list is not meant to be complete, especially as many modifications on the basic geometry occur. If a reader desires to study the use of the reflex discharge as a gage, he is referred to any modern book on vacuum techniques.

2. Vacuum pump (*147–164*). The discharge is used as a pump in one of two basic modes of operation: either the ions are accelerated through a hollow cathode region to a mechanical pump or they are absorbed in the cathode. The latter operation has proved extremely successful, especially when used with titanium cathodes or filaments that act as getters. Such "sputter ion" pumps have evolved considerably from the basic Penning configuration; as in the case of the gage, the reader is referred to a modern book on vacuum techniques for discussion beyond that provided under the special heading in the references.

3. Ion source (*81, 177–189, 191–205, 207*). Because of the large ion flux striking the cathode (Backus, *9*), numerous groups have constructed ion sources by placing holes in the cathode centers. The discharges have operated in several of the modes discussed previously, both with hot and with cold cathodes. The highest ion currents result in the high pressure mode, which is partly or wholly stabilized by proper design and choice of parameters. Typical results are 1 to 100 mA in cw operation and several amperes pulsed. The reader is referred to the special section in the references.

4. Electron source. An electron beam may be extracted through anode holes (*93, 94, 190, 206*). The discharge appears to offer no advantage over more conventional electron sources.

5. Plasma source. By placing a discharge with a large hole in one cathode at the end of a drift space, a plasma column may be formed (*62, 5, 72*). Such a column, having densities of 10^9 to 10^{13} cm^{-3}, electron temperatures of 5 to 10 eV, and background pressures of 10^{-4} torr, is used for plasma experiments.

6. Rocket. The thrust provided by the plasma emitted from the discharge is in the proper range for a possible interplanetary ion engine (*109, 44, 80*). The major difficulties are:

(*a*) The engine is relatively inefficient due to anomalous cross-field electron transport (*45*).

(*b*) Because of the magnetic field it is difficult to separate efficiently the ejected plasma from the engine (*80*).

7. Microwave amplifier. Chorney and Madore (*37*), Chorney and Fitzgerald (*36*), and Allen *et al.* (*3*) have shown that in the high pressure modes the discharge is highly unstable and can be used to amplify microwaves. Large amplifications have been reported, but because of the large discharge noise, commerical devices apparently have not been constructed.

8. VHF—microwave source. Because of the large efficiency for generating vhf in cold cathode discharges, the reflex arc holds promise as a wide band noise source (Agdur, private communication); the reader is referred to Section II-C. Practical development of such a source faces many practical difficulties and apparently is not complete. The high efficiency of emission from the discharge of Knauer *et al.* (*89*), utilizing the diocotron instability, also makes this a promising source.

9. High-voltage rectifier. As the low pressure discharge conducts electricity in one direction but not the other, it can be used as a high voltage rectifier (*91, 107*).

10. Amplifier. Because the I–V characteristics have a negative slope over part of the curve, the discharge may be used as an amplifier (Dukes, *50*). Such an amplifier appears to have no advantage over more conventional techniques.

APPENDIX A. IMPORTANT INSTABILITIES IN THE REFLEX DISCHARGE

A number of instabilities have been observed in or predicted for the reflex discharge and are described below.

1. Neutral Drag Instability

In the presence of a radial electric field a plasma rotates with a velocity E_r/B. The basic instability mechanism is a form of the drift instability (*132, 66, 67, 34*), and is illustrated in Fig. 19, where a region of increased plasma

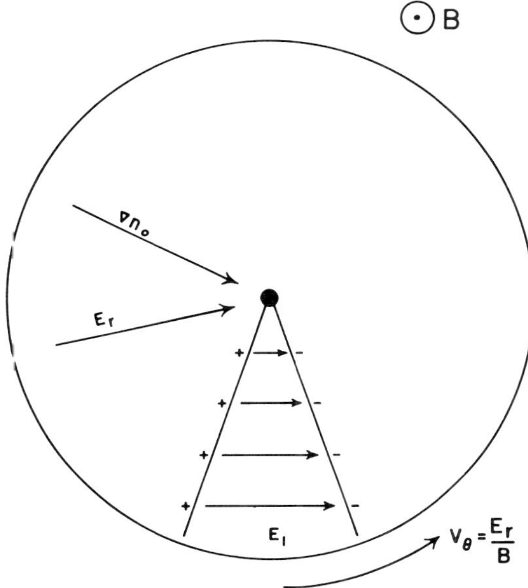

FIG. 19. The neutral drag instability.

density is shown. If the ions lag behind the electrons (say, owing to collisions with neutrals), an azimuthal field E_1 is established. The radial drift E_1/B then transports plasma outward from the dense central region, causing the original perturbation to grow. In an axially uniform plasma the fastest growing mode is axially independent, i.e., flutelike (*66, 67, 16*). Chen (*34*) has considered the importance of the ion centrifugal acceleration due to the rotation, and has showed that it increases the linear growth rate, although collisions are still necessary for instability. Although the theories generally assume that the instability has no radial variation in phase, unpublished experiments by the reviewer indicate that such a phase often exists. Such a variation may be more unstable than the assumed lack of variation (Morse, *111*).

2. Screw

In the presence of an axial field or of unequal streaming of electrons and ions to the ends of a discharge, a screw instability may develop (*61*) whose nature is similar to that of the screw instability in a positive column. The reader is referred to Hoh and Lehnert (68) for a physical description of the latter. It seems unlikely that this instability exists in its pure form in a reflex discharge, but the presence of an axial field may cause the neutral drag instability to take on a spiral form. Such an instability may have been observed by Kerr (79).

3. Diocotron Instability

The diocotron instability occurs in an electron sheet or sheath. A simple form is shown in Fig. 20, where the electric fields resulting from a deformed

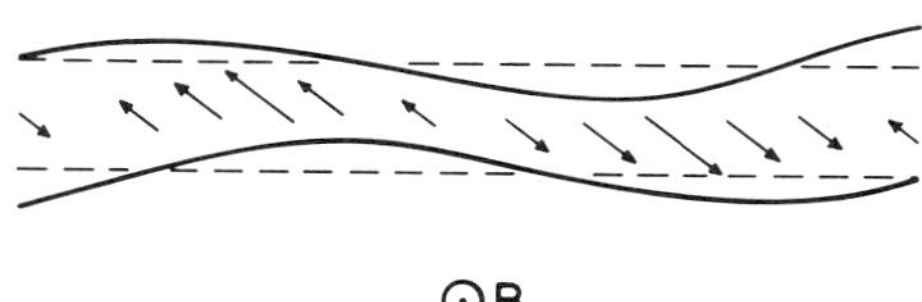

FIG. 20. The diocotron instability. A charge sheet (dotted lines) is deformed into the solid form. The arrows show the perturbed part of the electric field; the resulting $E \times B$ drifts cause the perturbation to grow.

sheet are shown. The directions of these fields are such that the $E \times B$ forces cause the deformation to grow. The frequency of the instability is purely imaginary in a frame stationary with respect to the average velocity of the sheet, and consequently corresponds to the rotational velocity in a rotating discharge. The instability criteria (88, 86) depend on the magnetic field, the discharge length and radius, and the anode sheath thickness in a complicated way; the reader is referred to the references for the results. For a very rough calculation, one may expect instability for

$$T \gtrsim L/5B$$

where T is the anode sheath thickness in centimeters, L the discharge length in centimeters, and B the magnetic field in kilogauss.

4. Beam-Plasma Instability

An electron beam passing through a plasma can excite many types of waves. The condition for maximum growth is, roughly speaking, that the beam velocity equal the wave phase velocity. The important waves that are excited in the reflex discharge appear to be the ion acoustic wave and a high frequency wave which is suspected to be the whistler (helicon) wave.

5. *Electron Cyclotron Harmonic Instability*

A nonthermal electron distribution function can excite many harmonics of the electron cyclotron frequency. These are electrostatic waves (the Bernstein modes) propagating essentially perpendicularly to the magnetic field and have been observed in a large variety of discharges; see Crawford (*40*) for a review of experiments and theory.

6. *"Argon" Instability*

Gas in a closed discharge will be pumped, owing to the absorption of gas by the electrodes. However, if argon is pumped, a pressure instability is observed below pressure of 2×10^{-4} torr, of the nature shown in Fig. 21 (*149*,

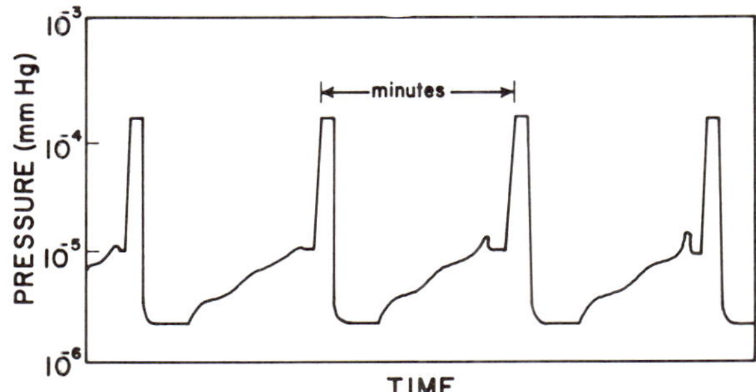

FIG. 21. Typical pattern of pressure versus time for a getter-ion pump exhibiting the argon instability. (From Jepsen *et al.*, *77*.)

77, *155*, *161*). This instability limits the minimum pressure attainable by a reflex discharge in argon (unless a special design is used; see Ref. *161*). Because air contains about 1% argon, a sputter-ion discharge pumping against a small leak may eventually go unstable. The mechanism for the instability is not clear, but it is apparently connected with a change in the pumping mechanism at about 3×10^{-5} torr (*161*) (Rutherford *et al.*, 1960). At pressures above this, high frequency instability (of uncertain character) permits argon ions to reach the anode and be buried there, whereas at low pressures, pumping occurs only at the cathode. It is not clear why this change should lead to the observed instability.

APPENDIX B. ELECTRIC FIELDS PARALLEL TO MAGNETIC FIELDS IN DISCHARGES

Although in a thermal plasma large electric fields can exist only over distances comparable to a Debye length, they can exist over much longer distances in nonthermal plasmas. The phenomena was first pointed out by Rose (*128*). To see the mechanism, consider plasma streaming from the left into a region of dc potential (Fig. 22). For a reasonably dense plasma, even large

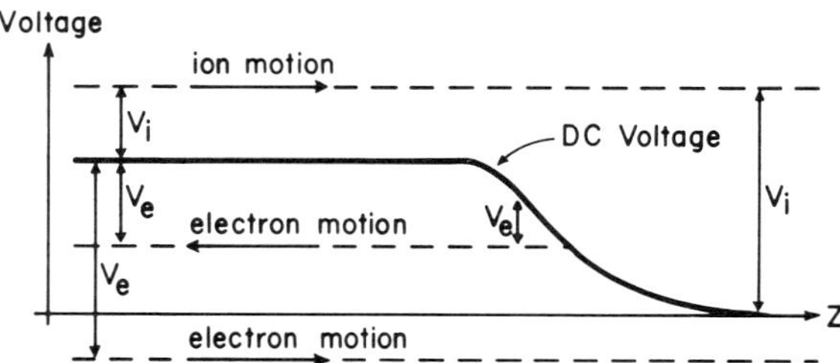

FIG. 22. Electric fields in a nonthermal plasma. Beams of electrons and ions enter from the left with energies V_e and V_i. A spatially varying voltage transmits the ions and some electrons and reflects some electrons. As discussed in the text, it is possible to obtain values of the fluxes such that the plasma is everywhere quasineutral, even though the variation in dc voltage is much larger than expected for a thermal plasma.

potentials represent only a small charge separation, so we may consider the plasma to be quasineutral. Because the ions are accelerated and the electrons decelerated by the potential, the plasma to the right of the diagram will have an excess of electrons unless some of the electron current is reflected from the potential. The equation for quasineutrality may be immediately written down as

$$\int_0^\infty \frac{1}{v_+} \frac{d\Gamma_+}{dV} \, dV = 2 \int_0^{V_0} \frac{1}{v_-} \frac{d\Gamma_-}{dV} \, dV + \int_{V_0}^\infty \frac{1}{v_-} \frac{d\Gamma_-}{dV} \, dV \tag{16}$$

Here, the velocities are

$$v_\pm = \left[\frac{2e(V \pm V_0)}{m_\pm} \right]^{1/2}$$

where V is the particle energy, V_0 the local value of the dc voltage, and $\Gamma_\pm$ the particle fluxes. Given any monotonic potential, values of $\Gamma_\pm$ on the left of the figure may be found such that Eq. (16) is satisfied.

The reflex discharge is more complicated, of course, as electrons are entering the discharge from the cathodes and ionization is occurring. Electrons born somewhere in the discharge oscillate in a potential well until they cause further ionization or are lost radially. Salz *et al.* (*110, 129*) have established a self-consistent scheme for determining the potential; this is clearly a difficult calculation, since ionization phenomena are strongly connected to the potential determination.

REFERENCES

A. General References

1. B. Agdur and U. Ternström (1964), *Phys. Rev. Letters* **13**, 5. Instabilities in Penning Discharges.

2. M. A. Allen, P. Chorney, and H. S. Maddix (1963), *Appl. Phys. Letters* **3**, 30. Stabilization of a Hot-Cathode PIG Discharge in a Magnetic Field.

3. M. A. Allen, C. S. Biechler, and P. Chorney (1964), *Cong. Intern. Tubes Hyperfréquences, 5th, Paris.* Beam-Plasma Amplification for High Power Density Applications.

4. B. Angerth, J. Ehrensvärd, and H. Persson (1965), *IAEA Conf. Plasma Phys. and Controlled Nuclear Fusion Res., 2nd,* Vol. II, p. 901. Production of a Hot Plasma by Discharge in a Strongly Inhomogeneous Magnetic Field.

5. J. R. Apel and A. M. Stone (1965), *Proc. Seventh Intern. Conf. Ionization Phenomena in Gases, 7th, Vol.* II, p. 405. Experiments on Wave Interactions between Plasma and an Electron Stream in a Magnetic Field.

6. M. C. Ayphassorho and M. P. Loup (1963), *Proc. Intern. Conf. Ionization Phenomena in Gases, 6th,* Vol. 2, p. 401. Rayons X emis par Bombardement Electronique de l'Anode, dans une Décharge Gaseuse Haute Tension, a Configuration de Penning.

7. J. Backus (1949), in "The Transuranium Element Research Papers." McGraw-Hill, New York, 1949. Theory and Operation of a Phillips Ionization Gauge Type Discharge.

8. J. Backus (1949), in "Characteristics of Electrical Discharges in Magnetic Fields" A. Guthrie and R. K. Wakerling eds, Chap. 11. McGraw-Hill, New York, 1949.

9. J. Backus (1959), *J. Appl. Phys.* **30**, 1866. Studies of Cold Cathode Discharges in Magnetic Fields.

10. J. Backus and N. E. Huston (1960), *J. Appl. Phys.* **31**, 400. Ion Energies in a Cold Cathode Discharge in a Magnetic Field.

11. E. P. Barbian, C. E. Rasmusson, and J. Kistemaker (1967), *Intern. Conf. Phenomena in Ionized Gases, 8th,* p. 176. Springer, Vienna. Ionization and Current Growth in an $E \times B$ Discharge.

12. H. W. Batten, H. L. Smith, and H. C. Early (1956), *J. Franklin Inst.* **262**, 17 (1956). Plasma Fluctuations in Crossed Electric and Magnetic Fields.

13. N. Bayet and G. Dumas (1953), *Compt. Rend. Acad. Sci.* **236**, 1648. Sur le Fonctionnement de la Gauge de Penning.

14. A. B. Berezin, D. G. Buliginsky, M. I. Viljunas, and S. S. Kalmykov (1965), *Proc. Intern. Conf. Ionized Gases, Belgrad, 7th,* Vol. I, p. 455. Gradvinska, Beograd, 1966. Investigation of Pulsed Toroidal Penning Discharge.

15. A. B. Berezin, D. G. Buliginsky, M. I. Viljunas, V. A. Rodichkin, and A. F. Ioffe (1967), *Intern. Conf. Phenomena Ionized Gases, 8th.* Springer, Vienna. Investigation of Turbulent Oscillations in PIG-Type Discharge Plasma.

16. R. Bingham (1964), *Phys. Fluids* **7**, 1001. Long Wavelength Instability in a Reflex Discharge.

17. R. Bingham and F. F. Chen (1962), *Bull. Am. Phys. Soc.* **7**, 401. Helical Instability in the Reflex Arc.

18. R. Bingham, F. F. Chen, and W. Harries (1962), Princeton Rept. MATT-63 (unpublished). Preliminary Studies of a Reflex Arc.

19. S. Bliman (1963), *Compt. Rend.* **256**, 3035. Mise en Evidence dans une Décharge Reflex a Cathodes Froides d'une Instabilite du Type "Deux Faisceaux."

20. S. Bliman, A. Bouchoule, and A. Septier (1965), *Compt. Rend.* **260**, 2751. Oscillations Haute Fréquence et Acceleration d'Electrons dans une Décharge du Type PIG a Cathodes Froides.

21. S. Bliman, A. Bouchoule, and A. Septier (1965), *Proc. Intern. Conf. Ionized Phenomena Gases, Belgrad, 7th*, Vol. I, p. 448. Gradvinska, Beograd 1966. Existence d'Oscillations H. F. Intenses dues á un Phenomene d'Interaction Faisceau Plasma dans une Décharge Reflex a Cathodes Froides.

22. J. F. Bonnal, G. Briffod, and C. Manus (1961), *Phys. Rev. Letters* **6**, 665. Enhanced Diffusion and Oscillations in Weakly Ionized Plasmas. (1962); *Nucl. Fusion Suppl.*, p. 995. Étude due la Diffusion Perpendiculaire au Champ Magnetique.

23. F. Bottiglioni, M. Chapet, E. Fischoff, and F. Prevot (1961), *Proc. Intern. Conf. Ionized Phenomena Gases, 5th*, Vol. II, 2100. Colonne de Plasma Fortement Ionisee dans un Champ Magnetique.

24. F. Bottiglioni, M. Fumelli, and F. Prevot (1963), *Proc. Intern. Conf. Phenomena Gases, 6th*, Vol. 2, p. 427. Fluctuations et Rotations d'Une Colonne de Plasma.

25. C. M. F. Brachet and P. L. Vasseur (1961), *Proc. Fifth Intern. Conf. Ionized Phenomena Gases, 5th*, Vol. 2, p. 1370. Entretien et Confinement d'un Arc Axial dans une Decharge Reflex a Tres Basse Pression.

26. L. V. Brezechko, Yu. K. Kuznetsov, and O. S. Pavlichenko (1967), *Intern. Conf. Phenomena Ionized Gases, 8th*, p. 187. Springer, Vienna. Resistive Drift Instability in a Cold Cathode Penning-Type Discharge.

27. G. Briffod (1964), *Plasma Phys. (J. Nucl. Energy, Pt. C.)* **6**, 329. Instability in a Cold Cathode Reflex Arc.

28. G. Briffod, M. Gregoire, and C. Manus (1961), *Phys. Letters*, **2**, 201. Diffusion Anomale d'un Plasma Perpendiculairement au Champ Magnetique.

29. G. Briffod, M. Gregoire, and S. Gruber (1963), *Proc. Intern. Conf. Ionized Phenomena Gases, 6th*. Instabilites dans les Decharges Reflex.

30. F. F. Chen (1962), *Phys. Rev. Letters* **8**, 234. Radial Electric Field in a Reflex Discharge.

31. F. F. Chen (1963), *Proc. Intern. Conf. Ionized Phenomena Gases, 6th*. Low Frequency Instabilities of a Fully Ionized Gas.

32. F. F. Chen (1964), Princeton Rept. MATT-249 (unpublished). Preliminary Results on the Interpretation of Low-frequency Fluctuations Observed in the Reflex Arc.

33. F. F. Chen (1965), *Phys. Fluids* **8**, 912. Resistive Overstabilities and Anomalous Diffusion.

34. F. F. Chen and R. Bingham (1961), *Bull. Am. Phys. Soc.* **6**, 189. A Hot Cathode Reflex Discharge.

35. F. F. Chen and A. W. Cooper (1962), *Phys. Rev. Letters* **9**, 333. Electrostatic Turbulence in a Reflex Discharge.

36. P. Chorney and D. J. Fitzgerald (1964), *Congr. Intern. Tubes Hyperfréquences, 5th, Paris*. Beam plasma Amplification at 11 Gc.

37. P. Chorney and R. J. Madore (1964), *Cong. Intern. Tubes Hyperfréquences, 5th, Paris*. Generation of Plasmas for Beam-Plasma Amplifiers.

38. G. T. K. Conn and H. N. Daglish (1953), *Vacuum* **3**, 24. Cold Cathode Ionization Gauges for the Measurement of Low Pressures.

39. T. Consoli, R. Geller, and R. Le Gardeur (1961), *Proc. Intern. Conf. Ionized Phenomena Gases, 5th, Munich*. Decharge Pulsee Reflex du Type PIG a Forte Densite.

40. F. W. Crawford (1965), *Nucl. Fusion* **5**, 73. A Review of Cyclotron Harmonic Phen. in Gases.

41. F. R. Crownfield (1963), *Proc. Intern. Conf. Ionized Phenomena Gases, 6th, Paris*, Vol. 2, p. 451. Interactions of Plasma Oscillations with Conduction in a Penning Gauge.

42. F. R. Crownfield, D. Raiford, and M. D. Holt (1965), *Proc. Intern. Conf. Phenomena Ionized Gases, 7th, Belgrad*, Vol. I, p. 451. Gradvinska, Beograd, 1966. Non-linear Response of a Penning Ionization Gauge Discharge to Applied Very High Frequency Signals.

43. J. Ďatlov, J. Musil and F. Žáček (1968), *Czech. J. Phys.* **B18**, 75. Characteristics of a Reflex Discharge in Hydrogen in a Magnetic Field.

44. J. W. Davis, A. P. Walch, R. G. Meyerand, Jr., F. Salz, and E. C. Lary (1961), IAS–ARS Joint Meeting, Los Angeles, Preprint 61-103-1797. Theoretical and Experimental Description of the Oscillating Electron Ion Engines.

45. J. W. Davis, A. W. Angelbeck, and E. A. Pinsley (1963), *AIAA. J.* **1**, 2497. Plasma Behavior in an Oscillating-Electron Ion Engine.

46. H. Doucet (1963), *Proc. Intern. Conf. Ionized Phenomena Gases, 6th*, Vol. 2, p. 387. Décharge Reflex a Basse Pression en Champ Magnetique Croissant vers la Peripherie.

47. D. G. Dow (1963), *J. Appl. Phys.* **34**, 2395. Electron-Beam Probing of a Penning Discharge.

48. W. E. Drummond and M. N. Rosenbluth (1962), *Phys. Fluids* **5**, 1507. Anomalous Diffusion Arising from Microinstabilities in a Plasma.

49. L. V. Dubovoĭ and P. G. Popov (1965), *J. Exp. Theoret. Phys. (USSR)* **49**, 27; (1966), *Soviet Phys. JETP (English Transl.)* **22**. Stabilization of a Powerful Penning Discharge by a Radially Increasing Magnetic Field.

50. J. N. Dukes (1959), Univ. of California (Electron. Res. Lab.), Ser. 60, **240**. A Gaseous-Conduction Linear Amplifier with Magnetic Focusing. Quoted in Lundgren *et al.* (1961).

51. G. Dumas (1955), *Rev. Gen. Elec.* **64**, 331. Study of Gyromagnetic Resonance in a Penning Gauge.

52. T. Ya. Fishkova, E. V. Shpak, and S. Ya. Yavor (1964), *Zh. Tekhn. Fiz.* **34**, 53; (1964), *Soviet Phys.-Tech. Phys. (English Transl.)* **9**, 40. Investigation of the Leakage of Charged Particles from a Discharge with Reflected Electrons.

53. M. D. Gabovich, O. A. Bartnovskii, and Z. P. Fedorus (1960), *Zh. Tekhn. Fiz.* **30**, 345; (1960), *Soviet Phys.-Tech. Phys. (English Transl.)* **5**, 320. Droop in the Axial Potential of a Discharge with Electron Oscillations in a Magnetic Field.

54. R. Geller (1963a), *Appl. Phys. Letters* **2**, 218. A Pulsed Toroidal PIG Discharge.

55. R. Geller (1963b), *Phys. Rev. Letters* **10**, 463. Experimental Results of a Bumpy Torus Discharge.

56. R. Geller and J. Hosea (1963), *Appl. Phys. Letters* **2**, 215. A Highly Ionized Pulsed PIG Reflex Discharge.

57. R. Geller and D. Pigache (1961), *Proc. Intern. Conf. Ionization Phenomena Gases, 5th*, Vol. 1, p. 1111. Mechanisme de la Decharge Reflex a l'Equilibre.

58. R. Geller and D. Pigache (1962), *Plasma Phys. (J. Nucl. Energy, Pt. C)* **4**, 229. Mecanisme d'Equilibre d'une Decharge PIG Reflex et Determination Experimentale de $D_\perp$ dans un Plasma.

59. J. D. Gow and J. S. Foster, Jr. (1953), *Rev. Sci. Instr.* **24**, 606. A High-Intensity Pulsed Ion Source.

60. V. G. Grigorenko, L. A. Dushin, O. S. Pavlichenko, and A. I. Skibenko (1966), *Zh. Tekhn. Fiz.* **36**, 1800; (1967), *Soviet Phys.-Tech. Phys. (English Transl.)* **11**, 1340. Anomalous Decay of the Plasma in a PIG Discharge in a Strong Magnetic Field.

61. G. Guest and A. Simon (1962), *Phys. Fluids* **5**, 503. Instability in Low Pressure Plasma Diffusion Experiments.

62. L. S. Hall and A. L. Gardner (1962), *Phys. Fluids* **5**, 788. Highly Ionized Steady State Plasma System.

63. J. C. Helmer and R. L. Jepsen (1961), *Proc. IRE* **49**, 1920. Electrical Characteristics of a Penning Discharge.

64. E. H. Hirsch (1964), *Brit. J. Appl. Phys.* **15**, 909. Excess Energy Electrons.

65. E. H. Hirsch (1964), *Brit. J. Appl. Phys.* **15**, 1535. On the Mechanism of The Penning Discharge.

66. F. C. Hoh (1963), *Phys. Fluids* **6**, 1184. Instability of Penning-Type Discharges.

67. F. C. Hoh (1963), *Arkiv Fysik* **24**, 285. Instability of Gas Discharges in a Magnetic Field.

68. F. C. Hoh and B. Lehnert (1961), *Phys. Rev. Letters* **7**, 75. Screw Instability of a Positive Column.

69. W. Honig and P. Parzen (1955). *IRE Intern. Conv. Rec.* **3**, P. 3, *Electron Devices and Components Parts*, p. 3. A Gas Discharge Noise Source.

70. E. B. Hooper, Jr. (1968), *Bull. Am. Phys. Soc.*, **13**, 1499. Correlation Measurements in an Unstable Reflex Arc.

71. E. B. Hooper, Jr. and G. Bekefi (1965), *Proc. Intern. Conf. Ionized Phenomena Gases*, *7th*, Vol. 3, p. 209. Electron Density Measurements with a Laser Interferometer.

72. G. Horikoshi, T. Kuroda, K. Matsuura, A. Miyahara, M. Masuzaki, and S. Nagao (1967), *Intern. Conf. Ionized Phenomena Gases*, *8th*, p. 550. Springer, Vienna. Pig Plasma Source with Divertor Type Magnetic Field Configuration.

73. M. Hoyaux (1954), *J. Phys. Radium* **15**, 264. Theorie des Sources d'Ions a Electrons Oscillants.

74. M. Hoyaux, R. Lemaitre, and P. Gans (1955), *J. Appl. Phys.* **26**, 110. Theory and Probe Measurements in a Magnetic Ion Source.

75. T. I. Ianova, G. E. Pustovalov, and E. M. Reikhrudel (1966), *Soviet Phys.-Tech. Phys.* (*English Transl.*) **11**, 688. Solution of Laplace's Equation in a Penning Discharge.

76. R. L. Jepson (1961), *J. Appl. Phys.* **32**, 2619. Magnetically Confined Cold-Cathode Gas Discharges at Low Pressures.

77. R. L. Jepsen, A. B. Francis, S. L. Rutherford, and B. E. Kietzmann (1960), in " Vacuum Symposium Proceedings," p. 45. Pergamon Press, New York. Stabilized Air Pumping with Diode Type Getter-Ion Pumps.

78. M. G. Kaganskii, D. L. Kaminskii, and A. N. Klyucharev, (1964), *Zh. Tekhn. Fiz.* **34**, 1050; (1964), *Soviet Phys.-Tech. Phys.* (*English Transl.*) **9**, 815. Coherent Oscillations in a High-Voltage Penning Discharge.

79. D. M. Kerr, Jr. (1966), *Phys. Fluids* **9**, 2531. Helical Instability in a Penning Discharge.

80. W. D. Kilpatrick, J. H. Mullins, and J. M. Feen (1963) *AIAA. J.* **1**, 806. Propulsion Application of the Modified Penning Arc Plasma Ejector.

81. J. Kistemaker and H. L. Douwes Dekker (1950), *Physica* **16**, 198, 209. Investigations on a Magnetic Ion Source, I, II.

82. J. Kistemaker and J. Snieder (1953), *Physica* **19**, 950. Some Measurements on a not Selfsustaining Gas Discharge with an Axial Magnetic Field.

83. F. Klan (1967), *Proc. Intern. Conf. Ionized Phenomena Gases*, *8th*, p. 190. Rotational Instability and Cyclotron Harmonic Radiation of a Cold Cathode Reflex Discharge.

84. W. Knauer (1961), *Proc. Intern. Conf. Ionized Phenomena Gases*, *5th*, Vol. 2, p. 1350. Mechanisms of the Low Pressure Penning Discharge

85. W. Knauer (1962), *J. Appl. Phys.* **33**, 2093. Mechanism of the Penning Discharge at Low Pressures.

86. W. Knauer (1966), *J. Appl. Phys.* **37**, 602. Diocotron Instability in Plasmas and Gas Discharges.

87. W. Knauer and M. A. Lutz (1963), *Appl. Phys. Letters* **2**, 109. Measurement of the Radial Field Distribution in a Penning Discharge by Means of the Stark Effect.

88. W. Knauer and R. L. Poeschel (1965), *Proc. Intern. Conf. on Ionized Phenomena Gases, 7th,* Vol. 2, p. 719. The Diocotron Effect in Plasmas and Gas Discharges.

89. W. Knauer, A. Fafarman, and R. L. Poeschel (1963), *Appl. Phys. Letters* **3**, 111. Instability of Plasma Sheath Rotation and Associated Microwave Generation in a Penning Discharge.

90. H. C. Koons and G. Fiocco (1968), *Phys. Letters* **26A**, 614. Anisotropy of the Electron Velocity Distribution in a Reflex Discharge Measured by Continuous-Wave Laser Scattering.

91. O. H. Krause (1952), M. S. Thesis, Univ. of California, Berkeley. A Study of Plasma Discharge in a Magnetic Field. Quoted in Lundgren *et al.* (*108*).

92. Yu. E. Kreindel (1963), *Zh. Tekhn. Fiz.* **33**, 883; (1964), *Soviet Phys.-Tech. Phys.* (*English Transl.*) **8**, 662. Electron Current at the Exit of a Penning-Type Tube.

93. Yu. E. Kreindel (1966), *Zh. Tekhn. Fiz.* **36**, 903; (1966) *Soviet Phys.-Tech. Phys.* (*English Transl.*) **11**, 665. Efficient Extraction of Electrons from a Modified Penning Discharge.

94. Yu. E. Kreindel and E. N. Fakhrutdinov (1965), *Zh. Tekhn. Fiz.* **35**, 312; (1965), *Soviet Phys.-Tech. Phys.* (*English Transl.*) **10**, 249. Pulse Characteristics of Modified Penning Discharge Tubes.

95. Yu. E. Kreindel and A. S. Ionov (1964), *Zh. Tekhn. Fiz.* **34**, 1199; (1965), *Soviet Phys.-Tech. Phys.* (*English Transl.*) **9**, 930. Some Characteristic Features of Low-Pressure Pressure Discharges in Penning Tubes.

96. Ye. T. Kucherenko and V. P. Ignatko (1964), *Radio Eng. Electron* **9**, 140. One Form of an Inhibited Discharge in a Magnetic Field.

97. Ye. T. Kucherenko and O. K. Nazarenko (1959), *Radio Eng. Electron.* **4**, 8, 1253. Characteristics of a Discharge with Oscillating Electrons in a Magnetic Field.

98. Ye. T. Kucherenko and V. A. Saenko (1964), *Uks. Fiz. Zh.* **9**, 194. Reflected Discharge in a Weak Magnetic Field.

99. Ye. T. Kucherenko and V. A. Saenko (1967). *Czech. J. Phys.* **B17**, 67. Reflex Discharges at Low Pressure.

100. Ye. T. Kucherenko and V. A. Saenko (1967), *Zh. Tekhn. Fiz.* **37**, 112; (1967), *Soviet Phys-Tech. Phys.* (*English Transl.*) **12**, 76. Reflex Discharge at Low Magnetic Field.

101. Ye. T. Kucherenko, V. A. Saenko, and Ye. V. Akhtyrskaya (1965). *Radio Eng. Electron.* **10**, 1601. Properties of a Reflex Discharge in a Magnetic Field.

102. O. K. Kurbatov (1966), *Zh. Tekhn. Fiz.* **36**, 1655; (1967), *Soviet Phys-Tech. Phys.* (*English Transl.*) **11**, 1240. Energy and Angle Distributions of Ions Arriving at Various Parts of the Cathode in a High Voltage Penning Discharge.

103. H. Kühn (1957), *Z. Physik* **149**, 267. Untersuchungen uber das Auftreten von Hochfrequenten Schwingungen an einer Ionenquelle mit Magnetischen Führungsfeld.

104. M. Laffineur and C. Pecker (1950), *Compt. Rend.* **231**, 1446. Emission radioélectrique due á l'effet Gyromagnétique dans une Décharge.

105. G. Landauer (1961), *Proc. Intern. Conf. Ionization Phenomena Gases, 5th,* Vol. I, p. 389. Magnetfeldabhängige Mikrowellenstrahlung aus einer He-Gasentladung.

106. G. Landauer (1962), *Plasma Phys.* (*J. Nucl. Energy, Pt. C*) **4**, 395. Generation of Harmonics of the Electron-Gyrofrequency in a Penning Discharge.

107. E. G. Linder and J. H. Coleman (1952), *Proc. IRE* **40**, 818. A High-Voltage Cold Cathode Rectifier.

108. R. E. Lundgren, C. Susskind, and J. R. Woogyard (1961), *IRE Trans. Electron Devices* **ED8**, 489. Applications of Cold-Cathode Parallel-Field Devices.

109. R. G. Meyerand Jr. (1961), *Prog. Astron. Rocketry, Electrostatic Propulation* **5**, 81. The Oscillating-Electron Plasma Source.

110. R. G. Meyerand, Jr., F. Salz, E. C. Lary, and A. P. Walch (1961), *Proc. Intern. Conf. Ionized Phenomena Gases, 5th, Munich.* Electrostatic Potential Gradients in a Non-Thermal Plasma.

111. D. L. Morse (1965), *Phys. Fluids* **8**, 1339. Low Frequency Instability of Partially Ionized Plasma.

112. S. Nagao, A. Miyahara, T. Kuroda, T. Sato, and K. Matsuura (1963), *J. Phys. Soc. Japan* **18**, 738. Oscillations in a PIG Discharge and a Type of Instability of Plasma.

113. O. S. Pavlichenko, L. A. Dushin, I. K. Nikol'skii, and L. V. Brzhechko (1964), *Zh. Tekhn. Fiz.* **34**, 590; (1964), *Soviet Phys.-Tech. Phys. (English Transl.)* **9**, 459. Macroscopic Instability of Plasma in a Negative Discharge.

114. O. S. Pavlichenko, L. A. Dushin, Yu. K. Kuznetsov, and I. Yu. Adamov (1965), *Zh. Tekhn. Fiz.* **35**, 1394; (1966), *Soviet Phys.-Tech. Phys. (English Transl.)* **10**, 1082. Instability in an Oscillating-Electron Discharge. I. Microwave Radiation.

115. O. S. Pavlichenko, L. A. Dushin, Yu. K. Kuznetsov, and I. Yu. Adamov. (1965), *Zh. Tekhn. Fiz.* **35**, 1401; (1966), *Soviet Phys.-Tech. Phys. (English Transl.)* **10**, 1088. Instability in an Oscillating-Electron Discharge. II. Anomalous Diffusion of the Plasma.

116. F. M. Penning, (1937), *Physica* **4**, 71. Ein Neues Manometer fer Niedrige Gasdrucke, Instesondere Zwischen 10^{-3} und 10^{-5} mm.

117. K. Phillips (1954), *J. Sci. Instr.* **31**, 110. Some Experiments with a Cold Cathode Vacuum Gauge.

118. C. E. S. Phillips (1898), *Proc. Roy. Soc. (London)* **A64**, 172. The Action of Magnetized Electrodes upon Electrical Discharge Phenomena in Rarefied Gases. Preliminary Note.

119. C. E. S. Phillips (1901), *Phil. Trans. Roy. Soc. London Ser. A* **A197**, 135. The Action of Magnetized Electrodes upon Electrical Discharge Phenomena in Rarefied Gases.

120. I. M. Poliak (1961), *Radio. Electron.* **6**, 3, 395; (1961), *Radio Eng. Electron.* **6**, 73, The Electron Current in Penning Tubes.

121. Ya. S. Popov (1967), *Zh. Tekhn. Fiz.* **37**, 118; (1967), *Soviet Phys.-Tech. Phys. (English Transl.)* **12**, 81. Low Pressure Cold-Cathode Penning Discharge.

122. E. M. Reikhrudel (1956), *Radio. Electron.* **1**, 253. Measurements in a Penning-Type Discharge.

123. E. M. Reikhrudel and E. Kh. Isakaev (1966), *Zh. Tekhn. Fiz.* **36**, 653; (1966), *Soviet Phys.-Tech. Phys. (English Transl.)* **11**, 486. The Striking of a Discharge in a Penning Gauge at Low Pressures.

124. E. M. Reikhrudel, G. V. Smirnitskaya, and E. P. Sheretov (1962), *Radio Eng. Electron.* **7**, 1672. High Frequency Discharge Radiation in an Ion Pump with Cold Cathodes.

125. E. M. Reikhrudel, G. V. Smirnitskaya, and R. P. Babertsian (1965), *Proc. Conf. Phenomena Ionized Gases, 7th,* Vol. I, p. 443. Gradevinska, Belgrad, 1966. A New Method for Determination of the Potential Distribution in a Penning Type Discharge.

126. E. M. Reikhrudel, G. V. Smirnitskaya, and R. P. Babertsyan (1966), *Zh. Tekhn. Fiz.* **36**, 1226; (1967), *Soviet Phys.-Tech. Phys. (English Transl.)* **11**, 909. A New Method for the Determination of the Potential Distribution in a Penning Discharge.

127. E. M. Reikhrudel, G. V. Smirnitskaya and Y. Nguyen-Khin-Tee (1967), *Intern. Conf. Phenomena Ionized Gases, 8th,* p. 187. Springer, Vienna. Dependence of the Current on the Parameters in Discharge with Oscillating Electrons in a Magnetic Field.

128. D. J. Rose (1959), Mass. Inst. of Technol. Res. Lab. of Electron. Quart. Progr. Rept. No. 53 (unpublished). Acceleration of a Neutral Ion Beam.

129. F. Salz, R. G. Meyerand Jr., E. C. Lary, and A. P. Walch (1961), *Phys. Rev. Letters* **6**, 523. Electrostatic Potential Gradients in a Penning Discharge.

130. W. Schuurman (1966), Rijnhuizen Rept. 66–28, FOM-Institute voor Plasma-Fysica, Rijnhuizen, Jutphaas, Nederland. Investigation of a Low-Pressure Penning Discharge.

131. W. Schuurman (1967), *Physica* **36**, 136. Investigation of a Low-Pressure Penning Discharge.

132. A. Simon (1963), *Phys. Fluids* **6**, 382. Instability of a Partially Ionized Plasma in Crossed Electric and Magnetic Fields.

133. G. V. Smirnitskaya and R. P. Babertsyan (1966), *Zh. Tekhn. Fiz.* **36**, 1217; (1966) *Soviet Phys.-Tech. Phys.* (*English Transl.*) **11**. On the Kinetics of Positive Ions in a Penning Type Discharge.

134. G. V. Smirnitskaya and E. M. Reickhrudel (1959), *Soviet Phys. Tech. Phys.* (*English Transl.*) **4**, 131; Cold-Cathode Electric Discharge at Low Pressures in a Magnetic Field [(1957), *Radio. Electron.* **2**, 10, 1303].

135. T. H. Stix (1967), *Phys. Fluids* **10**, 1601. Resonant Diffusion of Plasma across a Magnetic Field.

136. R. J. Strutt (1913), *Proc. Roy. Soc.* (*London*) **A89**, 68. A Peculiar Form of Low Potential Discharge in the Highest Vacua.

137. G. A. Swartz and L. S. Napoli (1965), *Phys. Fluids* **8**, 1550. Highly Ionized Dense Plasma in a Cesium Penning Arc.

138. T. Tanaka and K. Yamamoto (1963), *J. Phys. Soc. Japan* **18**, 735; (1962), **17**, 715. A Characteristic of the PIG Type Discharge Tube.

139. K. I. Thomassen (1965), *Phys. Rev. Letters* **14**, 587. Rotational Instability in a Penning-Type Discharge.

140. K. I. Thomassen (1966), *Phys. Fluids* **9**, 626. Measurement of an Anomalous Diffusion Coefficient.

141. K. I. Thomassen (1966), *Phys. Fluids* **9**, 1836. Turbulent Diffusion in a Penning Type Discharge.

142. K. I. Thomassen (1967), *Intern. Conf. Phenomena Ionized Gases, 8th*, p. 327. Springer, Vienna. Oscillations and Transport in the Reflex Discharge.

143. M. N. Vasil'eva and E. M. Reickhrudel (1962), *Zh. Tekhn. Fiz.* **32**, 725; (1962), *Soviet Phys.-Tech. Phys.* (*English Transl.*) **7**, 528. Influence of Space Charge on the Kinetics of Electrons in Penning-Type Tubes.

144. J. Verwell (1957), *Le Vide* **67**, 32.

145. R. W. Waniek, R. T. Grannan, and D. G. Swanson (1964), *Appl. Phys. Letters* **5**, 89. Anomalous Cyclotron Radiation from a Plasma Discharge.

146. M. Wehrli (1922), *Ann. Physik* **69**, 285. Funkpotentiale in Transversalen Magnetfeld.

B. Pump

147. I. Ames and R. L. Christensen (1959), *in* " Vacuum Symposium Transactions," p. 311. Pergamon Press, New York. Some Studies of Getter-Ion Pumped Vacuum Systems.

148. G. F. Brothers, T. Tom, and D. F. Munro (1963). *Trans. Natl. Vacuum Symp., 10th*, p. 202. Macmillan, New York. Design and Performance of a 50,000 L/sec Pump Combining Cold Cathode Ion Pumping and Active Film Gettering.

149. W. M. Brubaker (1959), *in* " Vacuum Symposium Transactions," p. 302. Pergamon Press, New York. A Method for Greatly Enhancing the Pumping Action of a Penning Discharge.

150. G. Conisa and B. Iosifescu (1961), in "Vacuum Symposium Transactions," Vol. I, p. 413. Pergamon Press, New York. Pressure-versus-Time Variation in Closed Vessels Exhausted by Ionic Pumping.

151. J. S. Foster, Jr., E. O. Lawrence, and E. J. Lofgren (1953), *Rev. Sci. Instr.* **24**, 388. A High Vacuum High Speed Ion Pump.

152. A. J. Gale (1956), in "Vacuum Symposium Transactions," p. 12. Pergamon Press, New York. Cold Sealed Getter/Ion Pumped Supervoltage X-ray Tubes.

153. A. M. Gurewitsch and W. F. Westendorp (1954), *Rev. Sci. Instr.* **25**, 389. Ionic Pump.

154. L. D. Hall (1958), *Rev. Sci. Instr.* **29**, 367. Electronic Ultra-High Vacuum Pump.

155. R. L. Jepsen (1959), *Le Vide* **80**, 80. Important Characteristics of a New Type Getter-Ion Pump.

156. A. Klopfer (1961), *Vacuum Tech.* **10**, 113. Die Erzengung Von Hockstvakua mit Getter-Ionespumpen und das Messen von sehr Tiefen Drucken.

157. W. Knauer and E. R. Stack (1963) in "Vacuum Symposium Transactions," p. 180. Macmillan, New York. Alternative Ion Pump Configurations Derived from a More Thorough Understanding of the Penning Discharge.

158. N. Milleson (1961), in "Vacuum Symposium Transactions," Vol. I, p. 365. Pergamon Press, New York. Preliminary Pumping Results with Penning Type Discharges Supported on Condensible Metal Vapors.

159. E. M. Reickhrudel and G. V. Smirnitskaya (1963), *Zh. Tekhn. Fiz.* **33**, 1405; (1964), *Soviet Phys.-Tech. Phys.* (*English Transl.*) **8**, 1045. Modern Methods of Obtaining Ultrahigh Vacuum.

160. S. L. Rutherford (1963), in "Vacuum Symposium Transactions," p. 185. Macmillan New York, Sputter-Ion Pumps for Low Pressure Operation.

161. S. L. Rutherford, S. L. Mercer, and R. L. Jepsen (1960), in "Vacuum Symposium Transactions," p. 380. Pergamon Press, New York. On Pumping Mechanisms in Getter-Ion Pumps Employing Cold-Cathode Gas Discharges.

162. H. Schwarz (1953), *Rev. Sci. Instr.* **24**, 371. Methods of Obtaining High Vacuum by Ionization. Construction of an Electronic Pump.

163. H. Schwarz (1954) in "CVT Vacuum Symposium Transactions," p. 46. Pergamon Press, New York. Production of High Vacua by Ionization.

164. R. Zaphiropoulos and W. A. Lloyd (1959), in "Vacuum Symposium Transactions," p. 307. Pergamon Press, New York. Design Considerations for High Speed Getter-Ion Pumps.

C. Gage

165. H. I. J. Allwood (1948), *J. Sci. Instr.* **25**, 207. Vacuum Protection Systems for Oil Diffusion Pumps and Thermionic Filaments.

166. E. C. Evans and K. Burmaster (1950), *Proc. IRE* **38**, 651. A Philips-Type Ionization Gauge for Measuring of Vacuum from 10^{-7} to 10^{-1} mm of Mercury.

167. R. I. Gerrod and K. A. Gross (1948). *J. Sci. Instr.* **25**, 378. Combined Thermocouple and Cold-Cathode Vacuum Gauge.

168. A. M. Grigoriev (1960), *Proc. Intern. Conf. Vacuum Tech. 1st*, p. 308. Pergamon Press, New York. Enlarging the Range of the Pressures Measured by Cold-Cathode Ionization Gauges.

169. C. Hayashi, K. Hashimoto, K. Kaneko, K. Okamoto, and R. Sagane (1949), *Rev. Sci. Instr.* **20**, 524. Several Improvements on the Philips Gauges.

171. J. H. Leck and A. Riddoch (1956), *Brit. J. Appl. Phys.* **7**, 153. Observations on the Characteristics of the Cold Cathode Ionization Gauge.

170. J. H. Leck (1953), *J. Sci. Instr.* **30**, 271. Sorption and Desorption of Gas in the Cold-Cathode Ionization Gauge.

172. H. Maesta (1956), *Z. Angew. Phys.* **8**, 598. Uber das Verbatten der Gasentladung beim Vakuumeter Nach Penning.

173. F. M. Penring and K. Nienhuis (1949), *Philips Tech. Rev.* **11**, 116. Construction and Application of a New Design of the Philips Vacuum Gauge.

174. R. G. Picard, P. C. Smith, and S. M. Zollers (1946), *Rev. Sci. Instr.* **17**, 125. A Reliable High Vacuum Gauge and Control System.

175. N. Varičak and B. Vošicki (1955), *J. Sci. Instr.* **32**, 346. Oscillograph Measurement of the Penning-Gauge Characteristics.

176. J. J. Vermande (1952), *Le Vide* **7**, 1145. Utilisation de la Jauge de Penning.

D. Charged Particle Sources

177. E. J. Anderson and K. W. Ehlers (1956), *Rev. Sci. Instr.* **27**, 809. Ion Source for the Production of Multiply Charged Heavy Ions.

178. C. Bailey, D. L. Druky, and F. Oppenheimer (1949), *Rev. Sci. Instr.* **20**, 189. A Magnetic Ion Source.

179. A. Bariand, R. Becheres, J. Druaux, and F. Prevot (1965), *Proc. Intern. Conf. Ionized Phenomena Gases*, 7th, Vol. 3, p. 340. Source d'Ions Annulaire de Grande Intensite a Decharge de Penning.

180. C. F. Barnett (1953), *Rev. Sci. Instr.* **24**, 394. PIG Ion Source.

181. R. Basile and J. M. Legrange (1963), *Proc. Intern. Conf. Ionized Phenomena Gases*, 6th, Vol. 3. p. 203. Etude du Bilan Energique d'une Decharge Reflex.

182. H. Baumann, K. Bethge, and E. Keinkicke (1967). *Nucl. Instr. Methods.* **46**, 43. Production of Negative Lithium Ions in a Penning Discharge.

183. B. Cheon, H. Akimune and T. Suita (1966), *J. Appl. Phys. (Japan)* **5**, 628. Beam Controllable PIG Ion Source.

184. K. W. Ehlers, B. F. Gavin, and E. L. Hubbard (1963), *Nucl. Instr. Methods* **22**, 87. High Intensity Negative Ion Sources.

185. A. T. Finkelstein (1940), *Rev. Sci. Instr.* **11**, 94. A High Efficiency Ion Source.

186. J. Flinta (1958), *Nucl. Instr.* **2**, 219. Pulsed High Intensity Ion Source.

187. S. V. Goler and H. Wagner (1963), *Proc. Intern. Conf. Ionized Phenomena Gases*, 6th, Vol. 2, p. 405. Uber die Penning-Entladung mit heisser Kathode und ihre verwendung als Ionenquelle.

188. H. Heil (1943), *Z. für Phys.* **120**, 212. Uber Eine Neue Ionenquelle.

189. R. J. Jones and A. Zucker (1954), *Rev. Sci. Inter.* **25**, 562. Two Ion Sources for the Production of Multiply Charged Nitrogen Ions.

190. R. Keller (1948), *Helv. Phys. Acta* **21**, 170. Une Nouvelle Source d'Electrons et Son Inversion Comme Source d'Ions.

191. R. Keller (1949), *Helv. Phys. Acta* **22**, 78. Etude d'une Source d'Ions du Type Penning.

192. Ye. T. Kucherenko and V. A. Saenko (1965), *Radio Eng. Electron* **10**, 1639. Study of the Energy Spectrum of Ions from a Magnetic Source.

193. M. Legentil and J. L. Delcroix (1961), *Proc. Intern. Conf. Ionized Phenomena Gases*, 5th, Vol. **2**, p. 1987. Magnetic Ion Source.

194. P. Lorrain (1947), *Can. J. Res.* **A25**. 338. A Low-Pressure Glow-Discharge Proton Source.

195. P. Lorrain (1948), *Helv. Phys. Acta* **21**, 9497. Une Nouvelle Source d'Electrons et Son Inversion Comme Source d'Ions.

196. E. Lutand, C. Etievant, and M. Perulli (1965), *Proc. Intern. Conf. Ionized Phenomena Gases, 7th*, Vol. 3, p. 345. Investigation of a Penning Ion Source.

197. G. S. Mavrogenes, W. J. Rabler, and G. B. Turner (1695), *IEEE Trans. Nucl. Sci.* **NS-12**, 769. A Source for Multiply Charged Ions.

198. R. G. Meyerand Jr., and S. C. Brown (1959), *Rev. Sci. Inst.* **30**, 110. High-Current Ion Source.

199. C. B. Mills and C. F. Barnett (1954), *Rev. Sci. Instr.* **25**, 1200. High-Intensity Ion Source.

200. J. L. Nagy (1963), *Nucl. Instr. Methods* **32**, 229. The Energy Spectrum of Ion Beam emitted by a Penning-Type Ion Source.

201. R. Pauli and J. Flinta (1958), *Nucl. Instr.* **2**, 227. Pulsed High-Intensity Ion Source. Part II.

202. F. M. Penning and J. H. A. Moubis (1937), *Physica* **4**, 1190. Eine Neutronneurohre Ohne Pumpvorrichtung.

203. J. M. Sautter, M. Bariband, and J. M. Dolique (1966), *Compt. Rend.* **263**, 1211. Conception et Realisation d'une Source Tres Intense d'Ions d'Hydrogene Negativ H^-.

204. R. B. Setlow (1949), *Rev. Sci. Instr.* **20**, 558. A High Current Ion Source.

205. A. Svanheden (1961), *Nucl. Instr. Methods* **10**, 125. A Cold Cathode Ion Source for a Synchrocyclotron.

206. K. Tsukada (1951), *J. Phys. Soc. Japan* **6**, 415. The Ion Source of Electron Oscillation Type.

207. P. C. Veenstra and J. M. Milatz (1950), *Physica* **16**, 528. The Development of a Magnetic Ion-Source with High Ionization Efficiency.

E. Additional References

208. A. Bottreau and C. Marzat (1968), *Compt. Rend. (Paris)* **266B**, 703. Décharge réflex continue produisant un plasma de section quasi rectangulaire en vue de mesures interférométriques dans la bande des 8 mm.

209. G. G. Cloutier, C. Beaudry, and Z. Zakrzewski (1968), *Bull. Am. Phys. Soc.* **13**, 1524. Fine Structure of Rotating Instabilities in a Reflex Discharge.

210. H. W. Darwin and M. Fumelli (1965), *Proc. Phys. Soc. (London)* **85**, 997. Spectroscopic Measurement of Plasma Rotation in a Penning Discharge.

211. N. Hopfgarten, R. B. Johansson, B. Nilsson, and H. Persson (1968), *Phys. Fluids* **11**, 2272. Penning Discharge in a Strongly Inhomogeneous Magnetic Field.

212. J. J. Kim and A. Simon (1968), *Bull. Am. Phys. Soc.* **13**, 1530. Nonlinear Theory of the $E \times B$ Instability.

213. V. Kopecký and J. Václavïk (1966), *Plasma Phys.* **8**, 645. On Strong Turbulence of an Inhomogeneous Weakly Ionized Plasma in Crossed Electric and Magnetic Fields.

214. Yu. E. Kreindel (1968), *Zh. Tekh. Fiz.* **38**, 1675. Production and Investigation of Non-Homogeneous Plasma in a Penning Discharge.

215. W. G. Lange, J. H. Singleton, and D. P. Eriksen (1966), *Vac. Sci. Tech.* **3**, 338. Calibration of Low Pressure Penning Discharge Type Gauges.

216. J. R. Roth (1966), *Rev. Sci. Instr.* **37**, 1100. Modification of Penning Discharge Useful in Plasma Physics Research.

217. J. R. Roth (1967), *Phys. Fluids* **10**, 2712. A New Mechanism for Low-Frequency Oscillations in Partially Ionized Gases.

218. E. M. Rudnitskii (1968), *Zh. Tekh. Fiz.* **38**, 830, translated in *Soviet Phys.–Tech. Phys.* **13**, 623 (1968). Power Distribution between the Electrodes in a Penning Discharge.

219. T. Sato and T. Tsuda (1967), *Phys. Fluids* **10**, 1262; Erratum, *Phys. Fluids* **11**, 259 (1968). Computer Study of Nonlinear Cross-Field Instability.

220. R. F. Stetson and J. L. Hanisch (1968), *Bull. Am. Phys. Soc.* **13**, 1524, Study of a Reflex Discharge.

221. T. H. Stix (1968), *Bull. Am. Phys. Soc.* **13**, 1561. A Proposed Source for Highly Stripped Ions.

222. K. I. Thomassen (1968), *J. Appl. Phys.* **39**, 5017. Instabilities and Anomalous Diffusion in the Reflex Discharge.

223. Y. Tanaka and K. Yamamoto (1966), *Japan. J. Appl. Phys.* **5**, 1240. On Modes of a PIG Discharge.

224. Y. Tanaka and K. Yamamoto (1967), *Japan. J. Appl. Phys.* **6**, 520. Resonant Oscillation in a PIG Discharge

225. N. S. Wolf, B. S. Newberger, W. M. Hooke, and T. H. Stix (1968), *Bull. Am. Phys. Soc.* **13**, 1534. Standing Whistler in a Reflex Discharge.

Author Index

Numbers in parentheses are reference numbers and indicate that an author's work is referred to, although his name is not cited in the text. Numbers in italics show the page on which the complete reference is listed.

A

Adamov, I. Yu., 312(115), 314(115, 116), 316(114), 317(114), *338*
Agdur, B., 312, 315, *333*
Akhtyrskaya, Ye. V., 305(101), *337*
Akimune, H., 327(183), *341*
Alexandrov, E. B., 39, *57*
Allen, M. A., 321, 323, 324, 328, *333*
Allwood, H. I. J., 327(165), *340*
Ames, I., 327(147), *339*
Anderson, E. J., 327(177), *341*
Angelbeck, A. W , 328(45), *335*
Angerth, B., 325, *333*
Apel, J. R., 328(5), *333*
Ayphassorho, M. C., 325, *333*

B

Babertsian, R. P., 309, 318(125, 126), *338, 339*
Bachynski, M. P., 191(62a), 287(62a), *293*
Backus, J., 311, 312, 327, *333*
Badan, P., 44(21). *57*
Bailey, A. D., 102(29), *183*
Bailey, C., 327(178), *341*
Baird, J. R., 189, 190, 235, 237, 238, 239, *292*
Barbian, E. P., 325, *333*
Bariand, A., 327(179), *341*
Bariband, M., 327(203), *342*
Barnett, C. F., 327(180, 199), *341, 342*
Barrat, J. P., 20(1), 22(1), *56*
Barrett, E. B., 189, 190(60), *292, 293*
Bartnovskii, O. A., 324(53), *335*
Basile, R., 327(181), *341*
Batten, H. W., 310, *333*
Baumann, H., 327(182), *341*
Bayet, N., 311, *333*
Beaudry, C., *342*
Becheres, R., 327(179), *341*

Bekefi, G., 326, *336*
Belford, R. L., 119(55, 56), *184*
Bell, M. J., 6, *18*
Béné, G. J., 25, 26, 36, 46(23), 51(23), 54(37), 55(38), *56, 57*
Bennewitz, H. G., 119, *184*
Berezin, A. B., 324, 326, 327, *333*
Berk, H. L., 286(84), *291, 293*
Berkling, K., 62(9), 71, 73, 143, 147(9), *183*
Bethge, K., 327(182), *341*
Biechler, C. S., 328(3), *333*
Bieri, R., 91(24), *183*
Bierrum, N. R., 190, *292*
Bingham, R., 320(16, 17), 322(34), 323, 329(16, 34), *333, 334*
Birmingham, T. J., 286(84), 291, *293*
Bitter, F., 38, *56*
Blachier, B., 189, *292*
Blauth, E. W., 60(1), 180, *181*
Blewett, J. P., 61, *183*
Bliman, S., 312, 315, *333, 334*
Blum, P., 102, 104, *183*
Böhm, H., 102, *183*
Bonnal, J. F., 312, 313, *334*
Bottiglioni, F., 325, *334*
Bottreau, A., *342*
Bouchoule, A., 315(20, 21), *334*
Brachet, C. M. F., 311, *334*
Brezechko, L. V., 317(26), *334*
Briffod, G., 312(22), 313(22), 314, *334*
Brink, G. O., 109(44), *184*
Brossel, J., 38, *56*
Brothers, G. F., 327(148), *339*
Brown, S. C., 327(198), *342*
Brubaker, W. M., 85, 93, 94(20), 95(20), 96(20), 97, 98(27), 99(27), 100(28), 101, 102(28), 107, 108(42, 43), *183, 184*, 327(149), 331, *339*

Brzhechko. L. V., 312(113), 314(113), 317(113), 326(113), *338*
Buliginsky, D. G., 324(15), 326(14), 327(14), *333*
Bullis, R. H., 2(2), *18*
Burch, D. S., 8(12), *18*
Burmaster, K., 327(166), *340*
Burnham, D. C., 144, 169, *184*
Burt, J. A., 109, *184*

C

Callaway, J., 10, 11, *18*
Carr, H. Y., 53, 55, *57*
Cavalleri, G., 4(4), 6, *18*
Chapet, M., 325(23), *334*
Chen, F. F., 319, 320, 321(33), 322, 323, 329, *333, 334*
Cheon, B., 327(183), *341*
Chiyoda, K., 189, 199, 222, 227, 237, 286(82), 291, *291, 292, 293*
Chorney, P., 321(2), 323(2), 324(2), 328, *333 334*
Christensen, R. L., 327(147), *339*
Christophorou, L. G., 16(27), *18*
Churchill, T. L., 2(2), *18*
Cloutier, G. G., *342*
Cohen-Tannoudji, C., 39, 40, 41(19), 43(19), *57*
Colegrove, F. D., 46, *57*
Coleman, J. H., 328(107), *337*
Coleman, P. D., 189, 190, 235, 237, 238, 239, *292*
Compton, R. N., 16(27), *18*
Conisa, G., 327(150), *340*
Conn, G. T. K., 309, *334*
Consoli, T., 325, *334*
Cooper, A. W., 320(35), 322, 323, *334*
Cottet, H., *57*
Courant, E. D., 61, *181*
Crawford, C. K., 119, *184*
Crawford, F. W., 316(40), 331, *334*
Crettenand, E., 39(15), *57*
Crompton, R. W., 4(3, 5), 6(9), 7(11), 8(3, 5, 16), 10(3), 13(21, 24), 14(25), 15, 16, *18*
Crownfield, F. R., 326, *335*

D

Daglish, H. N., 309, *334*
Dalgarno, A., 13(22), 17, *18*
Darwin, C. G., 222, *293*
Darwin, H. W., *342*
Ďatlov, J., 320(43), 322, 323, *335*
Davis, F. J., 8(15), *18*
Davis, J. W., 328(44, 45), *335*
Dawson, P. H., 73(17), 77, 78(17), 81, 121(64, 65), 126(64), 127(64, 65), 128(64), 129(64), 130(65), 131, 143, 147(17, 70, 73), 149(17), 150(17), 151(17), 152(17), 154(17), 155(17), 160, 161, 162, 163(70, 82), 164, 168(82), 170(72, 73, 90), 172(90), 174(72, 73, 90), 175(90), 176(73), 177(70, 72), 178(72), 179(72, 73), 180(72), *183, 184, 185*
Dayton, I. E., 82(19), 91(19), *183*
Dehmelt, H. G., 144, 161, 162, 164, 168(80), 169, 176(74), *184, 185*
Delcroix, J. L., 286(88), *293*, 327(193), *341*
Desloge, E. A., 189, 203, 278, *292, 293*
Dolique, J. M., 327(203), *342*
Doucet, H., 326, *335*
Douwes Dekker, H. L., 327(81), *336*
Dow, D. G., 298(47), 299, 302, 306, 307, 310(47), *335*
Dreicer, H., 190, *292*
Druaux, J., 327(179), *341*
Druky, D. L., 327(178), *341*
Drummond, W. E., 313(48), 321(48), *335*
Dubovoĭ, L. V., 312, 315, 321(49), 324, 325, *335*
Dukes, J. N., 328, *335*
Dumas, G., 311, *333, 335*
Dushin, L. A., 312(113, 115), 314(113, 115), 316(114), 317(113, 114), 326(60, 113), *335, 338*
Duxler, W. M., 10(19), 11(19), *18*

E

Early, H. C., 310(12), *333*
Eck, T. G., 47, *57*
Ehlers, K. W., 327(177, 184), *341*
Ehrensvärd, J., 325(4), *333*
Ehrhardt, H., 15, *18*
Elford, M. T., 4(3, 5), 8(3, 5, 16), 10(3), *18*
Ellis, R. A., 286(83), *291, 293*

SUBJECT INDEX

14 DAY USE

RETURN TO DESK FROM WHICH BORROWED

ENGINEERING LIBRARY

Tel. No. 642-3339

This book is due on the last date stamped below, or
on the date to which renewed.
Renewed books are subject to immediate recall.

AUG 26 1971